Medicinal Chemistry

ONE WEEK LOAN

Medicinal Chemistry
The Role of Organic Chemistry in Drug Research

Second edition

edited by
C. R. Ganellin
*Department of Chemistry, University College London,
London, U.K.*

and

S. M. Roberts
*Department of Chemistry, University of Exeter,
Exeter, U.K.*

Academic Press
Harcourt Brace Jovanovich, Publishers
London   San Diego   New York   Boston
Sydney   Tokyo   Toronto

ACADEMIC PRESS
24–28 Oval Road
London NW1 7DX

ACADEMIC PRESS
San Diego, CA 92101

Copyright © 1993, by
ACADEMIC PRESS LIMITED

Third printing 1999

A catalogue record for this book is available from the British Library

ISBN 0-12-274120-X

Cover Illustration
A computer graphics representation of the active site of a carbon–carbon bond breaking enzyme
(human aldolase) highlighting the residues important for catalysis and the substrate fructose-1,6-
bisphosphate in space filling mode. Courtesy of J.A. Littlechild and H.C. Watson (1991) *J. Mol.
Biol.* **219**, 573–576.

Typeset by Alden Multimedia Limited
Printed in the United Kingdom at the University Press, Cambridge

99 00 01 02 CU 9 8 7 6 5 4 3

Contents

Contributors

J. Bradshaw, Glaxo Group Research Ltd, Ware, Hertfordshire SG12 0DP, U.K.

A. G. Brown, SmithKline Beecham Pharmaceuticals, Betchworth, Surrey RH3 7AJ, U.K.

A. Collis, Pfizer Central Research, Sandwich, Kent CT13 9NJ, U.K.

M. G. Davis, Division of Biosciences, University of Hertfordshire, (Formerly Hatfield Polytechnic), Hatfield, Hertfordshire AL10 9AB, U.K.

J. M. Evans, SmithKline Beecham Pharmaceuticals, Medicinal Research Centre, Harlow, Essex CM19 5AD, U.K.

I. François, SmithKline Beecham Pharmaceuticals, Betchworth, Surrey RH3 7AJ, U.K.

C. R. Ganellin, Department of Chemistry, Christopher Ingold Laboratories, University College London, 20 Gordon Street, London WC1E 6BT, U.K.

A. J. Gibb, Department of Pharmacology, University College London, Gower Street, London WC1E 6BT, U.K.

T. J. R. Harris, Biotechnology, Glaxo Research Ltd, Greenford, Middlesex UB6 0HE, U.K.

L. H. C. Lunts, Glaxo Group Research Ltd, Ware, Hertfordshire SG12 0DP, U.K.

B. G. Main, Chemistry Department, ICI Pharmaceuticals PLC, Macclesfield SK10 4TG, U.K.

E. G. Maliski, Glaxo Research Institute, Research Triangle Park, NC 27709, U.S.A.

S. Redshaw, Roche Products Ltd, Welwyn Garden City, Hertfordshire Al7 3AY, U.K.

K. Richardson, Discovery Chemistry, Central Research, Pfizer Ltd, Sandwich, Kent CT13 9NJ, U.K.

S. M. Roberts, Department of Chemistry, University of Exeter, Exeter EX4 4QD, Devon, U.K.

G. Stemp, SmithKline Beecham Pharmaceuticals, Medicinal Research Centre, Harlow, Essex CM19 5AD, U.K.

H. Tucker, Chemistry Department, ICI Pharmaceuticals PLC, Macclesfield SK10 4TG, U.K.

Preface

Chemists play a key role in pharmaceutical research; they are required to synthesize, purify and analyse novel compounds for biological testing. The preparation of a selected compound is *relatively* straightforward: with the wealth of chemical literature available, a chemist should be able to construct any thermodynamically stable molecule (though, depending on the complexity of the molecule, the task may take many man-months to complete and a number of seemingly reasonable routes might fail owing to unforeseen and unexpected courses being taken by the reactions). The real difficulty is in deciding what to make. Which fundamental molecular framework will have the desired biological action? What are the optimum positions for substituents on the basic skeleton to maximize activity? Using chemistry, we must also try to understand the aetiology of the disease in order to be in a position to suggest molecules that might correct the underlying malfunction.

The first edition of the book *Medicinal Chemistry: the Role of Organic Chemistry in Drug Research* was published in 1985. The aim of the book was to identify important contributions made by organic chemists in the discovery of new ethical drugs. In order to illustrate how these contributions were made, eleven important drugs were selected for detailed discussion. The eleven chapters were written by chemists who were intimately involved with the projects in the discovery phase. These chapters followed three sections which discussed basic concepts concerning enzymes, receptors and the action of drugs.

This new book, the second edition, updates and expands the earlier text. Advances in the discovery and identification of new receptors has led to a complete revision of that section. The recent impact of molecular biology on the drug discovery process is acknowledged by a chapter on the topic. In addition, the increased understanding of the design and utilization of prodrugs has led to a discussion of the relevant issues in this text. Similarly there have been major advances in the employment of computers in structure–activity analysis and a discussion of the state of the art in this area is also included. At the end of the introductory part of the book, the present-day thinking regarding the best approaches to drug discovery are considered.

Once again the role played by the organic chemist is emphasized by a series of accounts of the discoveries of some important compounds that are being used widely as medicines. Four chapters from the earlier edition have been updated (i.e. the sections dealing with adrenergic β_2-stimulants, β_1-blockers, histamine H_2-antagonists, and β-lactam anti-biotics) while three others are completely new, featuring, naturally, drugs that have recently made very important contributions to medicinal chemistry. Thus angiotensin-converting enzyme inhibitors have become well-established compounds in the treatment of some forms of heart disease and the design of cilazapril is detailed in Chapter 9. Similarly the research into the action of compounds on potassium channels that led to the discovery, synthesis and use of cromakalim is described (Chapter 8). The discovery of fluconazole, a noteworthy orally active antifungal agent, is described in Chapter 13. The seven examples have been carefully chosen to cover a range of disease areas, different structural types of active compounds and different discovery processes.

We hope that the reader will be stimulated by the accounts; moreover we hope he or she will have the good fortune to be part of a team that discovers one or more drugs for the twenty-first century.

We thank all the contributing authors most sincerely for their hard work. In addition we acknowledge the work of postgraduate students in the Organic Section of the Chemistry Department at the University of Exeter in reading the proofs and deciding which terms should appear in the glossary.

Robin Ganellin
Stan Roberts

Glossary

Actinomycetes	non-acid-fast bacteria, some varieties used in the production of antibodies
Adipose	of or related to fat
Agonist	a drug that can interact with receptors (specific sites in certain cells) and initiate a drug response (e.g. acetylcholine)
Allosteric control	alteration of an enzyme's activity by molecules non-competitively bound to sites other than the active or catalytic site of the enzyme
Angina	a spasmodic oppressive severe pain
Angina pectoris	sudden intense pain in the chest caused by lack of blood supply to the heart muscle
Antacid	an agent which reduces the acidity of gastric juice or other secretions
Antagonist	a drug that opposes the effect of another by physiological or chemical action or by a competitive mechanism for the same receptor sites
Antihypertensive	agent that causes lowering of (high) blood pressure
Apnoea	the cessation or suspension of breathing
Arrhythmia	irregularity, especially of the heart beat
Atria	upper chambers of the heart
Autonomic	independent, self-controlling
Biliary	relating to bile and the bile ducts
Bronchospasm	contraction of the bronchi
Carboxypeptidase	an enzyme that cleaves the amido-link of peptides at the carboxylic acid terminus
Carcinoma	a malignant tumour causing eventual death (i.e. cancer)
Catecholamines	amine compounds which have sympathomimetic activity and are concerned with nervous transmission and many metabolic activities
Chimera	an organism consisting of at least two genetically different types of tissue
Chromatin	the portion of the cell nucleus, easily stained by dyes, consisting of nucleic acids and proteins
Corticosteroid	steroid produced by adrenal cortex
Cytochrome P-450	an enzyme that catalyses hydroxylation
Cytoplasm	the substance of a cell surrounding the nucleus, carrying structures within which most of the cell's life processes take place
Cytosol	the soluble portion of the cytoplasm after all particles (e.g. mitochondria, endoplasmic reticular components) are removed

Denaturation	modification of the enzyme's biological structure (secondary and tertiary) due to extremes of pH or temperature, organic solvents, high salt concentrations, oxidation or reduction
Diastolic	dilation of the chambers of the heart that follows each contraction during which they refill with blood
Disseminated	widely distributed throughout an organ, tissue or the body
Dopaminergic	neurotransmitting action by the substance dopamine
Double-blind study	experiment wherein both experimenters and subjects do not know the particulars of the test items
Endometrium	mucosal layer lining the cavity of the uterus, its structure changes with age or with menstrual cycle
Flexor	muscle that flexes a joint
Ganglia	(plural of ganglion) collection of nerve cell bodies located outside the brain and spinal cord
Genome	the complement of chromosomes in a nucleus
Granulocyte	a type of phagocytic white blood cell
Haemodynamic	blood movement
Hypertension	abnormally high blood pressure
Hypophysis	the pituitary gland, important to growth, maturity and reproduction
Infarction	formation of a dead area of tissue caused by an obstruction of the artery supplying the area
Intravenous infusion	introduction of a fluid into a vein
In vitro	made to occur outside of an organism in an artificial environment
In vivo	carried out in the living organism
Isoenzyme	one of a group of enzymes which catalyses the same chemical reaction but which has different physical properties
Juxtaposition	a side-by-side position
Lumen	the interior space of a tubular structure
Lymphokines	agents that cause migration of constituents of the lymph
Macrophage	any large phagocytic cell occurring in the blood or lymph
Mesentery	double layer of peritoneum supporting most of the small intestine
Micturition reflex	involuntary desire to urinate
Mitochondria	organelles present in nearly all living cells, generating energy by the formation of ATP (adenosine triphosphate)
Mycelium	vegetative body of fungi
Myocardial infarction	destruction of an area of heart muscle due to blockage of the coronary artery
Neoplasm	any abnormal new growth of tissue; tumour
Neuropsychiatry	the study of both organic and functional diseases of the nervous system
Oncogene	any of several genes that can cause cancer

Osmolarity	the osmotic concentration of a solution expressed as moles of the dissolved substance per litre of solution
Parietal	pertaining to the wall of a cavity
Pathogen	agent that causes or produces disease
Pellagra	a condition caused by vitamin deficiency, marked by skin lesions, gastrointestinal disturbances and nervous disorders
Pernicious anaemia	deficiency of vitamin B_{12} resulting in degeneration of posterior/lateral columns of the spinal cord
Pluripotent	more pronounced chemical effect
Potency	an expression of drug activity (i.e. the dose required to produce a specific effect of given intensity as compared to a standard of reference)
Prophylaxis	prevention of disease
Prostatic hypertrophy	enlargement of the prostrate gland
Proteolytic enzymes	enzymes which hydrolyse (break down) proteins into simpler more soluble forms
Psoriasis	chronic inflammatory skin disease characterized by red patches covered with silvery scales
Psychotomimetic	producing symptoms that resemble those of psychosis, a severe mental illness (i.e. loss of contact with reality, regressive behaviour, delusions)
Pulmonary endothelial cells	layer of cells that line the lungs
Purine	fundamental part of organic compounds found in nature (e.g. uric acid compounds, adenine, guanine and xanthines)
Renal	of, relating to, resembling or situated near the kidney
Reperfusion arrhythmias	irregular heart beat on restoration of blood flow to the heart
Secretagogue	a substance promoting secretion (e.g. in the stomach)
Serum	clear fluid of blood, devoid of fibrinogen and cells
Spheroplast	cell surrounded solely by its limiting membrane
Sterol	any of a group of natural steroid alcohols (e.g. cholesterol)
Sympathomimetic	producing effects similar to those caused by stimulation of the sympathetic nervous system
Systemic	relating to or affecting the whole body
Tachycardia	abnormally rapid beating of the heart
Teratogenic	producing malformation in a foetus
Thymic	relating to the thymus gland
Topical	for application to the body surface
Vagus	tenth cranial nerve
Vascular	consisting of, or provided with, vessels
Vascular endothetial cells	layer of cells that line the blood vessels
Vasorelaxant (vasodilator)	an agent that can cause dilation of the blood vessels
Viscera	(plural of viscerous) large internal organs, especially in the abdomen

–1–

Introduction to Receptors and the Action of Drugs

S.M. ROBERTS
Department of Chemistry,
University of Exeter,
Stocker Road,
Exeter EX4 4QD, Devon, U.K.

Professor Stan Roberts received B.Sc. and Ph.D. degrees from the University of Salford (Ph.D. supervisor, Professor Hans Suschitzky). After further study at the Universities of Zürich and Harvard, with Professors Andre Dreiding and Robert B. Woodward respectively, he returned to Salford University to take up a position as lecturer in organic chemistry and biochemistry. In 1980 Professor Roberts joined Glaxo Group Research (Greenford) as Head of Chemical Research and in 1986 moved to his present position as Head of Organic Chemistry in the Department of Chemistry, Exeter University. His research interests include medicinal chemistry, agricultural chemistry and the use of enzymes in organic synthesis.

I. Communication Between Cells: the Roles of Receptors and Enzymes

Some forms of life are composed of a single independent cell (the protozoa) while mammals are multicellular organisms. In between these two extremes there are life forms of varying complexity. All these organisms possess cell(s) to compartmentalize various chemical reactions in order to use available materials for energy and the maintenance of life's processes.

Cells of different life forms have different characteristics (Fig. 1.1) and, indeed, different cells from the same organism can be distinguished readily. For example, mammalian cells come in all shapes and sizes: compare the spheroidal leucocyte (the white blood cell), the flat epithelial cells found lining the mouth and the nerve cell (Fig. 1.2).

The cells are organized such that chemical transformations can be accomplished efficiently, the rate of these transformations being controlled by Nature's catalysts –

MEDICINAL CHEMISTRY 2nd Edition
ISBN 0-12-274120-X

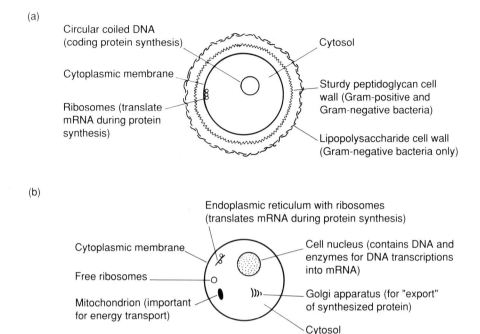

(a)

Circular coiled DNA
(coding protein synthesis)

Cytosol

Cytoplasmic membrane

Sturdy peptidoglycan cell
wall (Gram-positive and
Gram-negative bacteria)

Ribosomes (translate
mRNA during protein
synthesis)

Lipopolysaccharide cell wall
(Gram-negative bacteria only)

(b)

Endoplasmic reticulum with ribosomes
(translates mRNA during protein synthesis)

Cytoplasmic membrane

Cell nucleus (contains DNA and
enzymes for DNA transcriptions
into mRNA)

Free ribosomes

Mitochondrion (important
for energy transport)

Golgi apparatus (for "export"
of synthesized protein)

Cytosol

Fig. 1.1   A prokaryotic bacterial cell (a) and a eukaryotic (possessing a nucleus) human cell (b). Not all
the substructures that may occur in a cell are represented.

enzymes (see Chapter 2). Enzymes are high molecular weight compounds which catalyse anabolic (synthesis) and catabolic (degradation) reactions. The trivial name of the enzyme often gives a guide to its role (a more comprehensive list of enzyme activities is contained in Chapter 2):

$$RCO_2R^1 \quad \xrightarrow[\text{H}_2\text{O}]{\text{Esterase}\atop\text{enzyme}} \quad RCO_2H + R^1OH \qquad \text{Eq. (1)}$$

$$RCHR^1 \atop |\atop OH \quad \xrightarrow{\text{Dehydrogenase}\atop\text{enzyme}} \quad RCOR^1 \qquad \text{Eq. (2)}$$

$$A^+ \qquad AH + H^+$$

$$\begin{array}{c} R \qquad\qquad R^1 \\ CH-CH_2-CH \\ |\qquad\quad| \\ O-\!\!-P-\!\!-O \\ \parallel\quad| \\ O\quad O^- \end{array} \quad \xrightarrow[\text{H}_2\text{O}]{\text{Phosphodiesterase}\atop\text{enzyme}} \quad \begin{array}{c} RCH-CH_2-CHR^1 \\ |\qquad\qquad| \\ OPO_3H \qquad OH \end{array} \qquad \text{Eq. (3)}$$

In order to coordinate their activities the different cells in multicellular organisms need to communicate and this correspondence is accomplished mainly by small chemical molecules. For example, on receiving the appropriate signal, nerve terminals may release substances such as acetylcholine (**1**), noradrenalin (**2**) or dopamine (**3**), and these substances, known as neurotransmitters, can interact with the appropriate receptors.

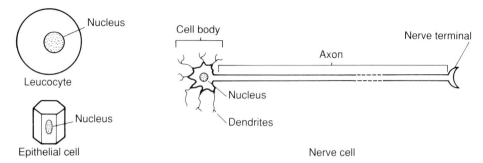

Fig. 1.2   Shapes and sizes of mammalian cells.

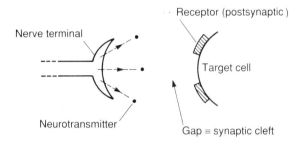

Fig. 1.3   A neuro-effector junction (synapse).

$$CH_3COOCH_2CH_2\overset{+}{N}(CH_3)_3$$

(1)

(2) R = H
(6) R = CH$_3$

(3)

The receptors lie on the surface of the cells opposite the nerve terminal (Fig. 1.3). The interaction of a neurotransmitter (agonist*) with its receptor usually effects a change in conformation of the macromolecular receptor, leading to a change in enzyme activity within the cell (Fig. 1.4), and/or movement of ions into or out of the cell (Fig. 1.5).

One specific example of the process diagrammatically illustrated in Fig. 1.4 is given by noradrenaline, which will act on a receptor (see later) to activate the intracellular enzyme adenylate cyclase to produce cyclic adenosine 3′,5′-monophosphate (5) (cyclic AMP) from adenosine triphosphate (ATP) (4) as shown in Eq. (4).

---

*Agonist is a name coined by Gaddum in 1937 and is now used to describe all physiological mediators and drugs which mimic their action by activating the same cellular reactions.

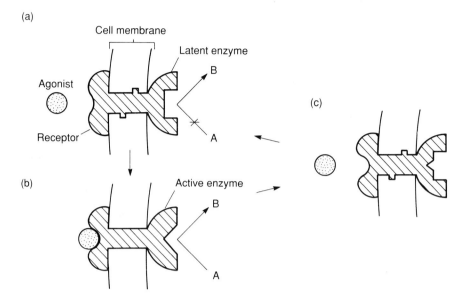

Fig. 1.4   Activation of an enzyme by occupation of a receptor by an agonist. (a) Receptor free, enzyme inactive. (b) Receptor occupied, enzyme triggered into action (allosteric activation of enzyme). (c) Agonist leaves receptor surface, and enzyme quickly returns to inactive form.

Cyclic AMP initiates a cascade of other enzyme activations leading to the observed biological response. Cyclic AMP is inactivated by a phosphodiesterase enzyme [Eq. (4)].

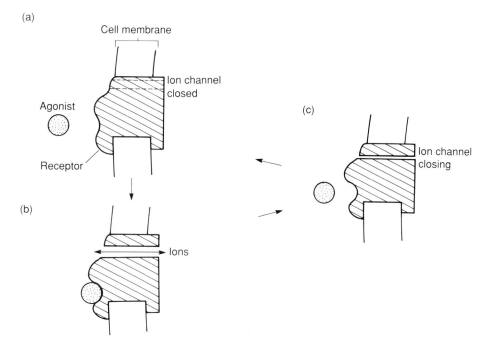

Fig. 1.5 Opening of an ion channel by the occupation of a receptor by an agonist. (a) Receptor free, channel closed. (b) Receptor occupied, channel opened, ion migration rapidly takes place down the electrochemical gradient. (c) Channel closes, neurotransmitter diffuses away. Difference in metal ion (e.g. Na^+) concentration across the membrane is re-established by exergonic metal-ion pump.

An example of the process represented in Fig. 1.5 is the interaction of acetylcholine with its receptor which causes opening of the receptor ion channel resulting in a movement of sodium ions into the cell. Figures 1.4 and 1.5 illustrate that receptor occupation may result in an effect (e.g. change in enzyme activity) that will last for the lifetime of the occupation, or the effect (e.g. ion movement) may be triggered and will not reoccur until disengagement of the agonist, repriming of the system and re-engagement of the agonist and the receptor.

Note that, if a neurotransmitter remains in the synaptic cleft, disengagement/re-engagement will continue and the receptor will be activated repeatedly; this will carry on until the chemical diffuses away from the site. To allow a faster return to the resting state after the neurotransmitter has ceased to be released from the nerve terminal, an enzyme is usually present which will convert the neurotransmitter into an inactive substance. For example, acetylcholine is deactivated by an esterase [Eq. (5)], while noradrenaline is rendered inactive by methylation of one of the phenolic groups [Eq. (6)] through the enzyme catechol O-methyltransferase (COMT).*

$$CH_3COOCH_2CH_2\overset{+}{N}(CH_3)_3 \xrightarrow[\text{H}_2\text{O}]{\text{Acetylcholinesterase}} CH_3CO_2H + HOCH_2CH_2\overset{+}{N}(CH_3)_3 \qquad \text{Eq. (5)}$$

*Some neurotransmitters including noradrenaline are removed from the synaptic cleft by reabsorption into the presynaptic terminal, and indeed the majority of the noradrenaline released from a nerve terminal is removed from the synapse by this method.

$$\text{HO} - \text{CHCH}_2\text{NH}_2 \xrightarrow[\substack{\overset{+}{\text{RSR}} \\ | \\ \text{Me}}]{\text{COMT}} \overset{\text{RSR}}{\underset{+}{H^+}} \quad \text{HO} - \text{CHCH}_2\text{NH}_2 \qquad \text{Eq. (6)}$$

These examples illustrate that enzymes are not always contained within cells. Equally, not all chemical messengers are released from nerve terminals to act on adjacent terminals before being degraded. For example, adrenaline (6), noradrenaline (2) and various steroids are released into the circulation from endocrine glands. Local hormones or autocoids such as histamine (a key component in the inflammatory response) are released from cells, and travel through extracellular fluid to act on nearby cells. The three types of intercellular communication processes are shown in Fig. 1.6.

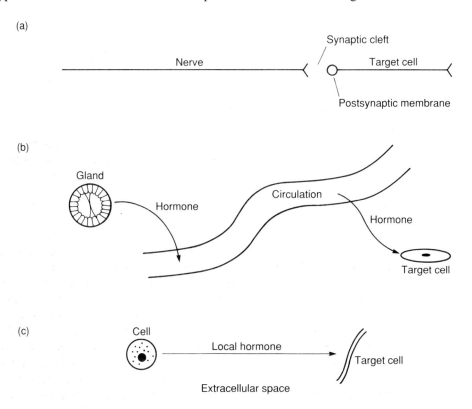

Fig. 1.6   Intercellular communication processes. (a) Nerve releases a neurotransmitter substance, e.g. acetylcholine, which diffuses across a synapse to act upon a postsynaptic membrane. (b) Endocrine gland releases a hormone, e.g. steroid, which is distributed throughout the body by the circulatory system. (c) Local hormone (autocoid) is released by cells and diffuses through the extracellular space to act locally, e.g. histamine released from mast cell or cells in the stomach.

It is important to understand that, while enzymes and receptors are both composed of amino acids condensed into high molecular weight polypeptide chains, and can be associated with ions and small molecules, there the likeness ends. One of the major differences is that enzymes catalyse bond-making and bond-breaking reactions while the receptors release the agonist unchanged.

In the normal healthy state, all cells are communicating, synthesizing and degrading molecules, and changing ion concentrations for the overall well-being of the organism. As a result of disease, damage or degeneration, cellular activities may become impaired and the correct dynamic equilibrium must then be reinstated by the means of a suitable drug. (Sometimes it may also be desirable to alter the normal physiology (in anaesthesia for example) by administration of a drug substance.)

It may be desirable to amplify the effect of a neurotransmitter. This can be accomplished by:

1. increasing the concentration of the natural neurotransmitter by (a) direct supplementation through introduction of the substance into the body; (b) inhibition of enzymes that degrade the transmitter; (c) inhibition of reuptake;
2. using a more potent and/or less readily metabolized surrogate of the natural substance (an unnatural agonist).

Alternatively it may be prudent to decrease the effect of a particular neurotransmitter at a given receptor. This can be done using an antagonist substance,* i.e. an unnatural compound which will bind strongly to a receptor without eliciting a response and which will prevent access of the neurotransmitter to the receptor.

Similarly, certain enzyme substrates and specific or highly selective enzyme inhibitors can prove useful drug substances.

Before amplification of these points, the central control of cell communication and a more detailed consideration of certain neurotransmitters and receptors is warranted.

II. Neurotransmitters, Receptors and the Nervous System

In complex organisms such as man there are a considerable number of receptors and a multitude of different enzymes. Actions are coordinated by the central nervous system (CNS) (the brain and spinal cord) and some actions are the result of sensory input (sight, sounds, touch, hearing, etc.) Output from the CNS is directed towards the autonomic nervous system (the sympathetic and parasympathetic systems) and nerves associated with voluntary motor functions as illustrated in Fig. 1.7. Voluntary motor function deals with the controlled movement of muscles (skeletal muscle) and the associated limbs; some of the organs controlled by the parasympathetic and sympathetic nerves are listed in Fig. 1.8 and the effects on selected organs from the two systems are listed in Table 1.1. In short, the sympathetic and parasympathetic systems operate in a complementary fashion. In response to a particular external influence, enhanced stimulation of the sympathetic nervous system occurs and leads to preparation for 'fight or flight' (Fig. 1.9). In the relaxed state (Fig. 1.10), stimulation of the parasympathetic nervous system predominates and deals with secretion and voidance of materials from the body.

Some important neurojunctions and the associated neurotransmitters in the nervous system are shown in Fig. 1.11.

---

*An antagonist diminishes or abolishes the effects of its corresponding agonist. Two types of antagonist are known: a competitive antagonist competes with the agonist for a binding site on the active site of the receptor while a non-competitive antagonist binds at a different site from that of the agonist. In the former situation the effect of an antagonist decreases in the presence of increasing concentrations of agonist while in the latter situation the effect of the antagonist is independent of agonist concentration.

8   S.M. ROBERTS

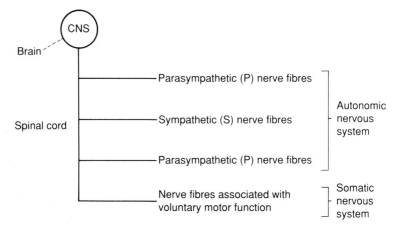

Fig. 1.7   The peripheral nervous system.

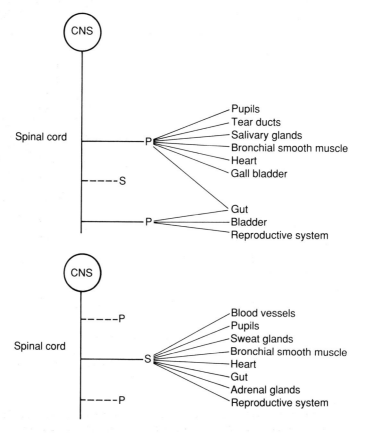

Fig. 1.8   Organs controlled by sympathetic (S) and parasympathetic (P) nerves.

TABLE 1.1
Response to Activation of Sympathetic and Parasympathetic Nervous System

Organ	Response to activation of nerve	
	Sympathetic	Parasympathetic
Pupil	Dilation	Constriction
Bronchi	Dilation	Constriction
Heart	↑ Acceleration	↓ Slowing
Digestive tract	↓ Slowing	↑ Increase
Sphincters of gut	Constriction	Relaxation
Bladder	Relaxation	Contraction
		Emptying
Blood vessels	Constriction	
Glands of alimentary canal (salivary, gut, pancreas)		↑ Increase in activity

Note that the acetylcholine receptors are divided into two categories; the classification is based on the actions of two drugs of plant origin.* The receptors which are activated by muscarine are termed muscarinic receptors, and those activated by binding nicotine are called nicotinic receptors. Thus nicotine and muscarine mimic the action of acetylcholine at two different distinct receptors. The effects on various end-organs of stimulation of the appropriate acetylcholine receptors are listed in Table 1.2. Inhibition of the action of acetylcholine at the neuromuscular junction leads to muscle relaxation.

Fig. 1.9 Stimulation of the sympathetic nervous system due to *fright* leads to preparation of the system for *fight* or *flight*: increase in heart rate, dilation of bronchi, dilation of pupils, constriction of peripheral blood vessels (pallor), etc. (From B.L.A.T. Booklet 'Action of Drugs', Centre for Health and Medical Education, London).

*The subdivision of the various receptors is mainly based on agonist and/or antagonist actions of substances not normally found in the mammalian system; it is known that the amino acid sequence of receptor subtypes is similar, but not the same.

Fig. 1.10 In the relaxed state stimulation of the parasympathetic nervous system is predominant and leads to (*inter alia*) slowing of the heart and increase in the activity of the gastrointestinal tract. (From B.L.A.T. Booklet 'Action of Drugs', Centre for Health and Medical Education, London).

Noradrenaline and adrenaline stimulate adrenoceptors. When noradrenaline is released from a presynaptic nerve terminal it crosses the synaptic cleft and initiates a response in the postsynaptic tissue by combining with one of two types of receptor called α-adrenoceptors and β-adrenoceptors.* The type of receptor found postsynaptically to noradrenergic nerves in the sympathetic system depends on the type of tissue; classification of adrenoceptors in different tissues is again based on the ability of agonists to initiate responses and antagonists to prevent responses. A further subclassification of β-adrenoceptors has been made: in man the majority of β-adrenoceptors in the heart are called β_1-adrenoceptors and these are distinct from other β-adrenoceptors (dubbed β_2-adrenoceptors*) found elsewhere in the periphery (outside the CNS). Many tissues have a mixed population of β_1- and β_2-receptors. Noradrenaline and adrenaline have different effects on α- and β-receptors; noradrenaline is potent at stimulating α-receptors but is less potent at activating β-receptors while adrenaline elicits activity from both α- and β-receptors at about the same level.

Some important sites of α- and β-adrenergic receptors are given in Table 1.3.

Note that activation of α-receptors generally results in a stimulant response (except in the gut) while activation of β-receptors leads to an inhibitory response, namely, relaxation of muscle (except in the heart). Blocking of α- and β-receptors causes, *inter alia*, relaxation of peripheral blood vessels and slowing of the heart respectively and can have beneficial effects in the treatment of angina and hypertension (Chapter 10), while β_2-stimulants can alleviate mild to moderate asthmatic attacks by relaxation of bronchiolar muscle and widening of airways (Chapter 11).

Some peptides act as neurotransmitters. For example, receptors for the enkephalins

---

*β_3-adrenoceptors have now been identified – see Chapter 3.

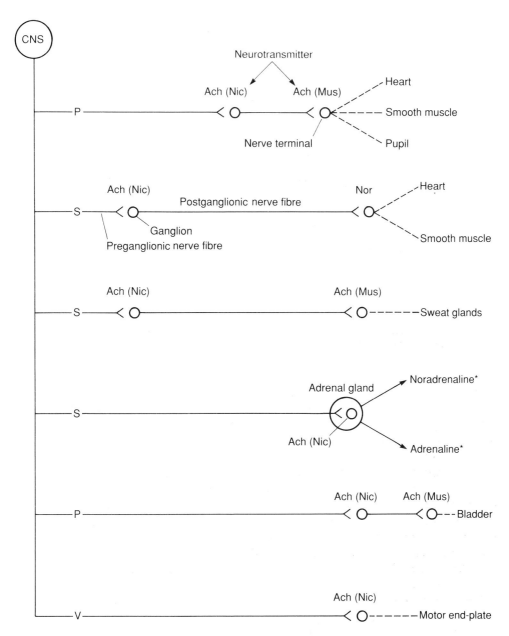

Fig. 1.11 Neurotransmitters in the peripheral nervous system. Asterisk indicates release into the circulation to act on distant receptors in the periphery. S, sympathetic nervous system; P, parasympathetic nervous system; V, voluntary motor function; Ach, acetylcholine; Nor, noradrenaline; Mus, muscarinic receptor; Nic, nicotinic receptor.

TABLE 1.2
Situation of Some Acetylcholine-Controlled Neuroeffector Junctions

Organ	Type of receptor	Control by voluntary motor (VM), parasympathetic (P), or sympathetic (S) nervous system	Effect of agonist
Heart	Muscarinic	P	Slowing rate of contraction, decrease force of contraction
Eye	Muscarinic	P	Constriction of pupil
Sweat glands	Muscarinic	S	Activation
Gastrointestinal smooth muscle	Muscarinic	P	Increase in tone, increase in contractions and peristalsis
Adrenal gland	Nicotinic	S	Release of adrenaline and noradrenaline
Bladder	Muscarinic	P	Contraction
Skeletal muscle	Nicotinic	VM	Muscle contraction
CNS	Muscarinic and Nicotinic	-	Various

(7) have been demonstrated within the CNS. It is believed that morphine (8) and the other opiates exert their analgesic action by interaction with these receptors. Thus morphine (8), heroin (9) and codeine (10) can be considered to be agonists at the enkephalin receptor; other compounds such as the pain-killing drug buprenorphin are partial agonists i.e. have mixed agonist and antagonist properties (see Chapter 3 for a fuller explanation). Opiate receptor blockers (antagonists), e.g. naloxane, are also known.

TABLE 1.3
Important Adrenaline and Noradrenaline Receptors

Organ	Major adrenoceptor present	Effect of stimulation
Eye	α	Pupillary dilation
Blood vessels (periphery including skin)	α	Constriction
Gastrointestinal tract	$\alpha + \beta$	Relaxation
Heart	β_1	Tachycardia (speeding of the heart) Increased force of contraction
Small intestine	β_1	Relaxation
Blood vessels (skeletal muscle)	β_2	Relaxation (vasodilation)
Bronchioles (lung)	β_2	Relaxation (bronchodilation)

H-Tyr-Gly-Gly-Phe-Met-OH   Methionine-enkephalin

H-Tyr-Gly-Gly-Phe-Leu-OH   Leucine-enkephalin

(7)

(8) $R^1 = R^2 = H$

(9) $R^1 = R^2 = COCH_3$

(10) $R^1 = CH_3; R^2 = H$

(11)

Histamine (11) is a neurotransmitter which also occurs in many tissues in the body. It is stored in mast cells and platelets and is released from these sites in response to stimuli such as allergic reactions and injury. Three main types of histamine receptor, termed H_1, H_2 and H_3, have been identified; they differ in sensitivity to various unnatural agonists and antagonists. Bronchiolar smooth muscle has H_1 receptors; activation by histamine causes contraction of the muscle and bronchoconstriction. The actions of histamine on vascular smooth muscle are complex, species dependent and mediated by both H_1 and H_2 receptors, while gastric acid secretion from parietal cells is stimulated by mucosally released histamine acting at H_2 receptors. H_2 receptor blockers are useful in the treatment of conditions in which there is excess acid secretion in the stomach, especially in duodenal ulceration (see Chapter 12). The H_3 receptors appear to be autoreceptors which inhibit further histamine synthesis or release; they occur especially in nervous tissue.

It is noteworthy that all the previously mentioned transmitters, i.e. acetylcholine, noradrenaline, enkephalins and dopamine, as well as histamine and adrenaline, have receptors in the CNS. Interaction with these receptors will elicit a response and if interaction of a potential drug substance with peripheral receptors (e.g. at the neuro-muscular junction, on bronchiolar smooth muscle, in parietal cells, in heart muscle, on blood vessels etc.) is beneficial, then it is often necessary to ensure that the compound does not cross the 'blood–brain barrier' to cause unwanted side effects through inter-action with receptors in the CNS.

III.   Enzymes and Enzyme Inhibitors

The active centres of enzymes are similar in many ways to the agonist binding sites of receptors, and enzyme inhibitors are identical in principle with receptor blockers. Simple inhibitors derange the active centre by engaging it directly (isosteric inhibition) or by inducing a conformational change affecting the active site through binding to a distant site (allosteric inhibition). Unnatural substrates for an enzyme which are slowly pro-cessed are also effective inhibitors of the physiological enzyme action provided that they have a substantially greater affinity than the natural substrate for the enzyme centre. Substantial inhibition of a key enzyme-controlled process in an organism will generally lead to the demise of the organism. If the enzyme in question is peculiar to a bacterium or fungus that has invaded the mammalian host then inhibition will eradicate the pathogen and leave the host unharmed (see for example Chapter 14).

IV. Other Types of Bioactive Molecules

Not all drugs act on discrete receptors or at active sites of particular enzymes. Others, such as some general anaesthetics, act by forming monomolecular layers on membranes, thereby modifying transport across the membrane. Some drugs have a stabilizing or a labilizing effect on the membranes of cells. Yet others, such as antacids or some diuretics, produce their effects by means of their physicochemical properties. Many steroid drugs and hormones must first pass into cells, becoming associated with specific cytosolic proteins which facilitate their transport into the cell nucleus where gene expression, and ultimately protein synthesis, is modified.

V. Factors Influencing Drug Action

The aim of administering a drug is to get it to the right place, in the right concentration and for the right period of time. Except for topical treatments (e.g. application of an anti-inflammatory agent to the skin), humans are usually dosed by a route which is remote from the intended site of action. Thus there will generally be a certain latent period before the action of the drug is initiated. This latent period will depend upon the route of administration, the formulation of the compound and the mode of distribution. The duration and intensity of action will, in turn, depend on the relative rates of arrival at, and removal from, the site of action. These rates depend primarily on the distribution, metabolism and excretion of the drug. The overall chronology of events between drug administration and elimination is summarized in Fig. 1.12.

Some drugs (prodrugs) will not exert pharmacological activity until they have undergone biotransformation within the body. For example the penicillin ester pivampicillin (12) is well absorbed when taken by mouth unlike the parent acid; the ester must be de-esterified by a blood-borne esterase enzyme before antibacterial activity is exhibited (for a more comprehensive discussion of prodrugs see Chapter 4).

(12)

Having succeeded in getting a drug into the biophase, binding to the appropriate site of an enzyme or receptor must take place. The binding of the small drug molecule to the macromolecule involves many complementary forces (electrostatic forces, hydrogen-bonding, hydrophobic bonding, van der Waals forces).

Electrostatic attraction, for example between the quaternary ammonium group of

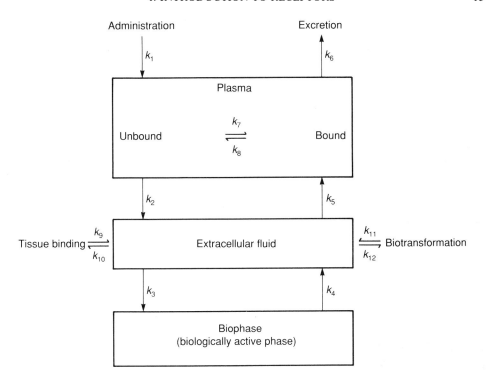

Fig. 1.12   Summary of the events between drug administration and elimination.

$$RCH_2CH_2\overset{+}{N}Me_3$$
$$|$$
$$^-O_2C$$
$$\cdot$$
E or R

acetylcholine analogues and a carboxylate residue in the macromolecule, is a powerful binding and stabilizing influence. Ion–dipole and dipole–dipole interactions are of less importance. Hydrophobic bonding is also very important and leads to good binding mainly due to the increase in entropy of the system ($\Delta G = \Delta H - T\Delta S$) through displacement of water molecules from 'uncomfortable' quasicrystalline arrangements adjacent to the hydrocarbon surfaces to a less ordered more favourable situation (Fig. 1.13). On the other hand the gain in free energy through hydrogen-bonding between enzyme or receptor and drug molecule is not as significant as may be imagined at first sight.

For example, the phenol moiety forming the part-structure of the drug molecule D—C_6H_4OH can be nicely accommodated at the active site of an enzyme (E) or receptor (R) as shown in Fig. 1.14a; strong hydrogen bonds are formed and good binding takes place. However, the hydrogen bonds between the drug and the enzyme or receptor can only be formed at the expense of the hydrogen bonds between the separate entities and associated water molecules (Fig. 1.14b). Hence the net gain in energy through hydrogen-bonding between the drug and macromolecule is marginal. Note that a closely related drug substance D—C_6H_5 lacking the phenolic hydroxy group would displace the water

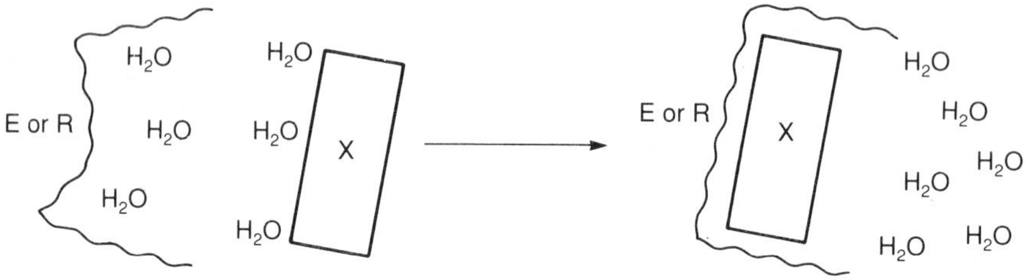

Fig. 1.13   Approach of hydrophobic faces of a drug molecule (X) and a receptor (R) or enzyme (E).

molecule(s) from the surface of the macromolecule without forming the compensating hydrogen bonds (Fig. 1.14c); in this case binding would be less secure.

The pharmacokinetic profile of a drug will vary from species to species and be influenced by many factors such as the route of administration, formulation, age, sex, disease, diet, environmental influences and the presence of other drugs.

Clearly the design of a drug molecule is hampered by the lack of detailed knowledge regarding pharmacological receptors, accessibility to target enzymes or other intended sites of action, and other undesirable and toxic effects. In addition, the development of a useful therapeutic agent is fraught with difficulties as the ideal structure to fit an enzyme or a receptor may not be suitable to allow that particular molecule to run the gauntlet between administration and elimination so as to be present near the enzyme/receptor at a concentration which would elicit the required physiological effect.

It can be appreciated that a drug molecule has somewhat greater difficulty in travelling to its intended site of action than, for example, a natural neurotransmitter which is released close to the appropriate receptor. The localized release and rapid disposal of

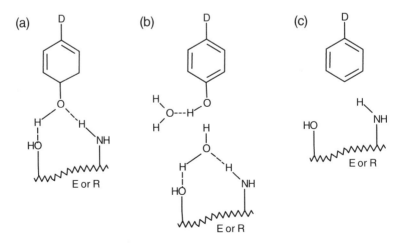

Fig. 1.14   Changes in hydrogen-bonding that take place on approach of a drug to the surface of an enzyme (E) or receptor (R).

these small molecules means that a complex organism can operate with the minimum number of different neurotransmitters; when an active substance such as histamine or adrenaline is released into a large part of the system, many different effects are seen owing to the stimulation of receptors at many sites. Drugs can be regarded as being more closely analogous to hormones and local hormones (autocoids) in that they are widely distributed in the body. Generally, these molecules are sterically more bulky and possess more functional groups; the extra features within the molecules restrict binding to relatively few receptors or enzymes.

Before a description of the case histories of drugs that act through interaction with specific enzymes or receptors, a more detailed account of 'receptorology' and enzymology is desirable and follows in the next two chapters.

Additional Reading

Czaky, T.Z. (1979). 'Cutting's Handbook of Pharmacology: The Actions and Uses of Drugs', 6th edn, Appleton-Century-Crofts, New York.

Fersht, A.R. (1984). Basis of Biological Specificity. *Trends in Biochemical Sciences*, **9**, 145.

Gill, E.W. (1965). In 'Progress in Medicinal Chemistry' (G.P. Ellis and G.B. West, eds), Vol. 4, p. 39, Butterworths, London.

Goodman, L.S. and Gilman, A. (1980). 'The Pharmacological Basis of Therapeutics', 6th edn, Macmillan, New York.

Kruk, Z.L. and Pycock, C.J. (1979). 'Neurotransmitters and Drugs', Croom Helm, London.

Lewis, P. and Rubenstein, D., (1979). 'The Human Body', Hamlyn, Feltham (U.K.).

Stryer, L. (1988). 'Biochemistry', 3rd edn, Freeman, San Francisco.

Wolff, M.E. (ed.) (1980). 'The Basis of Medicinal Chemistry: Burger's Medicinal Chemistry', 4th edn, Vols 1–3, Wiley (Interscience), New York.

−2−

Structure and Catalytic Properties of Enzymes

M.G. DAVIS
Division of Biosciences,
University of Hertfordshire,
(formerly Hatfield Polytechnic),
Hatfield, Herts., U.K.

After graduating in Chemistry from the University of Durham, *Michael Davis* specialized in Biochemistry via an M.Sc. course at Chelsea College, University of London. He then went on to a research post at St Bartholomew's Hospital, London, investigating metabolic disorders, where he obtained his Ph.D. in Biochemistry. His main work since then has been teaching biochemistry at Hatfield Polytechnic. Academic interests are mainly in the field of metabolic, membrane and nutritional biochemistry. Research work has involved collaboration with various hospitals and pharmaceutical companies.

I. Configuration of Enzymes

Although some metabolic reactions can occur spontaneously, by far the majority require *enzymes* as catalysts. Each cell requires over 500 different enzymes to enable it to carry out all its functions, although the types of enzymes required will depend on the nature of the cell. Some enzymes are membrane bound, some are found only in particular subcellular organelles such as mitochondria and others are cytoplasmic. Over 2000 enzymes have been described, the vast majority of which are proteins (though many proteins, such as haemoglobin and insulin, are not enzymes). Many enzymes also require additional non-protein components (known as *cofactors*) in order to be catalytically active; the role of cofactors will be discussed later.

MEDICINAL CHEMISTRY 2nd Edition
ISBN 0-12-274120-X

Like all proteins, enzymes have characteristic primary, secondary and tertiary structures. The *primary structure* is based on the sequence of amino acids linked by peptide bonds. Twenty-two amino acids are available for use in the primary sequence and each polypeptide chain may contain several hundred amino acid residues. An enzyme's *secondary and tertiary structure* give it its three-dimensional characteristics, as revealed by X-ray crystallography (Fig. 2.1). The secondary structure refers to the coiling and twisting of the polypeptide chain due to hydrogen-bonding, and tertiary structure to the way in which the enzyme folds into its characteristic globular form and is retained in this configuration by a number of types of bonding including disulphide bonds formed between adjacent thiol groups and ionic bonds between adjacent free carboxylic acid and amino groups. Hydrogen-bonding and hydrophobic bonding are also crucial in determining the overall shape of an enzyme. The digestive enzymes such as pepsin and trypsin consist of single polypeptide chains and have molecular weights varying from 15 000 to 35 000.*

Most enzymes catalysing metabolic reactions inside cells have what is called a *quarternary structure*, which means that they consist of more than one polypeptide subunit. Ultracentrifugation has shown that anything from two to 60 subunits can occur in a single enzyme, giving rise to molecular weights ranging from 35 000 to several hundred thousand. Most enzymes have from two to eight subunits. The enzyme may consist of aggregates of identical subunits, or of more than one type of subunit. For example, the enzyme lactate dehydrogenase, important for carbohydrate metabolism in muscle, consists of four subunits of which there are two types. Gel electrophoresis has shown that there are five different forms of this enzyme, depending on the combination of subunits. Enzymes like this which can occur in more than one molecular form are called *isoenzymes*. These have an important role in the regulation of metabolism, and are also valuable clinically in diagnostic enzymology, as elevated plasma levels of a particular isoenzyme may indicate that damage has occurred in a particular organ such as liver or heart.

Some metabolic systems involve even more complex levels of organization known as *multienzyme complexes* in which several different enzymes associate together physically to catalyse a series of sequential reactions. An example of this occurs in fatty acid synthesis in bacteria and plants in which seven different enzymes aggregate together to give a highly ordered and efficient system. In higher organisms such as mammals, *multifunctional enzymes* are found, in which one polypeptide can catalyse a sequence of reactions.

Because of their protein structure, enzymes are highly sensitive to their physical environment. Organic solvents, high salt concentrations, oxidation or reduction and extremes of pH or temperature will readily cause *denaturation*, precipitation due to breakdown of the tertiary and secondary structure. Loss of enzyme activity without denaturation can also occur as a result of relatively small changes in temperature and pH. Most enzymes have their optimal activity at physiological pH and temperature, that is, pH 7 and 37°C.

---

*These enzymes, incidentally, have an inactive precursor form called *zymogens* or proenzymes, which have to be converted to the active form by hydrolytic removal of a protective section of the peptide chain.

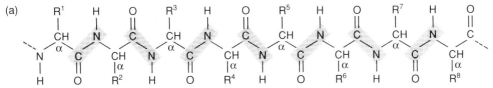

(a)

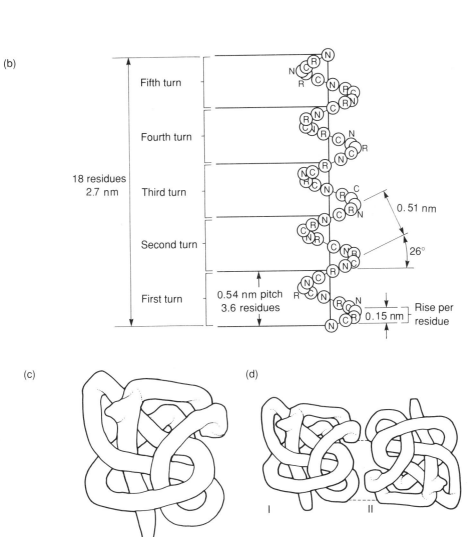

(b)

Fifth turn

Fourth turn

18 residues
2.7 nm

Third turn

0.51 nm

Second turn

26°

First turn

0.54 nm pitch
3.6 residues

0.15 nm

Rise per
residue

(c)

(d)

I   II

Fig. 2.1   (a) Primary structure. General formula of a polypeptide chain showing the linkage of adjacent amino acid residues through peptide bonds. (b) Secondary structure. Representation of a polypeptide chain as an α-helical configuration. N, NH; C, CO; R, CHR. [From L. Pauling and R.B. Corey, *Proc. Int. Wool Textile Res. Conf., B*, 249 (1955), as redrawn in C.B. Anfinsen, *The Molecular Basis of Evolution*, Wiley, New York, 1955, p. 101.] (c) Tertiary structure. (d) Quaternary structure. A protein dimer unit illustrating the quarternary structure of a complex globular protein. Modified, with permission, from E.E. Conn and P.K. Stumpf, *Outlines of Biochemistry*, 4th Edition, Wiley, New York, 1976.

II. Enzyme Specificity, Classification and Nomenclature

One of the properties of enzymes that most distinguishes them from inorganic catalysts is their high degree of specificity. This is due to the chemical and physical characteristics of an enzyme's protein structure. Enzymes are normally specific for one particular type of reaction and one particular type of substrate only. For example, a hydrolytic enzyme such as a peptidase or a lipase will not catalyse oxidation–reduction or group transfer reactions. Likewise the enzyme alcohol dehydrogenase, which catalyses the oxidation of ethanol to acetaldehyde, will also convert a variety of other alcohols to their corresponding aldehydes, with differing affinities for the substrates, but will have no effect on non-alcoholic substrates. Most enzymes will accept a range of chemically related compounds as their substrates; relatively few enzymes show absolute specificity for a single substrate.

Enzymes also display stereochemical specificity, and may be specific for particular optical or geometric isomers of a substrate. For example, D-amino acid oxidase will oxidize a variety of D-amino acids but has little or no activity with L-amino acids, and fumarase catalyses the hydration of the *trans* unsaturated dicarboxylic acid fumaric acid but not the *cis* isomer (maleic acid).

Enzymes are generally classified as belonging to one of the six categories shown in Table 2.1.

TABLE 2.1
Enzyme Classification

Enzyme category	Examples	Types of reaction catalysed
Oxidoreductases		Oxidation or reduction of substrate
	Dehydrogenases	Transfer of H from substrate to cofactor
	Reductases	Addition of H to substrate
	Oxidases	Transfer of H from substrate to oxygen
Transferases		Transfer of group from one molecule to another
	Aminotransferases	Transfer of amino groups
	Transacetylases	Transfer of acetyl groups
	Phosphorylases	Transfer of phosphate groups
Hydrolases		Hydrolysis of substrate (irreversible)
	Glycosidases	Hydrolysis of glycosidic bonds
	Esterases	Hydrolysis of ester bonds
	Peptidases	Hydrolysis of peptide bonds
Lyases		Elimination and addition reactions
	Hydratases	Addition of water to double bonds (reversible)
	Decarboxylases	Removal of CO_2 from substrate (reversible)
	Aldolases	Aldol condensations (reversible)
Isomerases		Molecular rearrangement
	Racemases	D- and L-isomer interconversion
	cis–trans Isomerases	Geometrical isomerization
	Mutases	Intramolecular group transfer
Ligases		Energy-dependent bond formation (generally irreversible)
	Synthetases	Condensation of two molecules
	Carboxylases	Addition of CO_2 to substrate

Enzymes have complex systematic names based on further subdivisions of the categories in Table 2.1. These are used for precise identification in technical literature, but trivial names are normally used in laboratories. These usually include the name of the substrate and type of reaction (e.g. alcohol dehydrogenase). Other common trivial names simply add the suffix -ase to the name of the substrate (e.g. ribonuclease), while many of the digestive enzymes are known by their historic names (e.g. amylase, lipase, pepsin and trypsin).

III. Characteristics of Enzyme Catalysis

Although enzymes show much greater specificity and sensitivity to temperature and pH than inorganic catalysts, they have the same function as all catalysts, accelerating the rate of reaction by lowering the activation energy required for a reaction to proceed. Enzymes also obey the normal rules of catalysis: (1) enzymes will not catalyse thermodynamically unfavourable reactions; (2) they will not change the direction of a reaction; (3) they will not change the equilibrium of a reaction; (4) they remain unchanged at the end of the reaction (though some enzymes undergo temporary covalent changes such as phosphorylation during the course of a reaction). Enzymes normally catalyse reactions much more efficiently than inorganic catalysts. For example, both platinum and the enzyme peroxidase (also known by its old name catalase) catalyse the decomposition of hydrogen peroxide into water and oxygen. The effects of the different catalysts on the activation energy (in kilojoules per mole) are as follows:

No catalyst	75
Platinum	50
Peroxidase	8

One mole of peroxidase can catalyse the decomposition of over one million moles of H_2O_2 per minute.

Enzymes are able to lower the activation energy for a reaction through the formation of an intermediate *enzyme–substrate complex*, which in turn can break down to enzyme and product as follows:

$$E + S \rightleftharpoons E\text{-}S \rightarrow E + P$$

S may consist of more than one substrate and P of more than one product. The effect of the formation of this complex is shown in Fig. 2.2.

The formation of this enzyme–substrate complex occurs at a small region on the surface of the enzyme called the *active site* or *active centre*, usually present as a crevice or pit. Only a few amino acids (5–10) are directly involved in the formation of this complex and subsequent catalysis at the active site. These amino acid residues are not normally consecutive, as different parts of the polypeptide chain come together at the active site as a result of the characteristic folding of the protein. The amino acids not directly involved at the active site are still important, however, as they are essential for maintaining the configuration of the protein required for the active site to function. This explains why some modifications to an enzyme's structure are more crucial than others. A modification or mutation leading to a change in amino acid sequence at the active

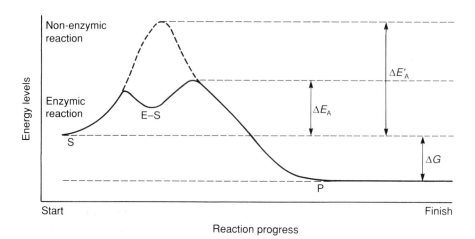

Fig. 2.2 The effect of enzyme catalysis on activation energy. ΔE_A is the activation energy for the enzyme-catalysed reaction and $\Delta E'_A$ for the non-enzymic reaction. ΔG is the free energy of the reaction.

centre can lead to complete loss of enzyme activity, while a change in sequence elsewhere may have far less effect.

The amino acids most frequently found at the active site are shown in Table 2.2; these amino acids possess groups on the side chain that are known to interact with the substrate.

As well as hydrogen-bonding and electrostatic interactions being involved between the substrate and complex, hydrophobic interactions are also important.

It has been shown that some of the amino acids at the active centre are only involved in binding the substrate to the enzyme. This part of the active site is called the *binding site*. Other amino acids at the active site are exclusively involved in catalysis, making up the *catalytic site*. The active site thus consists of binding and catalytic sites. Both binding and catalysis are essential for conversion of substrate(s) to product(s). The active site of an enzyme is shown diagrammatically in Fig. 2.3.

The presence of the binding and catalytic sites at the active site account for the specificity of enzyme reactions discussed earlier. Only the correct substrate (and closely related analogues) can bind to the enzyme, and only the appropriate type of reaction can be catalysed by a particular enzyme.

TABLE 2.2
Amino Acids Frequently Involved in the Active Centre

Amino acid	Side group	Interaction with substrate
Serine	—OH	Hydrogen-bonding
Cysteine	—SH	Hydrogen-bonding/disulphide bridging
Histidine	—Imidazole	Hydrogen-bonding/electrostatic
Lysine	—NH$_3^+$	Electrostatic
Arginine	—NH$_3^+$	Electrostatic
Aspartic acid	—COO$^-$	Electrostatic
Glutamic acid	—COO$^-$	Electrostatic

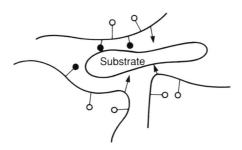

Fig. 2.3   Active site of an enzyme: —●, amino acids involved in binding site; —►, amino acids involved in catalytic site; —○, amino acids not directly involved in active site.

The old analogy of a rigid lock and key mechanism introduced in the nineteenth century by Emil Fischer has been largely superseded since the 1960s by Koshland's induced fit hypothesis, in which the substrate is believed to induce the required orientation of groups in the active site required for binding and catalysis.

Formation of the enzyme–substrate complex leads to a lowering of activation energy and subsequent catalysis due to the interaction between the substrate and the catalytic site. The binding between the substrate and amino acid groups at the active centre causes changes in bond energies and electron densities within the substrate, possibly accompanied by physical distortion and strain, resulting in a thermodynamically unstable conformation from which the reaction can readily proceed. When there is more than one substrate for a reaction, the enzyme catalyses the reaction by bringing the substrates together at the binding site in the juxtaposition required for the reaction to proceed. Provided the affinity of the product for the binding site is lower than that of the substrate, the product will dissociate from the active site, leaving it free for further reactions. As will be seen later, certain inhibitors have their effect by binding to the active site without further reaction and thus remain there, blocking the site.

Many enzymes require additional non-protein substances called cofactors for catalysis to occur. These substances can be either metal ion *activators* or relatively small organic compounds called *coenzymes*. These cofactors are normally attached to the enzyme by electrostatic bonds, but some coenzymes are linked covalently and are then called *prosthetic groups*.

Metal ion activators include Mg^{2+}, Ca^{2+}, Zn^{2+}, Fe^{2+}, Fe^{3+}, Cu^{2+}, Co^{2+}, Mo^+, K^+ and Na^+, and account for the trace metal elements required in the diet. Their requirement for enzyme activity also explains the inhibitory effect of chelating agents.

Coenzymes are derived from water-soluble vitamins and are involved in a variety of reactions, as shown in Table 2.3. They operate as second substrates, undergoing chemical modifications during the course of a reaction and reverting back to the original form by a further reaction. Diseases (such as scurvy) caused by the deficiency of a water-soluble vitamin are due to the resultant loss of enzyme activity.

IV.   Enzyme Reaction Rates

Enzyme activity is measured in units which indicate the rate of reaction catalysed by that enzyme expressed as μmol of substrate transformed (or product formed)/min. An

TABLE 2.3
Some Commonly Occurring Coenzymes

Coenzyme	Function	Vitamin processor	Deficiency disease
Nicotinamide adenine dinucleotide (NAD)	Hydrogen transfer	Nicotinamide (niacin)	Pellagra
Thiamine pyrophosphate	Decarboxylation	Thiamine (B_1)	Beriberi
Flavine mononucleotide	Hydrogen transfer	Riboflavin (B_2)	Skin lesions
Flavine adenine dinucleotide (FAD)	Hydrogen transfer	Riboflavin (B_2)	Skin lesions
Pyridoxal phosphate	Amino transfer	Pyridoxine (B_6)	Neurological disturbances
Ascorbic acid	Hydroxylation	Vitamin C	Scurvy
Cobalamine	Methylation	Vitamin B_{12}	Pernicious anaemia
Tetrahydrofolic acid (FH_4)	One-carbon transfer	Folic acid	Megaloblastic anaemia

enzyme unit is the amount of enzyme that will catalyse the transformation of 1μmol of substrate/min under specified conditions of pH and temperature. The specific activity of an enzyme is expressed as the number of units/mg of protein.

The rate of a biochemical reaction at a given temperature and pH depends on the enzyme concentration and the substrate concentration. Provided the substrate concentration remains in excess, the initial rate is directly proportional to enzyme concentration, as shown in Fig. 2.4.

When the enzyme concentration is kept constant and the substrate concentration varies, the effect of the substrate concentration on the rate of reaction is as shown in Fig. 2.5. Initially the reaction follows first-order kinetics with the rate proportional to substrate concentration, and eventually zero-order kinetics are followed with the velocity reaching a limiting value V_{max}. V_{max} is the reaction rate when the enzyme is fully saturated by substrate, indicating that all the binding sites are being constantly reoccupied. V_{max} is constant for a given amount of enzyme. The substrate concentration corresponding to half maximum velocity is known as K_m, the Michaelis constant, which is inversely

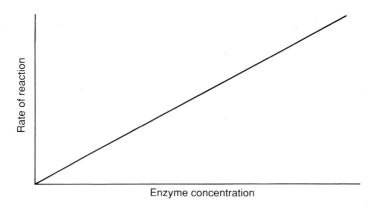

Fig. 2.4 The effect of enzyme concentration on the rate of reaction.

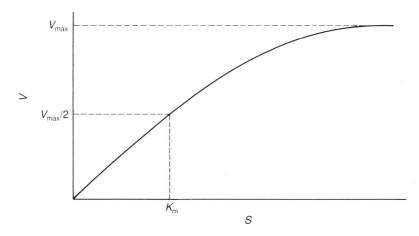

Fig. 2.5 The effect of substrate concentration (S on the rate of reaction (V).

proportional to the affinity of the enzyme for the substrate. K_m is constant for a particular enzyme and is independent of the amount of enzyme. A useful rule of thumb is that the substrate concentration has to be about $100 \times K_m$ to achieve maximum velocity. The relationship between reaction rate and substrate concentration is described by the Michaelis–Menten equation:

$$V = \frac{V_{max} S}{K_m + S}$$

This can also be expressed in the form

$$\frac{1}{V} = \frac{K_m}{V_{max} S} + \frac{1}{V_{max}}$$

A graph showing the reciprocal of the rate against the reciprocal of the substrate concentration, called the Lineweaver–Burk plot, will therefore give a straight line, as shown in Fig. 2.6. As the intercepts correspond to $1/V_{max}$ and $-1/K_m$, and the slope is K_m/V_{max}, this plot provides a useful way of determining the kinetic constants V_{max} and

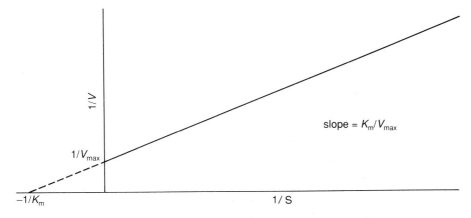

Fig. 2.6 A Lineweaver–Burk plot.

K_m from experimental data. As will be shown shortly, it is also a useful way of showing what type of inhibition may be occurring.

V. Enzyme Substrates as Drugs

Many of the basic building blocks (e.g. amino acids, sugars) for enzyme-controlled biosynthesis of complex natural products are provided in the diet. In contrast, there are few cases where an enzyme substrate is given to counter a disease, but one excellent example is in the treatment of Parkinson's disease.

Parkinson's disease is a disorder characterized by rigidity of the limbs, torso and face, and tremor, abnormal body posture and an inability to initiate voluntary motor activity (akinesia). It occurs mainly in elderly people and is a chronic and progressively degenerative disorder.

The disease is associated with decreased dopaminergic function in the brain, and many of the symptoms of the disease can be alleviated by oral administration of L-dopa (L-dihydroxyphenylalanine) (**1**). L-Dopa is given orally, absorbed from the gastro-intestinal tract, and carried to the brain by the bloodstream. In the brain the compound is a substrate for the enzyme dopa decarboxylase and dopamine (**2**) is produced, thus raising the activity of the dopaminergic neurones in the brain. Dopamine cannot be given itself because it does not have the lipophilic properties necessary to cross the blood–brain barrier (Fig. 2.7).

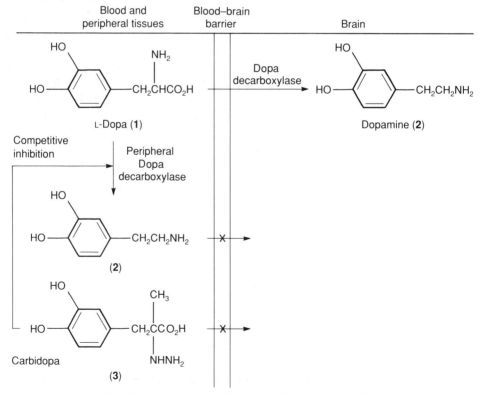

Fig. 2.7 Structural relationship of dopa, dopamine and carbidopa.

The enzyme dopa decarboxylase is also present in the liver and other tissues in the periphery. The action of peripheral dopa decarboxylase decreases the amount of L-dopa reaching the brain. To obtain a sufficient amount of compound at the necessary sites, large doses must be given with a consequent increase in side effects (nausea and vomiting). The administered dose of L-dopa can be reduced by inhibiting the peripheral dopa decarboxylase enzyme using a compound such as carbidopa (3) which does not cross the blood–brain barrier.

VI.   Enzyme Inhibition and Enzyme Inhibitors as Drugs

As was seen earlier, a variety of physical and chemical conditions such as extremes of temperature and pH and the presence of organic solvents can lead to a loss of enzyme activity. This, however, is due to gross damage to the protein's secondary and tertiary structure, resulting in denaturation rather than a specific effect on the enzyme's active site. In contrast, enzyme inhibitors are substances which bind to the enzyme with resulting loss of activity, without damaging the enzyme's protein structure. Inhibitors exert their effect by decreasing the affinity of the enzyme for the substrate, or by decreasing the amount of active enzyme available for catalysis, or by a combination of these effects. Different categories of inhibitor are described below.

A.   Irreversible Inhibitors

These are compounds which bind covalently to specific groups on the protein's surface, preventing binding and catalysis of the reaction. These compounds will inhibit a wide range of enzymes. Examples include iodoacetamide (4) and p-chloromercuribenzoate (5) (which bind to the sulphydryl group of cysteine residues), diisopropylfluorophosphate (6) (which binds to the hydroxy groups of serine residues) and 1-fluoro-2, 4-dinitrobenzene (7) (which binds to the amino groups in lysine residues and to the phenolic groups in tyrosine residues). Compounds such as these are mainly used for *in vitro* research studies, although some (e.g. mustards) are also used in anticancer therapy.

(4) 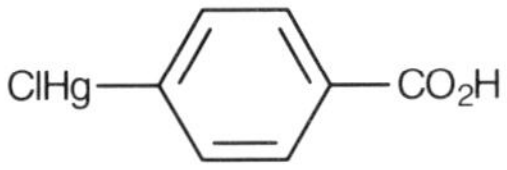   (5) 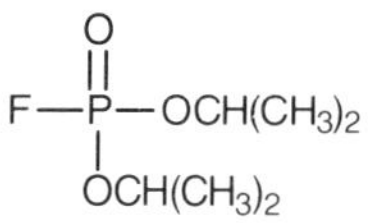   (6)

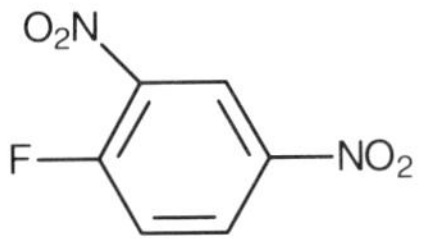

(7)

B.   Competitive Inhibitors

These compounds structurally resemble the substrate and can thus compete with the substrate for the enzyme's binding site. Such inhibitors are highly specific for a particular

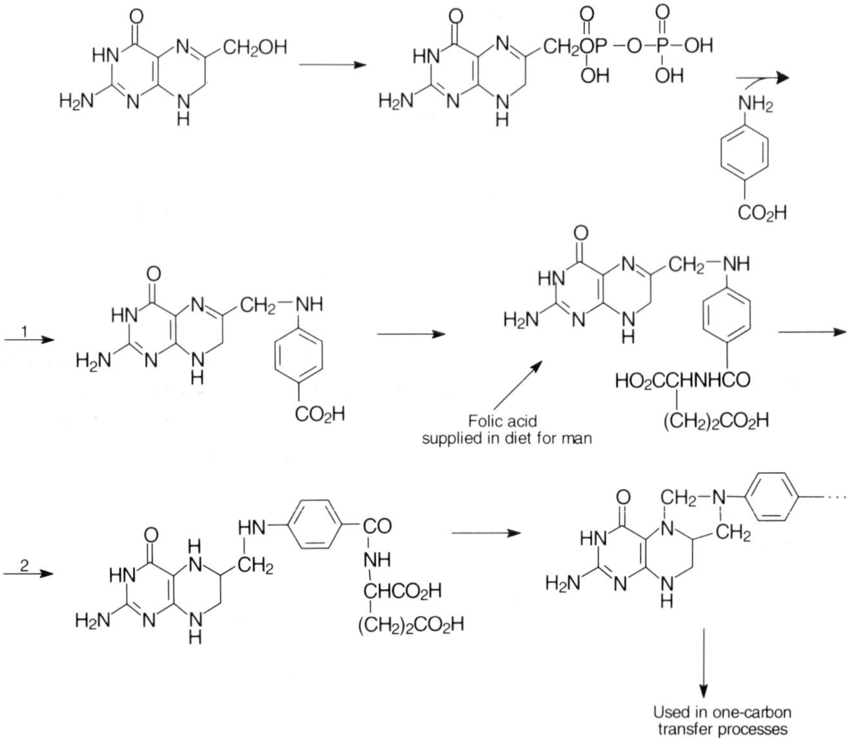

Scheme 2.1 Biosynthesis of the folates. Transformation 1 is catalysed by dihydropteroate synthetase. This enzyme is inhibited by sulphanilamide (8). Transformation 2 is catalysed by dihydrofolate reductase. The bacterial enzyme is inhibited by trimethoprim (10).

enzyme. Binding of the inhibitor is reversible, so the inhibitor can be displaced from the binding site by excess substrate. The affinity of the enzyme for the substrate is thus lowered in the presence of a competitive inhibitor (i.e. K_m is increased) but as catalysis is not directly affected, V_{max} can still be attained, even though a higher concentration of substrate will be required for this. The competitive inhibitor may be a naturally occurring alternative substrate for the enzyme which is also metabolized by it, though with a different degree of affinity, or it may be a chemical analogue of the substrate which binds to the enzyme without being further metabolized. A number of toxins and drugs operate in the latter manner. One of the earliest examples to be recognized was sulphanilamide (8), which inhibits a bacterial enzyme dihydropteroate synthetase (Scheme 2.1) required for folic acid synthesis, through its chemical similarity to p-aminobenzoic acid (9), a component of folic acid.

Since the existence of many bacteria depends on the production of folates by this process (whereas in humans folic acid is used and supplied in the diet as a preformed vitamin), sulphanilamide (and related sulphur drugs) will destroy the infecting bacteria but not the human host.

A different situation arises when it is required to inhibit an enzyme which is common to bacteria and man. Consider transformation 2 in Scheme 2.1. This step is essential for the well-being of both bacteria and man; in order to have an antibacterial agent that is

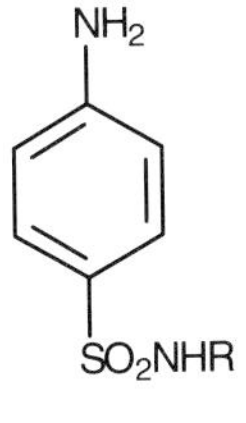

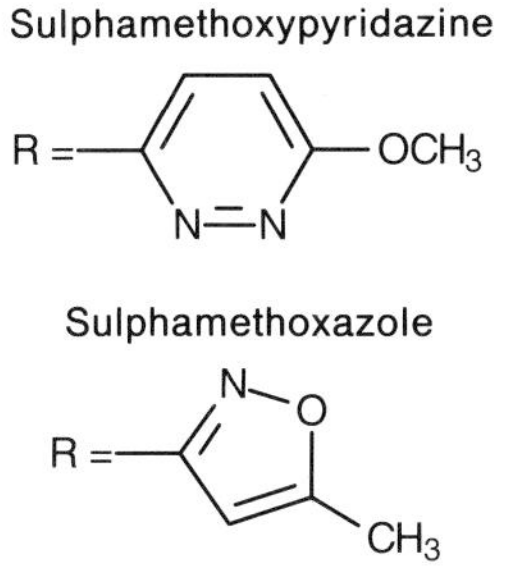

Sulphamethoxypyridazine

$R = $ ———OCH$_3$ (pyridazine ring, N=N)

(8) structure with NH$_2$ and SO$_2$NHR

Sulphamethoxazole

$R = $ (isoxazole ring, N—O, CH$_3$)

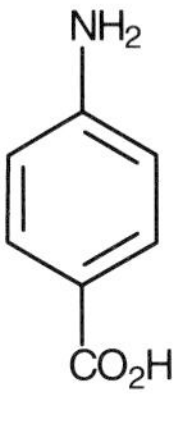

(9) structure with NH$_2$ and CO$_2$H

non-toxic to the host, advantage must be taken of the fact that the structures of bacterial dihydrofolate reductase and mammalian dihydrofolate reductase are different. Selective inhibitors of the bacterial enzyme have been found: for example the antibacterial drug trimethoprim (10) is bound to dihydrofolate reductase from the bacterium *Escherichia coli* about 10^4 times more strongly than to the same enzyme derived from rat liver.

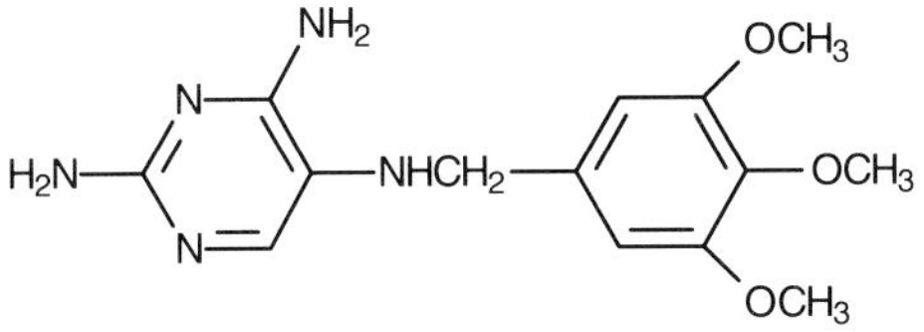

(10) structure: pyrimidine ring with NH$_2$, H$_2$N—, —NHCH$_2$— linked to trimethoxyphenyl (OCH$_3$, OCH$_3$, OCH$_3$)

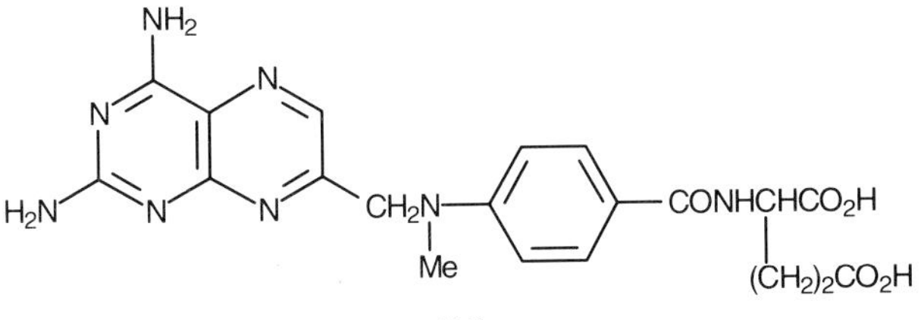

(11) methotrexate structure: pteridine ring with NH$_2$, H$_2$N—, —CH$_2$N(Me)— linked to phenyl —CONHCHCO$_2$H with (CH$_2$)$_2$CO$_2$H

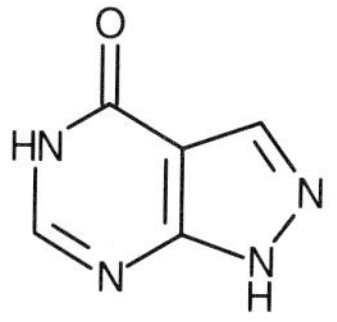

(12) allopurinol structure: pyrazolopyrimidine with O, HN, N, N—H

Methotrexate (11) is another dihydrofolate inhibitor that is in clinical use, in this case for the treatment of the widespread and sometimes crippling disease psoriasis, as well as in the treatment of some forms of cancer. The effect of reducing purine synthesis and cell division by blocking tetrahydrofolate production in psoriatic cells is beneficial; however, the drug must be used with caution since other, structurally similar and in many cases indistinguishable, dihydrofolate reductase enzymes in other tissues are affected, leading to severe toxicity on long-term treatment.

Another important enzyme inhibitor is allopurinol (12), which is used in the treatment of gout. This condition is due to a build-up in concentration of uric acid in joints. The enzyme controlling the production of urate from xanthine is xanthine oxidase (Scheme 2.2); on inhibition of this enzyme with allopurinol the cascades from guanine and hypoxanthine are stopped at xanthine and this compound is rapidly excreted. Such

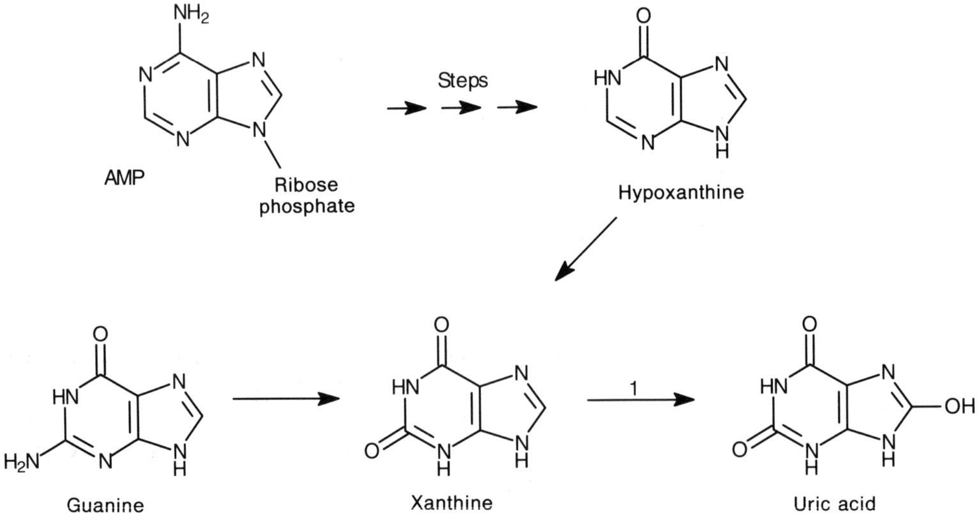

Scheme 2.2   Biosynthesis of uric acid. Transformation 1 is catalysed by xanthine oxidase. This enzyme is inhibited by allopurinol (12).

enzymes, having a very limited function in the body, may be safely and usefully inhibited by drugs.

   The above examples of drugs acting as enzyme inhibitors, and other examples of useful enzyme inhibitors described later in the book, are summarized in Table 2.4.

C.   Non-competitive Inhibitors

These substances, which are generally structurally unrelated to the substrate, bind reversibly to groups distant from the binding site of the enzyme, and are thus less specific than competitive inhibitors. The rate of reaction is decreased because the catalytic site is affected by the presence of the inhibitor. V_{max} is thus reduced, but because the binding site is not affected, the affinity of the enzyme for the substrate and therefore K_m remains unchanged. The effect of non-competitive inhibition is the same as that of less enzyme being present. Competitive and non-competitive inhibition can be distinguished from each other by the Lineweaver–Burk plot discussed earlier, as shown in Fig. 2.8.

TABLE 2.4
Drugs Acting as Enzyme Competitive Inhibitors

Drug	Enzyme inhibited	Disorder treatment
Sulphanilamides	Dihydropteroate synthetase	Bacterial infections
Trimethoprim	Bacterial dihydrofolate reductase	Bacterial infections
Penicillins and cephalosporins	Bacterial peptidoglycan transacylases	Bacterial infections
Ketoconazole	Steroid demethylase	Fungal infections
Allopurinol	Xanthine oxidase	Hyperuricaemia (primary metabolic gout)
Methotrexate	Dihydrofolate reductase	Psoriasis, cancers

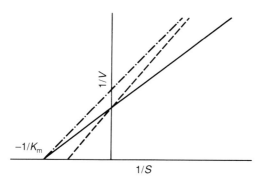

Fig. 2.8   Effects of competitive and non-competitive inhibition on enzyme kinetics: ———, no inhibitor present; – – –, competitive inhibitor present; —·—, non-competitive inhibitor present.

A number of poisons are harmful to cells because they are potent non-competitive inhibitors. Some examples are shown in Table 2.5.

VII.   Enzyme Regulation

Metabolic systems require regulation to ensure that adequate production output occurs (whether this be energy production from nutrients or biosynthesis of complex molecules) while avoiding the wasteful and potentially harmful consequences of overproduction. In general this fine control is achieved by a coordination of the regulation of enzyme synthesis and the regulation of enzyme activity. Enzyme activity is mainly controlled by the process of *allosteric regulation*, which can produce activation or, more commonly, inhibition of enzyme activity. The regulator or *effector* molecule is normally structurally unrelated to the substrate, but binds specifically and reversibly to the enzyme. However, this does not occur at the substrate binding site, but at a quite separate regulatory site (the name allosteric is derived from the Greek *allos steros*, meaning other space). Binding of the effector induces a conformational change which either increases or decreases the affinity of the enzyme for the substrate, depending on the nature of the effector.

Only certain enzymes, called regulatory or allosteric enzymes, are sensitive to this form of control, and all such regulatory enzymes have been shown to have quarternary structure, that is, two or more subunits. In many cases the regulatory site is on a different subunit from the active site, though in some enzymes both sites are on the same subunit.

TABLE 2.5
Poisons Acting as Non-competitive Enzyme Inhibitors

Compound	Mode of action	Biological effect
Organophosphate nerve gases and pesticides	Bind to serine —OH groups in cholinesterase active site	Paralysis
Mercuric salts and arsenic salts	Bind to —SH groups in many enzymes	Widespread cellular damage
Cyanide	Binds to cytochrome oxidase	Respiratory failure
Digitoxin	Binds to Na, K-ATPase	Inhibits sodium ion migration

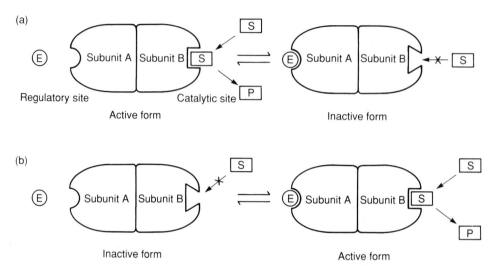

Fig. 2.9   Allosteric regulation of enzyme activity showing (a) inhibition and (b) activation. The effector is E, the substrate S and the product P.

Some enzymes are regulated by more than one allosteric effector; in such cases there are separate regulatory sites for each effector. Allosteric regulation is illustrated in Fig. 2.9.

In biosynthetic pathways a system of feedback inhibition frequently occurs whereby the end product of the pathway allosterically inhibits the first enzyme specific to that pathway. In branched-chain pathways, as occur in amino acid biosynthesis, the different products can separately inhibit the primary enzyme in the pathway and also the enzymes immediately after branching, as shown in Fig. 2.10.

This system of regulating biosynthesis ensures that the pathways operate normally when the products are required, but can be partially or completely shut down when

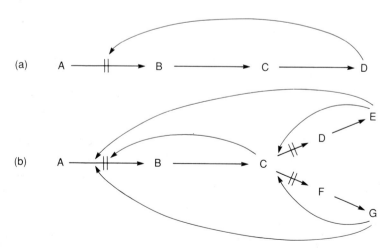

Fig. 2.10   Allosteric regulation of biosynthetic pathways by end product feedback inhibition illustrating (a) an unbranched pathway and (b) a branched pathway.

TABLE 2.6
Some Examples of Allosteric Regulation

Regulatory enzyme	Effector	Effect	Biochemical pathway
Phosphofructokinase	AMP	Activation	Glycolysis
	ATP	Inhibition	
Pyruvate carboxylase	Acetyl-CoA[a]	Activation	Gluconeogenesis
Mevalonate synthetase	Cholesterol	Inhibition	Steroid synthesis
Aspartokinase	Threonine, isoleucine, methionine, lysine	Inhibition	Amino acid synthesis

[a]CoA, coenzyme A.

adequate supplies of product have been formed. Some examples of allosteric regulation are shown in Table 2.6.

As well as the change in activity achieved through allosteric regulation, some enzymes undergo changes in activity due to *covalent modification*, normally phosphorylation. Well-studied examples of this form of regulation occur in the synthesis of the polysaccharide glycogen. Glycogen synthetase exists in an inactive phosphorylated form and an active dephosphorylated form. Through mechanisms not yet fully elucidated the protein–hormone insulin stimulates the dephosphorylation of the inactive form of the enzyme, thus increasing its activity in synthesizing glycogen. (Insulin also stimulates the uptake of the precursor glucose into the cells, further enhancing glycogen synthesis.) Glycogen breakdown (producing blood glucose in the liver and energy in muscle) requires the enzyme phosphorylase, which exists as a relatively inactive dimer or as a highly active phosphorylated dimer. Conversion to the active form requires intracellular cyclic AMP, which is produced from ATP by the membrane enzyme adenylate cyclase in response to the extracellular hormones adrenaline (Chapter 1) or glucagon binding to specific receptors on the membrane. Cyclic AMP also leads to phosphorylation and hence inactivation of glycogen synthetase. The opposite effects of the hormones adrenaline and glucagon from that of insulin can thus be explained by their indirect effects on enzyme activity.

Cyclic AMP is referred to as a secondary messenger as it is produced inside cells in response to plasma hormones binding to specific receptors on the outer surface of cell membranes. Many hormones other than adrenaline and glucagon operate in this way, including noradrenaline.

As well as this indirect effect on enzyme activity, many hormones help to control metabolism by their effect on enzyme synthesis. An increase in synthesis and subsequently the amount of enzyme is called *induction*, and a decrease in synthesis is called *repression*. In bacterial cells these processes frequently occur as a direct response to levels of nutrients present, and do not involve hormones. For example, when bacteria are grown in media rich in carbohydrates, the enzymes involved in carbohydrate absorption and breakdown are induced in order to optimize their utilization whereas when bacteria are grown in media rich in amino acids, enzymes required for their synthesis are repressed in order to avoid wasteful overproduction.

In animals, however, induction and repression of enzyme activity occur in response to hormones and some drugs. Most hormones cannot readily cross the cell membrane and will therefore achieve this effect by secondary messengers, but the steroid hormones and possibly the thyroid hormones do cross the membrane and exert a direct influence on protein synthesis.

The steroid hormones affect *transcription*, the transfer of genetic information required for protein synthesis from DNA to messenger RNA. Other hormones affect *translation*, the production of proteins at the ribosomes. In both cases metabolism is controlled through the regulation of enzyme synthesis. The mechanisms by which such hormones operate are not well understood at a molecular level, and the nature of these interactions will only be determined by further research in this area.

Additional Reading

Chaplin, M.F. and Bucke, C. (1990). 'Enzyme Technology', Cambridge University Press, Cambridge.

Cohen, P. (1983). 'Control of Enzyme Activity', 2nd edn, Chapman & Hall, London.

Bickerstaff, G.F. (1991). 'Enzymes in Industry & Medicine', Cambridge University Press, Cambridge.

Fersht, A. (1984). 'Enzyme Structure and Mechanism', 2nd edn, Freeman, New York.

Leadley, P.F. (1978). 'An Introduction to Enzyme Chemistry', The Chemical Society, London.

Palmer, T. (1991). 'Understanding Enzymes', 3rd edn, Ellis Horwood, Chichester.

Price, N.C. and Stevens, L. (1989). 'Fundamentals of Enzymology', 2nd edn, Oxford University Press, Oxford.

Stryer, L. (1988). 'Biochemistry', 3rd edn, Freeman, New York.

–3–

Receptor Pharmacology

A. J. GIBB
Department of Pharmacology,
University College London,
Gower Street,
London WC1E 6BT, U.K.

Dr Alasdair J. Gibb graduated in 1980 from the University of Strathclyde with a B.Sc. joint honours degree in Biochemistry and Pharmacology. He stayed on at the University of Strathclyde to do a Ph.D. studying the mechanisms of action of neuromuscular blocking drugs. After doing postdoctoral research at the Australian National University in Canberra, he came to the Pharmacology Department at UCL in 1986 to take up a Postdoctoral Fellowship with Professor Colquhoun. He was appointed as a New Blood Lecturer in the UCL Pharmacology Department in 1990. Dr Gibb's research interests are in the properties of glutamate receptors in the central nervous system (CNS). Within the various subtypes of glutamate receptors, his main interest is the NMDA receptor. The properties of NMDA receptors and how they are affected by drugs are investigated by electrophysiological methods, particularly patch-clamp techniques to record the single ion channel currents that occur when the NMDA receptor channel opens in response to glutamate binding to the receptor. The aim of the experiments is to provide a quantitative description of the NMDA receptor properties to improve understanding of central synaptic transmission.

MEDICINAL CHEMISTRY 2nd Edition
ISBN 0-12-274120-X

I. Introduction

Pharmacology is a relatively young science having developed almost entirely during this century. A crucial part of the establishment of pharmacology was the development of quantitative methods to describe the effects of drugs on living systems. Fundamental to these quantitative methods is the concept of receptors.

During the last 80 years or so, a succession of eminent pharmacologists such as Clark, Gaddum, Stephenson and Schild developed the receptor concept and laid the quantitative foundations of modern receptor pharmacology.

In principle, the term receptor could refer to any site to which drug molecules may bind. In practice, it is more useful in pharmacology to restrict the term receptor to those specific protein molecules whose functional role is to act as chemical sensors or transducers in the chemical communications systems that coordinate the activity of cells and organs within the body.

It is interesting to note that until about 20 years ago little information was available concerning the structural properties of receptors. Most pharmacologists may have supposed that receptors were proteins but none had been isolated to confirm this. However, the receptor concept was by then well established.

Several of the most significant advances in pharmacology and the treatment of human diseases have occurred as a result of using the receptor concept. A good example of this is the development of the β-adrenoceptor-blocking drugs now widely used in the treatment of high blood pressure. In 1948 Ahlquist suggested classifying receptors for adrenaline and noradrenaline into α- and β-adrenoceptors on the basis of their sensitivity to adrenaline and synthetic agonists like isoprenaline. α-Adrenoceptor stimulation was suggested to cause contraction of smooth muscle with adrenaline being much more effective than isoprenaline. In contrast, β-adrenoceptor stimulation was found to cause relaxation of smooth muscle, and stimulation of the heart. Isoprenaline was found to be more effective than adrenaline in stimulating β-adrenoceptors. The selective β-adrenoceptor-blocking drug propranolol was developed by Sir James Black and co-workers at ICI by using studies in a range of smooth muscle and heart preparations as a means of selecting drugs with specific and potent β-adrenoceptor-blocking activity.

Some years later Black and colleagues at the SmithKline & French Labs used a similar approach in developing the histamine H_2 receptor blocker cimetidine. This drug was the first selective H_2 receptor antagonist to be used successfully in the treatment of peptic ulcer disease.

The development of these two classes of drugs illustrated clearly how collaboration between pharmacologists and medicinal chemists, using the study of structure–activity relationships among a range of closely related chemicals, could allow a picture to be obtained of the chemical structure needed for action at a particular receptor. At the same time, the development of drugs with selective actions allowed receptors to be classified and advanced understanding of the normal physiology of the body.

This chapter gives a brief introduction to receptor structure and the quantitative methods used to describe drug–receptor binding. It then goes on to highlight how results from recent technical innovations have advanced our ideas on drug–receptor interactions. Most general textbooks of pharmacology cover parts of this subject area (e.g. Rang and Dale, 1991). In addition, the journal *Trends in Pharmacological Sciences*

(*TIPS*) regularly carries articles reviewing recent advances in receptor research. *TIPS* also publishes an excellent summary of receptor nomenclature each January.

II. Bioassay and the Measurement of Drug Effects

A. Bioassay

Traditionally bioassay involved the use of a biological response to a drug as a means of identifying an unknown substance, or estimating the concentration of a known substance (often when the substance being assayed had been released and collected from other biological tissue). The techniques of bioassay are also applied when comparing different drugs in a single tissue. In addition, bioassay techniques are used in *clinical trials* to measure the effectiveness of a drug treatment (for example when testing whether a new drug is better than an established treatment). It was largely through the early work of Gaddum and later Schild that the techniques of bioassay became widely used in pharmacology.

One of the earliest examples of the use of bioassay is the experiments of Otto Loewi who in 1920 demonstrated the chemical nature of neurotransmission using two perfused frog hearts. Loewi showed that chemical transmitters [later found to be acetylcholine (ACh) and noradrenaline (NorA)] were released from the nerve endings in the heart on stimulation of the vagus nerve. Depending on the season of the year, vagal stimulation caused either slowing (mediated by ACh) or speeding up (mediated by NorA) of the frog heart rate.

Historically, bioassay techniques were also crucial to the identification by Dale and co-workers of ACh as the neurotransmitter at the skeletal muscle neuromuscular junction. More recently, these techniques have been particularly useful in the identification of local hormones which are too labile to be studied by chemical means. This was particularly so in the identification of the prostaglandins and prostacyclins by Vane, Moncada and their colleagues at the Welcome Research Labs which led to the discovery that aspirin-like drugs act by inhibiting the synthesis of these local hormones.

A recent example of the use of bioassay techniques is the work which led to the identification of endothelium-derived relaxant factor (EDRF) as nitric oxide (NO) by Moncada and co-workers (Fig. 3.1). They studied the relaxation of arterial smooth muscle by bradykinin, and demonstrated that bradykinin did not relax strips of aorta which had been denuded of endothelium (Gryglewski *et al.*, 1986). When the superfusate from cultures of aortic endothelial cells was perfused over the aortic strips, these relaxed only when bradykinin was in the superfusate to cause release of NO from the endothelial cells. Bioassay techiques were crucial to these experiments because the half-life of NO is only about 6 s.

The receptor for NO is the cytosolic form of the enzyme guanylate cyclase which catalyses the conversion of guanosine triphosphate (GTP) to cyclic guanosine $3',5'$-monophosphate (cGMP). This forms part of the transduction mechanism mediating vasodilatation due to a range of vasodilators such as bradykinin, serotonin (5-hydroxytryptamine; 5-HT) or substance P which act on receptors on the endothelial cells to cause generation of NO which then diffuses into the neighbouring smooth muscle cell to cause

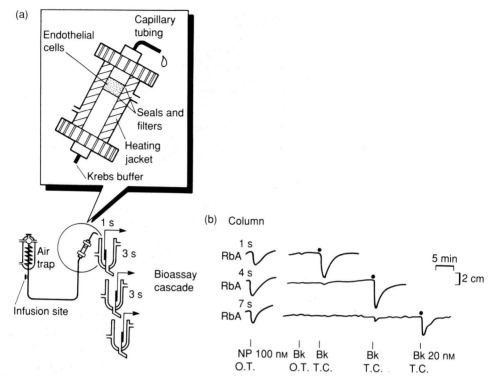

Fig. 3.1 (a) Diagrammatic representation of the bioassay cascade and the column containing cultured endothelial cells which was used by Moncada and collaborators to determine the nature of the endothelium-derived relaxant factor (EDRF). The principle of the bioassay is that perfusate from the preparation under investigation (in this case endothelial cells) is passed over a series of test preparations in the bioassay cascade (the responses of the test preparations could be contraction, relaxation, light emission, secretion, etc). The effects of the test solution are then compared with the effects of known concentrations of hormones suspected of being the unknown. Owing to the exquisite sensitivity of some biological preparations, they are often preferred to chemical techniques as a means of detecting physiological mediators. (b) Relaxation of contracted rabbit aortic strips (RbAs) in the bioassay cascade. There was a 1 s delay in perfusion between the column of endothelial cells and the first tissue in the cascade and a 3 s delay in perfusion between each bioassay tissue. The effect of nitroprusside over the bioassay tissues (NP, 100 nM O.T.) is compared with that of bradykinin (Bk, 20 nM) which has no effect when given O.T. but when given through the column (T.C.) it induces the release of a labile relaxing substance. The second and third Bk infusions T.C. are given when the effluent from the column directly superfuses the second and the third tissues in the cascade. The tissues were precontracted with a prostaglandin analogue U46619 (15 nM). [From Gryglewski *et al.* (1986) with permission.]

relaxation. NO is also an important mediator in platelet activation and immunological reactions.

B. Biological Variation

Studies of the effects of drugs on living systems require careful design because living systems display *biological variation*. No two people are alike, and no two preparations of biological tissue can be considered identical either. For example, it is likely that the β_2-adrenoceptors of all laboratory rats have exactly the same amino acid sequence.

However, owing to possible variations in experience, nutritional state, hormonal balance, etc., different preparations will respond in a slightly different way to a β_2-adrenoceptor agonist. On top of this, variations in the experimenter's technique must be taken into account. Repetition of pharmacological experiments is therefore essential, not only to avoid being misled by variation between different preparations, but also to avoid the effects of day-to-day variation in experimental technique. In designing pharmacological experiments it is important to bear this in mind. For example, if control responses and test responses are taken in the same tissue, the test response can be expressed as a percentage of the control response. Where the absolute magnitude of responses in each tissue may be quite different, results can be normalized in this way and combined. If during each experiment control responses are taken before *and* after each test response, then each preparation can be checked for changes in drug sensitivity during the experiment.

III. Quantifying Drug–Receptor Interactions

The theory of drug–receptor interactions was developed using the Law of Mass Action. This law states that the rate of any reaction is proportional to the concentrations of the reactants. It can be used to derive a relationship that, in many cases, describes drug binding to receptors.

A drug molecule D is assumed to combine with a receptor, R, to form a drug–receptor complex, DR. By the Law of Mass Action, the rate of this reaction is proportional to the product of the concentrations of D and R. Thus

$$\text{Rate} \ \alpha \ [D][R] \tag{1}$$

If we call the proportionality constant k_{+1} this means

$$\text{Rate} \ = \ k_{+1}[D][R] \tag{2}$$

Where [D] represents the concentration of drug molecules in solution and [R] is the concentration of free (unoccupied) receptors. The constant k_{+1} is known as the *microscopic association rate constant* (this is a first-order rate constant with dimensions of $M^{-1}s^{-1}$). The reverse reaction (*dissociation* of drug from receptor) is described by the microscopic dissociation constant k_{-1} (dimensions s^{-1}). In short this reaction is written

$$D + R \underset{k_{-1}}{\overset{k_{+1}}{\rightleftharpoons}} DR \tag{3}$$

At equilibrium the rate of the forward reaction ($k_{+1}[D][R]$) will be equal to the rate of the reverse reaction $k_{-1}[DR]$ so

$$k_{+1}[D][R] \ = \ k_{-1}[DR] \tag{4}$$

Drugs are characterized by their *dissociation equilibrium constant* K_D which is the ratio of dissociation to association rate constants. Alternatively, drugs are sometimes characterized by their *affinity* or *association constant*, which is the reciprocal of K_D. Rearranging Eq. (4) gives

$$K_D \ = \ \frac{k_{-1}}{k_{+1}} = \frac{[D][R]}{[DR]} \tag{5}$$

Note that K_D has units of concentration.

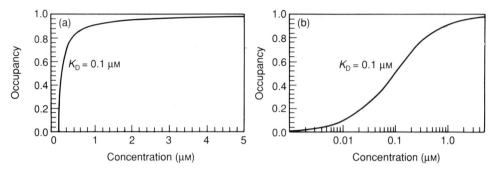

Fig. 3.2 Characteristic shape of the rectangular hyperbola. The curves show the relationship between drug concentration and receptor occupancy for a hypothetical drug–receptor interaction with a $K_D = 0.1\,\mu M$. (a) Hyperbola plotted on linear scales illustrates the rapid rise in occupancy at low drug concentration and the slow approach to saturation at high concentrations. (b) With a logarithmic concentration scale the hyperbola takes on the familiar symmetrical sigmoid shape. For many drug-mediated effects, the relationship between concentration and response often has the sigmoid shape. Nevertheless, the relationship between occupancy and response is usually complex.

It is not very useful to consider the 'concentration of receptors' since receptors are not free in solution but embedded in the cell membrane in most cases. It is more useful to consider this situation in a different way. If the total *number* of receptors, N_{tot}, reacts with a concentration, x_D of drug D to produce N_D drug–receptor complexes, then the fraction of receptors occupied by the drug will be

$$p_D = \frac{N_D}{N_{tot}} \tag{6}$$

which is known as the *occupancy*. The number of free receptors is therefore $N_{tot} - N_D$. Substitution into Eq. (5) gives

$$K_D = \frac{k_{-1}}{k_{+1}} = \frac{x_D(N_{tot} - N_D)}{N_D} \tag{7}$$

It is assumed here that because the number of drug molecules in solution is vastly greater than the number of receptors, x_D is not significantly changed by binding of drug to the receptor. Rearranging Eq. (7) gives the occupancy

$$p_D = \frac{N_D}{N_{tot}} = \frac{x_D}{K_D + x_D} \tag{8}$$

Eq. (8) is known as the *Hill–Langmuir equation*. It was derived by the physiologist A.V. Hill to describe the binding of nicotine in muscle (Hill, 1909), and later by the physical chemist Langmuir to describe the adsorption of gases on to metal surfaces. Notice that when $p_D = 0.5$, $x_D = K_D$. Thus the dissociation equilibrium constant is that concentration of drug that will produce 50% receptor occupancy.

This relationship between drug concentration and occupancy given by Eq. (8) generates a curve which is (part of) a rectangular hyperbola (Fig. 3.2a). In pharmacology, drug concentration is generally plotted on a logarithmic scale which converts the rectangular hyperbola to a sigmoid shape (Fig. 3.2b).

IV. The Hill Equation

The Hill equation is an empirical generalization of the Hill–Langmuir equation [Eq. (8)].

$$p_D = \frac{x_D^n}{K_D^n + x_D^n} \tag{9}$$

where the *Hill coefficient*, n, is a measure of the steepness of the curve [$n = 1$ in Eq. (8)]. For graphical purposes Eq. (9) is often rearranged in the form

$$\log \frac{p_D}{1 - p_D} = n \log x_D - n \log K_D \tag{10}$$

A log–log plot of $p_D/(1 - p_D)$ versus x_D (a *Hill plot*) will be linear, with a slope (n) of 1.0 if there is no cooperativity in drug binding to a receptor. If the Hill slope is greater than 1.0 this indicates positive cooperativity and if less than 1.0 suggests that possibly negative cooperativity is occurring or the presence of multiple receptor types or that an interfering process such as desensitization (see below) is occurring (*desensitization* is the tendency of an agonist response to wane despite the continued presence of the agonist).

A specific use of the Hill plot is in analysing the molecularity of the agonist–receptor reaction for ligand-gated ion channels. Where the response of this class of receptors (formation of open ion channels) can be measured as a current or conductance (in voltage clamp experiments), then this will be directly proportional to receptor occupancy. In this situation the slope of the Hill plot {a plot of $\log[y/(max - y)]$ versus $\log x_D$, where y is the response produced by concentration x and *max* is the maximum response} can suggest a lower limit on the number of agonist-binding sites on the receptor. In practice, desensitization often prevents measurement of a true maximum and so an approximation is used: $\log y$ versus $\log x_D$ is plotted assuming that the occupancy (p_D) is small [$p_D/(1 - p_D)$ is close to p_D] so that $y/(max - y)$ approximates y/max. Since $\log(y/max) = \log y - \log max$, the slope of this log–log plot is the same as the Hill plot. It turns out that, at low agonist concentrations (where the occupancy is small), the slope of this plot will be close to the expected number of agonist-binding sites on the receptor for quite a wide range of agonist mechanisms (see, for example, Section XA).

V. Radioligand-binding Studies – A Direct Measure of Occupancy

The theoretical basis for radioligand-binding studies is the same as used to derive the Hill–Langmuir equation [Eq. (8)]. In principle, receptor occupancy can be measured using a radiolabelled drug which can be either agonist or antagonist (usually labelled with ^3H, ^{35}S or ^{125}I). At equilibrium, the number of drug molecules bound (B) is a function of the drug concentration x_D

$$B = B_{max} \frac{x_D}{x_D + K_D} \tag{11}$$

and B_{max}, the total number of binding sites in the tissue. This relationship is often used in the linearized form

$$\frac{B}{x_D} = \frac{B_{max}}{K_D} - \frac{B}{K_D} \tag{12}$$

which, when plotted as B/x_D versus B, is known as a *Scatchard plot*. This has a slope of $1/K_D$ and an x-axis intercept of B_{max}. Although the Scatchard plot is a convenient way to display the results of radioligand-binding studies, linearizing transformations are statistically an unreliable procedure (Colquhoun, 1971). A weighted least-squares curve-fitting method is preferable for this procedure.

The main requirement for ligand-binding studies is to have a ligand that binds with sufficiently high affinity ($K_D < 10$ nM) and specificity to identify clearly the receptors of interest. Non-specific binding (drug binding to anything other than the receptor of interest) is measured by performing the binding experiment in the presence of an excess of non-radioactive ligand which will hopefully prevent the radioligand binding to the receptor but not prevent the radioligand binding to the non-specific sites. Specific binding is then calculated by subtracting the non-specific from total binding. It is the relationship between ligand concentration and specific binding which is often plotted as a Scatchard plot to estimate K_D and B_{max}. It should be born in mind that because of possible receptor conformational changes subsequent to ligand binding, binding studies will not necessarily give a true measure of K_D (see Section X). Nevertheless radioligand-binding studies are now very widely used to study many different types of receptor. When combined with functional studies, radioligand binding becomes a very powerful technique because it is a direct measure of occupancy.

Sometimes binding studies can identify a binding site that does not immediately have a functional correlate. An interesting example of this is in the 5-HT receptor field (Bradley *et al.*, 1986). Gaddum originally designated 5-HT receptors as D or M types based on studies in the guinea pig ileum. More recently various radioligand-binding sites for 5-HT have been identified in brain tissue based on relative affinities for several new drug tools. Two sites have high affinity for lysergic acid diethylamide (LSD). At one site (termed the 5-HT$_1$ receptor) 5-HT also has a high affinity while at the other (5-HT$_2$ receptor) spiperone has a high affinity. Subsequent studies demonstrated that 5-HT$_2$ receptors correlate well with effects mediated by the D receptor.

The 5-HT$_1$-binding site was found to be heterogeneous. Initial studies based on the use of spiperone and 8-hydroxy-2-(di-*n*-propylamino)tetralin (8-OH-DPAT) and mesulergine suggested there should be 5-HT$_{1A}$-, 5-HT$_{1B}$- and 5-HT$_{1C}$-binding sites, and subsequently the use of sumatripan identified a 5-HT$_{1D}$-binding site.

None of the 5-HT$_1$-binding sites correlated with the functional D and M receptors and so the M receptor was designated 5-HT$_3$ on the basis of results obtained with the selective agonist 2-methyl-5-HT and the antagonist ondansetron. Interestingly, a binding site equivalent to the 5-HT$_3$ receptor was not identified. In addition, a 5-HT$_4$ receptor was recently identified.

For several years during the 1980s functional correlates of the four 5-HT$_1$-binding sites were not found. However, functional correlates for these receptors have now been discovered, particularly in the cerebral arteries (Parsons, 1991), and with the use of functional studies, an additional 5-HT$_1$ receptor has been identified for which there is no clear binding correlate. This receptor termed '5-HT$_1$-like' has been found in both coronary and cerebral arteries (Parsons, 1991; Saxena and Villaton, 1991).

This fascinating story of 5-HT receptor discovery and classification continues with several 5-HT receptors having now been cloned (Table 3.1) and identified using binding and functional data. The extensive diversity of 5-HT receptors, first identified using

TABLE 3.1
Some Examples of Ion Channel and G-protein-coupled Receptors

Agonist	Receptor	Effector mechanism	Cloned?
Acetylcholine	Nicotinic (muscle/neuronal)	Ion channel (Na^+, K^+, Ca^{2+})	Yes
	Muscarinic-m_1, m_3, m_5	G-proteins couple via PLC and IP_3 to Ca^{2+} regulation	Yes
	$-m_2$, m_4	G_o (inhibits Ca^{2+}/K^+ channels)	Yes
Glutamate	NMDA	Ion channel (Na^+, K^+, Ca^{2+})	Yes
	Non-NMDA	Ion channel (Na^+, K^+)	Yes
	Metabotropic	G-proteins (IP_3 and Ca^{2+})	Yes
GABA	$GABA_A$	Ion channel (Cl^-)	Yes
	$GABA_B$	G_o (inhibits Ca^{2+}/K^+ channels)	Not yet
Glycine	Glycine	Ion channel (Cl^-)	Yes
5-HT	5-HT_{1a-d}, 5-HT_2	G-proteins, IP_3 or cAMP	Yes
	5-HT_4	(G_s) cAMP	Not yet
	5-HT_3	Ion channel (cation)	Yes
Dopamine	D_{1a}, D_{1b}, D_{2-5}	G-proteins	Yes
Noradrenaline and adrenaline	α_1, α_2	(G_o) Ca^{2+} and K^+ channel inhibition	Yes
	β_1, β_2, β_3	(G_s) cAMP and PKA	Yes
Histamine	H_1, H_2	(G_s) IP_3 or cAMP	Yes
	H_3	(G_s) and cAMP	Not yet
Adenosine	A_1	(G_i) Adenylate cyclase inhibition	Yes
	A_2	(G_s) Adenylate cyclase activation	Yes
ATP	P_{2X}, P_{2Z}	Ion channel (Na^+, K^+)	Not yet
	P_{2Y}, P_{2T}	IP_3 or cAMP	Not yet

Abbreviations: G_i, inhibitory G-protein (couples to adenylate cyclase); PLC, phospholipase C; IP_3, 1,4,5-inositol trisphosphate; NMDA, N-methyl-D-aspartate; GABA, γ-aminobutyric acid; 5-HT, 5-hydroxytryptamine (serotonin); G_s, stimulatory G-protein; cAMP, cyclic adenosine 3'-5'-monophosphate; ATP, adenosine triphosphate; PKA, protein kinase A (cAMP-dependent protein kinase). It should be noted that in different cell types, some G-protein receptors may couple to different G-proteins and hence link to different second messenger systems. The cloning information was compiled in October 1991.

radioligand-binding studies, has therefore now been confirmed by molecular genetic investigations.

VI. Receptor Structure

The idea that drugs interact with specific receptive substances, present in biological tissues, was developed long before there was any evidence regarding the physical nature of receptors. In his classical paper in 1956 (A modification of receptor theory) Stephenson said, "What a drug combines with to produce its effect is a subject for speculation." By the 1970s the difficult and time-consuming procedures of protein chemistry had allowed the structure of a few receptors to be studied, particularly the nicotinic acetyl-choline receptor (AChR). However, recent applications of biochemical and molecular genetic techniques and X-ray crystallographic studies have resulted in spectacular advances in knowledge of receptor structure (Unwin, 1989).

Receptors are often classified into two groups: the ligand-gated ion channels or the G-protein-coupled receptors (Table 3.1). These two groups constitute by far the majority of receptors which are important pharmacological targets. There are, however, other receptor classes such as the receptors for insulin and epidermal growth factor which have intrinsic tyrosine kinase activity, receptors that transport ligands across the cell membrane such as the low-density lipoprotein receptor or the transferrin receptor, and the steroid receptors which influence DNA transcription.

A.   Nicotinic AChR Structure: A Ligand-gated Ion Channel

Of the ligand-gated ion channels, by far the best studied is the nicotinic AChR (Fig. 3.3). Most studies have utilized the AChR from the electric organ of the Californian ray, *Torpedo californica*. This electric organ is an exceedingly rich source of receptors for isolation and biochemical study. By utilizing α-bungarotoxin (a component of cobra snake venom with extremely high affinity for the nicotinic AChR) it was possible in the late 1970s and early 1980s to isolate and characterize this receptor. The *Torpedo* AChR was found to be composed of four different protein subunits in a stoichiometry $\alpha_2\beta\gamma\delta$ arranged in a pseudosymmetrical fashion around a central ion channel pore [see Unwin (1989) for review] (Fig. 3.3). Each subunit is thought to cross the cell membrane four times giving four transmembrane (TM) domains which are numbered from the amino-terminus of the protein. Since five subunits make up each receptor, there are 20 trans-membrane domains per receptor (Fig. 3.3). The amino acid residues in TM2 of each subunit line the central ion channel and determine its conductance properties (Imoto *et al.*, 1988).

ACh, tubocurarine and other nicotinic receptor ligands were found to compete with α-bungarotoxin for its binding site on the α-subunit. The α-subunit was found to be unique in having two cysteine residues at amino acid positions 192 and 193. Subse-quently, molecular genetic techniques were used to isolate and sequence multiple subtypes of neuronal (found in the central nervous system and in autonomic ganglia) nicotinic receptor subunits (Deneris *et al.*, 1991), and all α-subunits were found to contain two cysteine residues at positions analogous to 192 and 193. These cysteine residues are located before the first transmembrane domain in the extracellular part of the protein and the agonist-binding site is thought to be close by, perhaps in a shallow cleft between the α- and adjacent subunits.

The neuronal nicotinic receptors are different from muscle receptors in a number of ways, including the fact that they have a higher sensitivity to nicotine and are not sensitive to α-bungarotoxin. In contrast with the α-, β-, γ- and δ-subunits of electric organ and muscle AChRs, neuronal nicotinic receptor subunits at present are divided into α- or β-subunits (Deneris *et al.*, 1991). The exact stoichiometry of neuronal nicotinic receptors is unknown but is probably $\alpha_2\beta_3$. Because seven different α- and five different β-subunits are now known, the possibilities for variation in neuronal nicotinic receptor structure are already enormous. A similar situation has arisen with the receptors for the inhibitory amino acid γ-aminobutyric acid (GABA) and with the excitatory amino acid

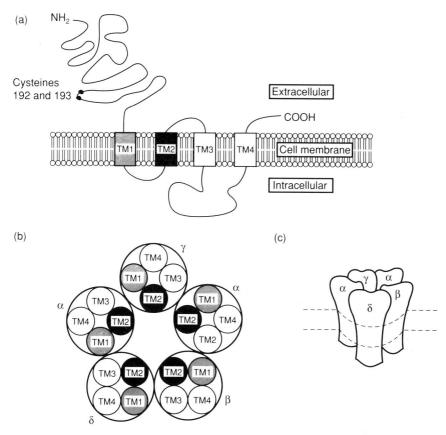

Fig. 3.3 Schematic diagram of the structure of the nicotinic AChR. This general structure is thought to hold for all ligand-gated receptors. The receptor is composed of five subunits each with the general structure shown in (a). (a) Schematic arrangement of a nicotinic AChR α-subunit. The four transmembrane domains (TM1–TM4) are shown crossing the cell membrane. The large extracellular domain at the amino-terminus of the α-subunit is where the main determinants of agonist and antagonist binding are located. (b) Arrangement of the AChR subunits as seen when viewed from above the cell membrane. Each subunit is thought to be arranged so that transmembrane segment TM2 lines the central ion channel of the receptor (Imoto et al., 1988). (c) Schematic of the three-dimensional structure of the nicotinic AChR as it sits in the cell membrane. [(a) and (b) adapted from Huganir and Greengard (1990) with permission.]

receptors of the non-*N*-methyl-D-aspartate (NMDA) type. Multiple subtypes of the subunits that make up the non-NMDA receptor and the $GABA_A$ subtype (Table 3.1) of GABA receptors have been cloned. In each case ten or more different subunits (based on amino acid sequence) are known.

B.   *β*-Adrenoceptor Structure: A G-protein-coupled Receptor

The *β*-adrenoceptor is one of the group of receptors that are coupled via G-proteins (GTP-binding proteins) to their effector system (Table 3.1). They are perhaps the largest and most diverse group of receptor proteins. Their effector systems include adenylate cyclase (stimulated by G_s and inhibited by G_i proteins), guanylate cyclase, phospholipase C (PLC), phospholipase A_2 (PLA_2) phosphodiesterases and Ca^{2+} and K^+ channels.

The 'prototype' for G-protein-coupled receptors is retinal rhodopsin. Rhodopsin is the most studied G-protein receptor because of its abundance in halobacteria. The ligand for this specialized receptor is retinal. Absorbtion of a photon of light activates rhodopsin which, via the G-protein transducin, activates a phosphodiesterase causing hydrolysis of cGMP and so indirectly causes the closure of the cGMP-activated channels in the photoreceptor membrane. Bacteriorhodopsin has been crystallized and its structure elucidated (Henderson and Unwin, 1975). It is a single polypeptide with seven transmembrane (TM)-spanning regions (numbered 1–7 from the amino-terminus of the protein). All G-protein-coupled receptors such as the β-adrenoceptors (Strader et al., 1989) have been proposed to have analogous structures on the basis of amino acid sequence homology (Fig. 3.4). They are single polypeptide receptors of approximately 400 amino acids. Other examples of this receptor type which have been cloned and sequenced include the muscarinic AChRs, the dopamine receptors, the histamine receptors and the 5-HT receptors (with the exception of the 5-HT$_3$ receptor which is a ligand-gated ion channel). In each case the intracellular domain between TM5 and TM6 is important for G-protein binding. Unlike the ligand-gated channels, agonist and antagonist binding occurs within the transmembrane domains involving particularly amino acids in TM3, TM5 and TM6 (Fig. 3.4). The receptor is thought to fold round on itself (Fig. 3.4c) so that a pocket is formed within the transmembrane domains. This deep binding within the receptor structure, where there may be scope for multiple interactions between agonist and receptor, perhaps accounts for the very high (nanomolar) affinity often found for agonist binding to G-protein-coupled receptors compared with the relatively low affinity (micromolar) of agonists for the ligand-gated receptors, where the agonist-binding site is thought to be more superficial.

The rapid advances made possible by molecular genetic techniques have resulted in the cloning and sequencing of a wide range of receptors (Table 3.1). This has made structural comparisons between receptors possible which should lead to a much more detailed understanding of the factors that determine why one receptor is, for example, selectively activated by muscarine and another by histamine, even although both receptors are of a similar overall structure.

VII.   Relating Occupancy to Response

Eq. (3) could be described as a *model* or a *mechanism* for drug binding to a receptor. However, it is rather a simple model since it gives no indication of how the occupied receptor might generate a response. Thus Eq. (3) may be quite sufficient to describe the action of an antagonist, but requires extension to describe agonist action adequately.

A.   Ligand-gated Ion Channels

Receptors like the nicotinic AChR or the GABA$_A$ receptor have both the agonist-binding site and the transduction mechanism (ion channel) as part of a single macromolecule. Relative to receptors which are linked to second messenger systems, the ligand-gated ion channels are simpler to study.

A simple extension to Eq. (3) can allow it to become a useful model of receptor

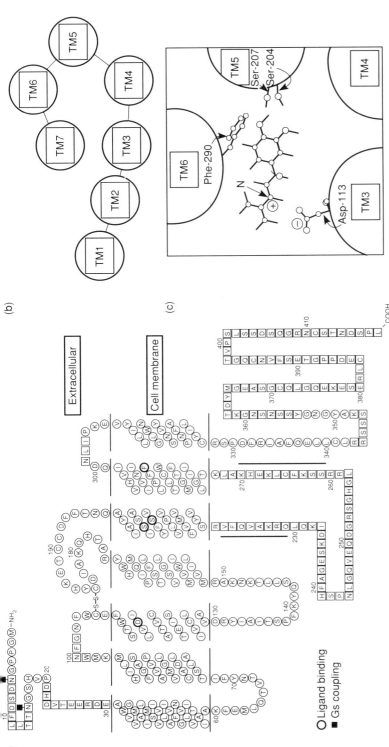

Fig. 3.4 Structure of the β-adrenoceptor (βAR). (a) Model of the transmembrane topology of the βAR. The protein amino-terminus is extracellular and the carboxy-terminus is intracellular. The horizontal lines indicate the likely limits of the plasma membrane. The receptor has seven transmembrane helices which are numbered from TM1 nearest to the amino-terminus to TM7 nearest to the carboxy-terminus. The amino acid residues represented by squares can be deleted without apparently affecting ligand binding or protein folding (Strader et al., 1989), whereas deletion of those in circles adversely affects folding or ligand binding. The residues shown in bold circles in transmembrane domains 3, 5 and 6 are proposed to be directly involved in interacting with βAR agonists (c). The regions of the third intracellular loop between TM5 and TM6 delineated by the solid black bar are critical for G-protein coupling. (b) Schematic of the arrangement of the seven transmembrane helices of the βAR as seen from above the cell membrane labelled TM1 to TM7 from the amino-terminus. (c) Detailed view of the arrangement of the transmembrane domains around the ligand-binding site of the βAR. The positioning of the helices is in accordance with the structure of bacteriorhodopsin (Henderson and Unwin, 1975). The residues that are suggested by mutagenesis studies to interact with the agonist isoprenaline are indicated by arrows. The structure of isoprenaline in the proposed binding pocket is shown with the amino nitrogen arrowed. It is almost certain that several other amino acids in the receptor will interact with agonists to a greater or lesser extent. [(a) and (c) adapted with permission from Strader et al. (1989).]

activation. This was first used to describe AChR activation by del Castillo and Katz (1957).

$$D + R \underset{k_{-1}}{\overset{k_{+1}}{\rightleftharpoons}} DR \underset{a}{\overset{b}{\rightleftharpoons}} DR* \tag{13}$$

Eq. (13) describes how the agonist drug, D, combines with receptor, R, to form the agonist–receptor complex (DR) which then undergoes a conformational change which opens (activates) the AChR ion channel (the open channel is denoted by DR*). Channel opening is described by a rate constant b and channel closing by a rate constant a. Eq. (13) describes a mechanism for the nicotinic AChR where the receptor can exist in only three *discrete* states. Two of these states are obviously going to be different: liganded and unliganded. The open state is also structurally different because it is where the receptor ion channel is open. If, for example, receptor desensitization were to be included in this mechanism, then an additional state (a desensitized state) would need to be added. In general, receptors are thought to exist in only a few discrete states.

The equilibrium fraction of occupied receptors in the closed state will be

$$p_{DR} = \frac{x_D/K_D}{1 + x_D/K_D(1 + b/a)} \tag{14}$$

However, the *response* of the receptor to agonist occupancy is p_{open}, the fraction of time for which an individual channel is open (DR*) or the fraction of a population of channels which are open at equilibrium. This is

$$p_{open} = \frac{x_D/K_D(b/a)}{1 + x_D/K_D(1 + b/a)} \tag{15}$$

This simple scheme provides a plausible mechanism for activation of the AChR (but see Section XA).

B.   Receptor Mechanisms that Involve Second Messengers

With receptors that couple to second messenger systems, the transduction mechanism is often much more complex than for the ligand-gated ion channels. For this reason it is generally not possible to postulate any specific mechanism linking receptor occupancy to response.

It is quite often the case that a long chain of events will separate occupation of the receptor from the actual response being measured. A good example of this might be the increase in heart rate caused by β-adrenoceptor agonists such as adrenaline which forms part of the 'fight or flight' response (see Chapter 1). The β-adrenoceptors in the heart are of the β_1-subtype. When adrenaline occupies the β_1-receptors of the heart, the G-protein (G_s) [a guanine-nucleotide-binding protein; see Neer and Clapham (1988) for review, or Birnbaumer *et al.* (1990) for a review of G-protein coupling systems] associated with the β_1-receptor releases guanosine diphosphate (GDP) and binds one molecule of GTP. This causes dissociation of the G-protein α-subunit (α_s) which can then bind to and activate adenylate cyclase (Fig. 3.5). Adenylate cyclase catalyses the conversion of ATP to cAMP. cAMP activates specific cAMP-dependent protein kinases which by phosphorylation of calcium channels in the heart sino-atrial (SA) node cells increases the calcium current into these cells during each action potential and so speeds up the rate of firing of the SA node cells. Because the SA node cell firing determines the heart rate, β-adrenoceptor agonists cause an increase in heart rate.

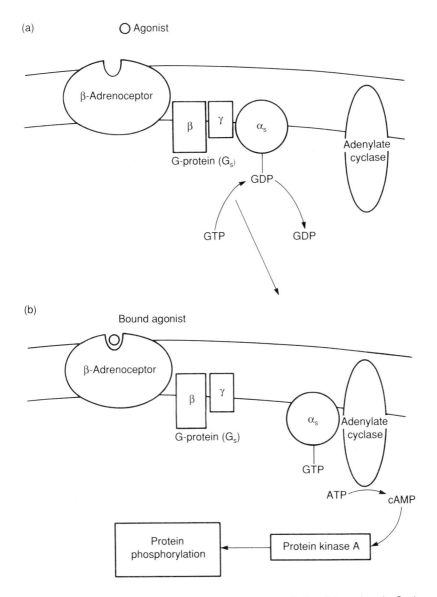

Fig. 3.5 Schematic representation of the mechanism of activation of adenylate cyclase by β-adrenoceptor agonists. (a) Resting state of the β-adrenoceptor and associated G-protein. The β-adrenoceptor couples to the G-protein, G_s. G_s is composed of three subunits denoted α_s, β and γ. On binding an agonist molecule the β-adrenoceptor conformation changes such that G_s loses one molecule of GDP and binds GTP (b). The α_s subunit of G_s then dissociates from the G-protein β and γ subunits and on collision with adenylate cyclase causes activation of the enzyme and conversion of ATP to cAMP. cAMP can then activate protein kinase A to cause phosphorylation of a variety of target proteins. Note that other receptors such as the muscarinic m_2 and m_4 receptors (Table 3.1), by coupling with G_i, can cause inhibition of adenylate cyclase whereas by coupling with G_o can cause inhibition of Ca^{2+} or K^+ channels.

The action of the β-adrenoceptor agonist is terminated because the GTP attached to the α_s-subunit is slowly hydrolysed back to GDP. With an attached GDP, α_s can then associate with the G-protein β- and γ-subunits again, until the next collision with an activated β-receptor occurs (Fig. 3.5).

Consider for a moment some of the points in this process where it might be possible to measure the response of the system to a β-adrenoceptor agonist: (i) heart rate, (ii) cell calcium current, (iii) cAMP level, (iv) adenylate cyclase activity. With receptors that couple to second messenger systems it cannot be assumed that in different preparations the relationship between receptor occupancy and response will be the same, even though the receptors expressed in each preparation are the same.

VIII. Competitive Antagonism and the Schild Equation

Drugs are often characterized as being either *agonists* or *antagonists*. Agonists have the ability to evoke a response on binding to the receptor while antagonists block the response of the receptor to an agonist, without evoking any response themselves. Acetylcholine and nicotine are agonists at the nicotinic AChR, whereas tubocurarine (one of the main components of the South American Indian arrow poisons) is a nicotinic antagonist used as a muscle relaxant during surgery. The β_2-adrenoceptor agonist salbutamol is used by asthmatics to dilate the airways while the β-adrenoceptor antagonist propranolol is widely used as an antihypertensive. The Schild equation provides a method of quantifying the actions of an antagonist. Many fundamental advances in pharmacology occurred as the result of the development of quantitative methods for the characterization of antagonist actions.

When an agonist drug (D) and antagonist (B) at concentrations of x_D and x_B are in equilibrium with a single population of receptors, and the binding of D and B are mutually exclusive then the fraction of receptors occupied by the agonist will be

$$p_D = \frac{x_D/K_D}{(x_D/K_D) + (x_B/K_B) + 1} \tag{16}$$

where K_B is the dissociation equilibrium constant of the antagonist. In effect, the antagonist reduces the occupancy of the receptor by the agonist. If the agonist concentration is increased to say x'_D such that agonist occupancy is restored to that level obtained by x_D in the absence of antagonist, then the factor by which the agonist concentration has to be increased

$$r = \frac{x'_D}{x_D} \tag{17}$$

is known as the *dose ratio*. For a competitive antagonist Schild demonstrated that

$$r = \frac{x_B}{K_B} + 1 \tag{18}$$

This simple result, known as the *Schild equation* (Arunlakshana and Schild, 1959), is obtained by assuming that equal agonist occupancies (whether in the presence or absence of antagonist) will always give equal responses. Equating the occupancies in the absence and presence of antagonist gives

$$\frac{x_D}{K_D + x_D} = \frac{x'_D/K_D}{(x'_D/K_D) + (x_B/K_B) + 1} \tag{19}$$

which can be rearranged to give Eq. (18). The importance of the Schild equation is that the dose ratio, r, is a characteristic *only* of the antagonist (neither the agonist concentration nor the agonist affinity appear in the Schild equation). The relationship between r and antagonist concentration means that a competitive antagonist will produce a parallel rightward shift of the log-concentration–response curves with no change in maximum response. The Schild equation is usually plotted on log–log scales giving

$$\log (r - 1) = \log x_B - \log K_B \tag{20}$$

Thus a plot of $\log(r - 1)$ versus $\log x_B$ (the *Schild plot*) for a competitive antagonist will give a straight line with slope of 1 and an intercept on the x-axis of $\log K_B$. The x-axis intercept is usually called the pA_2 value. Where a slope of 1 is found for experimental data, the pA_2 value will be the negative logarithm of the dissociation equilibrium constant (K_B) for the antagonist (see for discussion, Jenkinson, 1991). Verifying that the slope of a Schild plot for a particular antagonist is unity over a wide range of antagonist concentrations is by far the best way of demonstrating competitive antagonism. The Schild method is independent of agonist concentration and therefore a particular antagonist should give the same Schild plot for all agonists which act on the same population of receptors.

Where a Schild slope of greater than unity is encountered, this may be indicative of *non-competitive* antagonism: the situation where raising the agonist concentration cannot completely overcome the effect of the antagonist. This situation can arise when agonist and antagonist binding are not mutually exclusive. Figure 3.6 shows an example of the use of the Schild method to determine the K_B for tubocurarine block of nicotinic AChRs at the frog neuromuscular junction (Colquhoun *et al.*, 1979). The main action of tubocurarine is a competitive block of the AChR. However, as illustrated in Fig. 3.6, tubocurarine also blocks the AChR ion channel which is a non-competitive effect.

A. Drug Block of Open Ion Channels: A Non-competitive Antagonism

Apart from drug effects at receptors, the site of drug action that has received most attention in recent years has been the open ion channel. Several major classes of drugs mediate their effects by blocking specific ion channels in cell membranes. Prominent among these are the local anaesthetics such as lignocaine which block sodium channels, the dihydropyridine Ca^{2+} channel blockers such as nifedipine (Rampe and Triggle, 1990) and the K^+ channel blockers and openers (Castle *et al.*, 1989).

By blocking axonal Na^+ channels, the local anaesthetics inhibit conduction in pain fibres and hence cause a local anaesthesia. Lignocaine is also used to reduce arrhythmias of the heart where it blocks Na^+ channels in cardiac muscle causing a slowing of the cardiac action potential. The Ca^{2+} channel blockers on the other hand are used mainly in the treatment of angina and hypertension. Several K^+ channel openers are at present under development as potential antihypertensives because of their potent ability to relax vascular smooth muscle (see chapter 8).

IX. Desensitization and the Control of Receptor Number

Desensitization can be defined as the tendency of a drug response to become smaller with repeated doses, or during a single constant application of drug. It generally occurs with

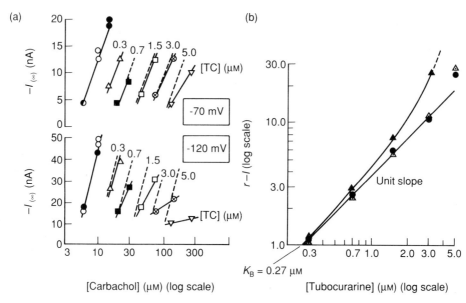

Fig. 3.6   Use of the Schild method to determine the K_D for tubocurarine (TC) blockade of nicotinic AChRs at the frog neuromuscular junction. (a) Partial log concentration–reponse curves for net inward current $(-I_{(\infty)})$ evoked by the agonist carbachol in the presence of increasing concentrations of tubocurarine. In the top panel the responses are recorded at a membrane potential of -70 mV and it can be seen that, except at the highest tubocurarine concentration (5 μM), the antagonist produces an approximately parallel shift in the dose–reponse curves as expected for competitive antagonism. The lower panel shows the same experiment but at a membrane potential of -120 mV. When the inside of the cell is made more negative with respect to the outside, the shift in dose–response curves is far from parallel. This is because the positively charged tubocurarine molecule is being attracted into the AChR ion channel by the negative membrane potential, increasing the channel-blocking action of tubocurarine: a non-competitive effect. The dashed lines in each panel show the responses predicted for pure competitive antagonism with $K_B = 0.27$ μM (b). Dose ratios were calculated at a response level of -8 nA at -70 mV and at a response level of -24 nA at -120 mV. (b) Schild plot of dose ratio (r) against tubocurarine concentration (x_B) according to Eq. (18). Filled circles: dose ratios from equilibrium response at -70 mV. Filled triangles: dose ratios from equilibrium response at -120 mV. Open triangles: dose ratios from peak responses at -120 mV. Because the non-competitive open channel block by tubocurarine is slow to develop, if the peak response is measured, then mainly competitive antagonism is seen and the Schild slope is close to unity. The fact that both curves coincide at low agonist concentrations (small dose ratios), suggests that the K_B for competitive binding to the receptor is independent of membrane potential as might be expected if the agonist-binding site is on the outermost part of the receptor molecule, outside the membrane potential field. [Adapted from Colquhoun et al. (1979), with permission.]

agonist responses and is absent with antagonists. Desensitization is probably a general phenomenon although it varies widely in extent and rate of appearance with different receptor systems. One example might be the tolerance that develops over a few days when repeated doses of opiate analgesics such as morphine are given to a patient: in order to obtain similar pain relief at each administration, the dose has to be stepped up. After cessation of treatment, several days may be needed to recover full opiate sensitivity. Another example occurs with responses to β-adrenoceptor agonists which desensitize over a period of a few minutes to an hour and then require a similar period for recovery. A much faster desensitization is observed with the ligand-gated receptors. The response of these receptors to a high agonist concentration can desensitize in less than one second, but recovery occurs over a period of seconds to minutes. The mechanisms involved in desensitization of any receptor type are not fully understood. However, it is clear that desensitization of the ligand-gated ion channel receptors involves conformational changes in the receptor-channel protein itself where the receptor enters a desensitized state. This can be expressed in a model as an extension to the simple agonist mechanism [Eq. (13)]:

$$D + R \underset{k_{-1}}{\overset{k_{+1}}{\rightleftharpoons}} DR \underset{a}{\overset{b}{\rightleftharpoons}} DR^* \underset{d_-}{\overset{d_+}{\rightleftharpoons}} DR_d \qquad (21)$$

In this mechanism the receptor forms the desensitized state DR_d from the open state with a rate constant d_+, and recovers from desensitization with a rate constant d_-. While in the desensitized state, the receptor cannot be activated by ACh and so desensitization reduces the size of the ACh response.

In fact, a more complex mechanism (the *cyclic model* of receptor desensitization) is necessary to describe nicotinic AChR desensitization. In the cyclic model, the desensitized state of the receptor can lose its agonist molecule and then revert to the resting state without going through the open state.

Desensitization of receptors which couple to second messenger systems generally involves both alterations in the transduction system (or coupling between receptor and transduction system), and changes in the number of receptors on the cell surface. In the case of the β-adrenoceptors this involves internalization of receptors which are subsequently re-inserted into the plasma membrane. With the β-adrenoceptor and probably many other G-protein-coupled receptors, desensitization involves phosphorylation of one of the internal domains of the receptor (Lefkowitz *et al.*, 1990).

X. Partial Agonists and Agonist Efficacy

The concept of *partial agonism* was introduced by Ariens ($\sim$ 1950) and used by Stephenson (1956) to provide a general method of dealing with the situation where the relationship between receptor occupancy and tissue response is different for different agonists. Stephenson distinguished between the ability of an agonist to *bind* to the receptor (measured by its equilibrium dissociation constant, or affinity constant), and its ability to elicit a response once bound (measured by an empirical constant termed the *efficacy* of the agonist). This concept conveniently accounts for the experimental observation that, in a single tissue, two different agonists, acting on the same receptors, will often generate different maximal responses from the tissue. Even at 100% receptor occupancy one agonist may generate a much smaller response than a different agonist. Associated with the idea of efficacy is the concept of *spare receptors*. A high efficacy agonist may be able to elicit a maximal response from the tissue when occupying only a small fraction of receptors (hence leaving *spare* receptors) whereas an agonist of low efficacy may be unable to generate a maximal tissue response, even at 100% receptor occupancy [see also Colquhoun (1987)].

The term *intrinsic activity* is sometimes used synonymously with efficacy. This term was introduced by Ariens along with the concept of partial agonism. The intrinsic activity of drugs was suggested to range from 0 to 1. However, this was a more restrictive term than efficacy as defined by Stephenson because the intrinsic activity of all full agonists was assumed to be 1 whereas efficacy may vary from zero to a large positive value.

The difference between agonists, partial agonists and antagonists becomes simply a matter of efficacy: antagonists have zero efficacy, partial agonists have low efficacy and full agonists have high efficacy. For example, different β-adrenoceptor agonists may, by occupying the receptor in slightly different ways, cause more or less G-protein activation. An interesting example of this occurs with the β-adrenoceptor agonist salbutamol. This agonist selectively activates the β_2-adrenoceptor subtype present in the airways relative to β_1-adrenoceptor activation in the heart. Yet there is less than a 2-fold difference in the affinity of salbutamol for β_1- and β_2-receptors. Salbutamol apparently has a much higher efficacy at β_2-adrenoceptors (see Chapter 11) leading to its development as an effective antiasthma agent.

It is a straightforward matter to include a term for efficacy in the simple receptor theory. So far we have the occupancy p_D expressed in terms of the agonist concentration x_D, and equilibrium dissociation constant K_D

$$p_D = \frac{x_D}{K_D + x_D} \tag{22}$$

The magnitude of the tissue response was written by Stephenson as some function, f, of the efficacy ε times the occupancy, i.e.

$$\text{Response} = f(\varepsilon p_D) \tag{23}$$

Notice that there is no explicit term here for the number of receptors in the tissue, although obviously for any particular occupancy, the larger the total number of receptors in the tissue, N_{tot}, then the larger the stimulus given to the tissue (and hence the response) is likely to be. This was taken into account by Furchgott (1966) who used the term *intrinsic efficacy* (ε_i say) to define the ability of the agonist to activate any individual receptor. The response of the tissue then becomes a function, f, of the intrinsic efficacy, ε_i, times the total number of receptors in the tissue, N_{tot}, times the occupancy p_D. Differences in the number of receptors in different tissues is one reason why a drug may show tissue selectivity without there being any difference between the receptors in each tissue [for example see Salles and Badia (1991)].

In the classical approach it is not necessary to know what the function f is. Usually efficacy can only be measured in relative terms by comparing one agonist with another [see for example Salles and Badia (1991)]. Unfortunately the classical methods used to estimate agonist affinities and efficacies may not be accurate because the Stephenson approach neglects the fact that the efficacy of the agonist will generally affect the binding of the agonist [see Colquhoun (1987) for details], as discussed above in relation to ligand-binding measurements. However, in the case of the nicotinic AChR, single channel recording experiments have allowed direct estimates of both receptor affinity and intrinsic efficacy for several different agonists (Colquhoun and Sakmann, 1985; Colquhoun, 1987; Colquhoun and Ogden, 1988). This is because the AChR system is simple enough that a precise receptor activation mechanism can be postulated. For the model described in Eq. (13) the ratio b/a is a direct measure of efficacy (i.e. it reflects the ability of the agonist once bound to the receptor) to open the receptor ion channel.

An interesting question arises here in relation to comparing agonists, at different receptors. Antagonists, it will be recalled, are supposed to have zero efficacy and so can be classified purely on the basis of their dissociation equilibrium constant (K_D) using the

Schild method. Comparing the selectivity of an *antagonist* between two different receptor types is done simply by comparing K_Ds to obtain a selectivity ratio. In contrast, the selectivity of an *agonist* for one receptor type compared with another will depend on both the K_D at each receptor and the efficacy of the agonist at each receptor. Indeed, the separation of these factors, which, in principle, is necessary to make full use of structure–activity relationships, was one of Stephenson's original aims.

When *agonist* and *antagonist* K_Ds are estimated using radioligand-binding studies, these will not necessarily give a direct measure of the microscopic equilibrium dissociation constant. Generally antagonist K_Ds estimated in this way (where either another antagonist or agonist can be used as the displacing agent) agree well with K_Ds estimated using the Schild method. In contrast, agonist K_Ds estimated from binding studies sometimes suggest a much higher affinity than predicted from functional studies, often because of problems with receptor desensitization (see below), but also because of the fundamental problem that the actual binding measured will generally reflect the efficacy of the agonist, as well as the K_D [e.g. for the mechanism in Eq. (13), the occupancy is $p_{DR} + p^*_{DR}$ which depends on b/a as well as K_D].

In practice it can be very difficult to separate the affinity (reciprocal K_D) of an agonist for its receptor from its efficacy. Any attempt to estimate the agonist K_D directly from the EC_{50} (concentration giving half-maximal response) of an agonist dose–response relationship is likely to fail, since conformational changes associated with receptor activation will prolong the lifetime of the agonist–receptor complex. This can be illustrated by considering the *rates* of the reactions involved in receptor activation. For simple drug binding to a receptor [Eq. (3)] the lifetime of the DR state is determined by k_{-1}, the dissociation rate constant. For example, if $k_{-1} = 10^3 \, \mathrm{s}^{-1}$, this predicts a mean lifetime for DR of $1/10^3 \, \mathrm{s}^{-1} = 1 \, \mathrm{ms}$. If $k_{+1} = 10^7 \, \mathrm{M}^{-1} \mathrm{s}^{-1}$, then $K_D = 100 \, \mu\mathrm{M}$. However, if, for example, receptor activation [transition to DR* in Eq. (13)] results in a mean life of 1 s for the occupied forms of the receptor, then this will give an apparent $K_D = 100 \, \mathrm{nM}$. Such a situation could arise because an agonist is of very high efficacy. It can also arise if receptor desensitization is occurring. For example, if 99% of each 1 s activation is spent in a desensitized state, then in addition to a low agonist affinity, the agonist efficacy may be quite low, but the agonist will still *appear* to have a high affinity for the receptor.

A. A More Complex Receptor Activation Scheme

The del Castillo and Katz model [Eq. (13)] provided a useful description of agonist activation of the nicotinic AChR. However, this model allows for only a single agonist-binding site and there is now considerable functional, biochemical and structural evidence that there are two agonist-binding sites on the nicotinic AChR (see also Section VI). For example, the Hill coefficient for AChR activation at low ACh concentrations approaches 2, suggesting that two ACh molecules must bind to the receptor to produce efficient receptor activation. Biochemically, there are found to be two ACh- and α-bungarotoxin-binding sites per receptor molecule and two agonist-binding subunits are present in each receptor (Unwin, 1989). This evidence suggests that the del Castillo and Katz model could be usefully extended to include a second agonist-binding reaction.

$$ D + R \underset{k_{-1}}{\overset{2k_{+1}}{\rightleftharpoons}} DR + D \underset{2k_{-2}}{\overset{k_{+2}}{\rightleftharpoons}} D_2R \underset{a}{\overset{b}{\rightleftharpoons}} D_2R^* \tag{24} $$

(The factor of 2 before k_{+1} and k_{-2} occurs because there are two agonist-binding sites and this mechanism presumes that either site can be occupied or vacated first.) The occupancy for this scheme of the doubly liganded closed state will be

$$p_{D_2R} = \frac{c_1 c_2}{c_1 c_2 (b/a + 1) + 2c_1 + 1} \tag{25}$$

where $c_1 = x_A/K_1$ and $c_2 = x_A/K_2$ and K_1 and K_2 are the microscopic dissociation equilibrium constants for the first and the second agonist-binding reactions. The occupancy of the open state (p_{open}) will be

$$p_{D_2R^*} = \frac{b}{a} p_{D_2R} \tag{26}$$

The Hill slope for this mechanism is complicated. The Hill plot is not a straight line and so the Hill slope depends on agonist concentration.

$$n = 2 \left(\frac{1 + c_1}{1 + 2c_1} \right) \tag{27}$$

At low agonist concentrations $n = 2$ but it declines to $n = 1$ at high concentrations (Colquhoun and Ogden, 1988). In their study of AChR activation at the frog end-plate, Colquhoun and Ogden estimated an EC_{50} of 15 μM, $K_D = 77\,\mu M$ and $n = 1.6$ at the EC_{50} concentration. This mechanism [Eq. (24)] has been found to be a good description of AChR activity in a wide range of experimental situations. Colquhoun and Sakmann (1985) used this mechanism to interpret data from single-channel recordings of AChR channel openings at the frog neuromuscular junction. They made the fundamental observation that for each occasion when a receptor is occupied by two agonist molecules, the ion channel opens several times in quick succession. These groups of openings are termed bursts. Colquhoun and Sakmann measured the mean open time [$\tau_{open} = 1/a$ in Eq. (24)], the mean closed time within bursts [$\tau_g = 1/(b + 2k_{-2})$] and the mean number of closures per burst ($N_g = b/2k_{-2}$). When ACh was the agonist these parameters were found to be $\tau_{open} = 1.4\,ms$ giving $a = 714\,s^{-1}$, $\tau_g = 20\,\mu s$ and $N_g = 1.9$ giving $b = 1.9 \times 2k_{-2} = 3.8k_{-2}$, $\tau_g = 1/3.8k_{-2} + 2k_{-2} = 1/5.8k_{-2}$ giving $k_{-2} = 8620\,s^{-1}$ and $b = 32600\,s^{-1}$. A $K_D = 77\,\mu M$ suggests that, if the two agonist-binding sites are identical ($k_{-1} = k_{-2}$ and $k_{+1} = k_{+2}$), then $k_{+1} = k_{+2} = 1.1 \times 10^8\,M^{-1}s^{-1}$. For each agonist tested the ratio $b/(a + b)$ was greater than 0.9. Thus agonists activate the AChR rapidly (k_{+1}, k_{+2} and b are large) and with high efficiency (b/a is large). In other words the *intrinsic efficacy* (Furchgott, 1966) for agonists at the AChR is high. This was confirmed in studies with high agonist concentrations (Colquhoun and Ogden, 1988) where it was observed that when the effects of receptor desensitization and agonist block of the AChR channel are taken into account, then ACh is capable of opening the receptor channel for more than 90% of the time.

XI. Conclusions

Receptor pharmacology grew out of the need to quantify the effects of drugs on tissues and living systems. Using quantitative methods allowed the classification of receptors according to agonist and antagonist potency and subsequently led to the development of receptor-selective therapeutic agents. This was achieved despite the lack of any detailed knowledge of receptor structure.

The wide structural diversity now becoming apparent among receptors suggests the possibility that traditional pharmacology and medicinal chemistry, when combined with detailed structural information concerning each receptor subtype, could lead to new receptor-selective drugs. If the structural diversity of drug receptors reflects functional diversity, then the possibilities for development of new specific therapeutic agents will be greatly enhanced.

Acknowledgements

I would like to thank Patrick Covernton, John Connolly, Angus Silver, Don Jenkinson and David Colquhoun for critically reading and discussing this manuscript.

References

Arunlakshana, O. and Schild, H.O. (1959). *Br. J. Pharmacol.* **14**, 48.
Birnbaumer, L., Abramowitz, J., Yatani, A., Okabe, K., Mattera, R., Graf, R., Sanford, J., Codina, J. and Brown, A.M. (1990). *Biochem. Mol. Biol.* **25**, 225.
Bradley, P.B., Engel, G., Feniuk, W., Fozard, J.R., Humphrey, P.P.A., Middlemiss, D.N., Mylecharane, E.J., Richardson, B.P. and Saxena, P.R. (1986). *Neuropharmacology* **25**, 563.
Castle, N.A., Haylett, D.G. and Jenkinson, D.H. (1989). *TINS* **12**, (2) 59.
Colquhoun, D. (1971). 'Lectures on Biostatistics', Clarendon Press, Oxford.
Colquhoun, D. (1987). In 'Perspectives on Receptor Classification' (J.W. Black, D.H. Jenkinson and V.P. Gerskowitch, eds), pp. 103–114, Alan R. Liss, New York.
Colquhoun, D. and Ogden, D.C. (1988). *J. Physiol.* **395**, 131.
Colquhoun, D. and Sakmann, B. (1985). *J. Physiol.* **369**, 501.
Colquhoun, D., Dreyer, F. and Sheridan, R.E. (1979). *J. Physiol.* **293**, 247.
Del Castillo, J. and Katz, B. (1957). *Proc. R. Soc. Lond. (Biol.)* **146**, 369.
Deneris, E.S., Connolly, J., Rogers, S.W. and Duvoisin, R. (1991). *TIPS* **12**, (1) 34.
Furchgott, R.F. (1966). *Adv. Drug. Res.* **3**, 21.
Gryglewski, R.J., Moncada, S. and Palmer, R.M.J. (1986). *Br. J. Pharmacol.* **87**, 685.
Henderson, R. and Unwin, P.N. (1975). *Nature (London)* **257**, 28.
Hill, A.V. (1909). *J. Physiol. (London)* **39**, 361.
Huganir, R.L. and Greengard, P. (1990). *Neuron* **5**, 555.
Imoto, K., Busch, C., Sakmann, B., Mishina, M., Konno, T., Nakai, J., Bunjo, H., Mori, Y., Fukuda, K. and Numa, S. (1988). *Nature (London)* **335**, 645.
Jenkinson, D.H. (1991). *TIPS* **12**, 53.
Lefkowitz, R.J., Hausdorff, W.P. and Caron, M.G. (1990). *TIPS* **11**, 190.
Neer, E.J. and Clapham, D.E. (1988). *Nature (London)* **333**, 129.
Parsons, A.A. (1991). *TIPS* **12**, 310.
Rampe, D. and Triggle, D.J. (1990). *TIPS* **11**, (3) 112.
Rang, H.P. and Dale, M.M. (1991). 'Pharmacology', 2nd edn, Churchill-Livingstone, Edinburgh.
Salles, J. and Badia, A. (1991). *Br. J. Pharmacol.* **102**, 439.
Saxena, P.R. and Villaton, C.M. (1991). *TIPS* **12**, 223.
Stephenson, R.P. (1956). *Br. J. Pharmacol.* **11**, 379.
Strader, C.D., Sigal, S.I. and Dixon, R.A.F. (1989). *FASEB J.* **3**, 1825.
Unwin, N. (1989). *Neuron* **3**, 665.

–4–

Drug Access and Prodrugs

A.J. COLLIS
Pfizer Central Research,
Sandwich,
Kent CT13 9NJ, U.K.

Dr Alan Collis was born in Derby in 1962. He received his degree from UMIST (1984) and completed his Ph.D. there in 1987, with Dr Brian Booth, on novel routes to isoquinolines using superacids. After his academic work, he joined the Discovery Chemistry department at Pfizer where he has enjoyed five years of varied research. Areas of interest include: inhibitors of proteolytic enzymes/ion channels; the design of prodrugs; the relationship of pharmacokinetics to physical properties; and organic synthesis (both asymmetric and heterocyclic). He dedicates his chapter to those who seek to achieve drug access by compound design.

I. Preface

Unless a biologically active compound reaches its site of action in the body in sufficient quantity and for a suitable duration, it will not be a clinically useful drug. Optimizing this aspect of performance is a vital part of drug design, and frequently provides a major stumbling block between a lead and developing a drug.

MEDICINAL CHEMISTRY 2nd Edition
ISBN 0-12-274120-X

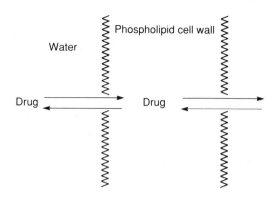

Fig. 4.1   Diffusion across a cell wall.

As a vast amount of literature relating to drug access and prodrugs has been published, this chapter will not attempt a comprehensive review. Instead, the key principles and considerations required to negotiate this area will be illustrated. Drug access will be discussed first, since it highlights the issues that prodrugs often have to address. The options available for improvement of drug access are then discussed, followed by principles and examples of prodrug design.

II.   Drug Access

Consideration of drug access requires an answer to three basic questions:

(i)   What is the preferred method(s) of introduction of the drug, or prodrug, into the patient?
(ii) Where does the drug need to be to exert its action?
(iii) How does it get from (i) to (ii)?

'What method' and 'where to' are usually defined by the disease area, e.g. bronchodilator–lung, or by the desire to improve on existing therapy, e.g. no oral agent currently available. The various aspects of drug access, and prodrug design, are embodied within the third question. Its answer depends upon both the physicochemical properties of the drug and its handling by physiological processes, i.e. pharmacokinetics. For example, following an oral dose of a bronchodilator, it must dissolve in gastrointestinal fluid, pass through the gut wall and liver, enter circulating blood and then into the lungs before it is eliminated. The first two steps of dissolution and absorption are dependent on physicochemical properties. Pharmacokinetics defines the rate of appearance of a drug in the circulation, its distribution and, ultimately, its rate of elimination. So how and why do these impact on drug access?

A.   Physicochemical Effects on Absorption and Dissolution

Drugs are usually dosed at a site remote from where they act, hence to gain access they must pass through tissue, i.e. transverse a matrix of cells. A large majority of drugs achieve this by passive diffusion through cell walls (phospholipid bilayers) (Fig. 4.1).

The rate of passive diffusion through a cell wall is dependent on the concentration gradient across it, and on the equilibrium between drug in the aqueous and cell wall phases (Yalkowsky and Morozowich, 1980). Higher concentration gradients result in faster diffusion, whereas the relationship to the equilibrium value reaches a maximum, beyond which rate declines. An initial concentration gradient, at a given dose, will depend upon how rapidly the drug dissolves (i.e. dissolution rate) or concentration when given in solution. Lipophilicity governs the equilibrium between aqueous and cell wall phases. Increasing molecular weight slows down passage through cell membranes, but drugs with molecular weights as high as 1200 [e.g. cyclosporin (Davies, 1989)] may still show good tissue penetration. As nearly all drugs have molecular weights less than cyclosporin, it is rarely a factor governing absorption.

B.   Determination of Lipophilicity

Lipophilicity is the preference of a drug to dissolve in a lipid phase (e.g. cell membrane) relative to an aqueous one. Hence, polar substituents (e.g. amides, alcohols, ethers, ketones or ionized groups) reduce lipophilicity, pushing equilibrium concentrations towards the aqueous phase, and non-polar groups (e.g. alkyl, aryl, chloro and bromo) have the opposite effect.

Ideally, lipophilicity would be measured directly as a partition coefficient between a phospholipid bilayer and water (e.g. Preston Mason et al., 1989). Experimentally, the fragile nature of phospholipid bilayers make them inconvenient for routine handling, so they are replaced by an organic solvent. This is an approximation: bulk phase organic solvent lacks the considerable structure and order associated with a cell membrane. Despite this approximation, this simple technique is predictive and broadly used. Values are usually reported as the $\log_{10}$ of the partition coefficient (P). Hence, a drug that distributes 10:1 in organic solvent/water has a log P of 1. The challenge is then to identify a solvent that closely mimics biological membranes.

Nearly a hundred different solvent systems have been proposed, including reversed-phase chromatographic systems. The most frequently employed system is octanol/water which accounts for about 25% of reported data, and from its popularity, octanol appears to parallel the properties of lipids. Other systems have been suggested when diffusion is particularly difficult, e.g. into the central nervous system (CNS). For this difficulty of penetrating the CNS (often referred to as the blood–brain barrier) the use of a combined figure from octanol/water and cyclohexane/water has been suggested (Young et al., 1988).

Ionization of a basic or acidic drug provides an additional factor to consider, since partitioning becomes pH dependent (Fig. 4.2).

Experimentally this may be measured with a buffered aqueous phase, usually at pH 7.4, the pH of many biological fluids, e.g. plasma. A value obtained this way is frequently referred to as a distribution coefficient, log D, rather than log P value and the two can be related by pK_a using the following formula:

$$\log D = \log P + \log (1 + 10^{pK_a - pH})$$

As a simple guide, a unit increase in pH will *increase* the log D of a base or *decrease* that of an acid by one unit. A limitation to this mathematical approach is that it assumes that ionized drug does not enter the organic phase. This assumption is an approximation as counter-ion effects from buffer can be observed with log D measurements. Similarly, the

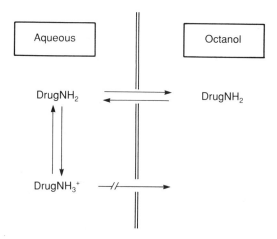

Fig. 4.2   Partition of a basic drug into octanol.

synthetic technique of phase transfer catalysis is known to take ion pairs from an aqueous phase into a non-polar organic solvent. Consequently, the lipophilicity of drugs may behave in a 'non-ideal' way with respect to pH. Qualitatively this effect is usually small, since the partitioning ability of an un-ionized drug is going to be far greater than for an ionized one, *unless* the drug is very highly ionized or lipophilic.

Calculated lipophilicities allow the routine and tedious measurement of partition coefficients to be reduced (Rekker, 1977). If one member of a related series of compounds has a measured lipophilicity, then that of the others may be calculated using fragmental values. This only requires arithmetical addition or subtraction for each 'functionality' introduced or deleted, although correction factors should be introduced if the altered functional groups are in close proximity. In principle, calculations may be extended to any structure without the need for measured values. However, these calculations do not account for the overall three-dimensional shape of a molecule. Shape may have large effects on lipophilicity since polar/lipophilic groups can be buried away from solvent (Dunn *et al.*, 1987), or subjected to intra-molecular hydrogen-bonding, which reduces polarity.

An ideal log D value for a series is dependent on the solubility and drug access properties required. If lipophilicity is too low, then passive diffusion will be slow. High lipophilicity may result in a majority of drug residing in lipid phases away from its site of action and may adversely affect aqueous solubility. However, as a broad guide, rapid tissue penetration is typically observed at a log D of 1 to 2. Exceptions do occur and the only absolutely certain measurement is to check tissue penetration or absorption *in vivo*.

Tissue penetration is occasionally observed for highly polar drugs, and consequently two mechanisms that do not rely on passive diffusion have been suggested. Either the drug enters through gaps between cells [the aqueous pore route (Ho *et al.*, 1977)] or by specific membrane transport systems. Access through the aqueous pore route is favoured by small molecular size (e.g. molecular weight < 200). The low molecular weight examples, *N*-hydroxyurea and 5-fluorouracil (Fig. 4.3), which are both well absorbed in man, are believed to enter by the aqueous pore route. Entry of larger compounds is restricted as the degree of absorption is inversely related to molecular weight.

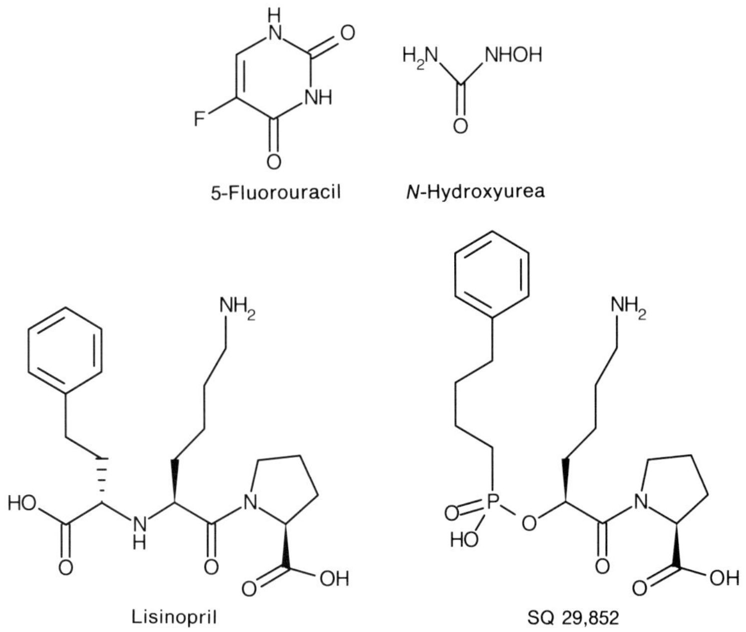

5-Fluorouracil   N-Hydroxyurea

Lisinopril   SQ 29,852

Fig. 4.3 Polar drugs believed to be absorbed by unusual routes.

Specific membrane transport systems require a high degree of structural recognition, therefore a structure/membrane transport relationship has to be fulfilled. Lisinopril and SQ29,852 (Fig. 4.3) are two examples of angiotensin-converting enzyme inhibitors that are reported (Amidon and Friedman, 1989a) to be absorbed after administration because of their similarity to dipeptides.

C. Dissolution Rate

As stated earlier, the rate at which a drug crosses a cell membrane is dependent on lipophilicity and concentration gradient. The concentration gradient of a solution dose is simply limited by aqueous solubility. When a drug is given in a pure or solid dose form, then the concentration gradient is determined by dissolution rate. Dissolution rate (R) and solubility (S) are related by the following formula, where K reflects the surface area of the drug particles and the degree of agitation.

$$R = KS$$

Predicting solubility is more complex than predicting lipophilicity. Liquids present the most straightforward example since solubility is approximately inversely proportional to lipophilicity. Therefore, any structural modification that increases lipophilicity may have to be balanced against a reduction in dissolution rate.

Crystalline drugs present more complex behaviour. The stabilizing forces present in the crystal lattice must be overcome during dissolution. These forces are reflected in melting point, but prediction from structure is complex. Quantification depends on calculating the interactions stabilized by the geometry of the crystal lattice and the preferred solid state molecular conformation. Consequently, introducing lipophilicity

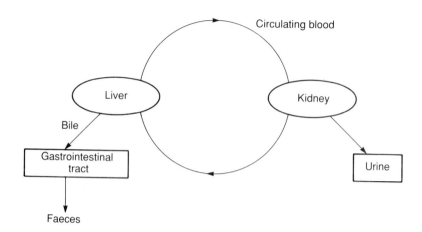

Fig. 4.4   Elimination of a metabolically stable drug.

into a high melting point crystalling series may have the opposite effect to a liquid and *increase* solubility due to reduction of intermolecular forces in the crystal.

Target solubilities are difficult to define since they are dose dependent, i.e. $10\,\mu g/ml$ may be fine for a potent drug whereas several mg/ml is required for others. Strategically it is best to aim high by designing good solubility into a series (i.e. in the mg/ml range) provided that loss of efficacy or low lipophilicity can be avoided.

D.   Systemic Pharmacokinetics and Drug Access

During systemic circulation, a drug will be eliminated either structurally intact or after chemical modification by enzymic processes (metabolism). Pharmacokinetics examines this balance of rates between absorption/distribution and elimination.

Elimination of structurally intact drug normally occurs in urine or faeces (occasionally sweating and expelled breath may contribute) (Fig. 4.4).

Circulation of the metabolically stable drug will continue until it is extracted by the kidneys to exit via the urine or is secreted into the gastrointestinal (GI) tract. Secretion into the GI tract via the bile follows extraction by the liver. Once a drug has entered the GI tract it may pass down it until it exits in the faeces or alternatively it may be absorbed through the gut wall back into the systemic blood (a process known as enterohepatic recirculation). If a drug fails to gain access to its site of action because of rapid excretion from the systemic circulation, then the use of prodrugs or formulation (discussed later) may do little to ease the problem.

The majority of drugs are not secreted intact as they are removed by metabolism. Metabolic modification may result in a loss of biological activity and/or a greater propensity to be extracted by the liver or kidneys. It may be very fast, highly efficient and proceed through several alternative routes, e.g. hydroxylation, heteroatom oxidation/de-alkylation, reduction, hydrolysis or conjugation (the introduction of a new unit of structure, e.g. an alcohol becomes sulphated). Most tissues have some metabolic capacity, but the gut wall, gut contents and the liver cause particular problems for drug access by the oral route.

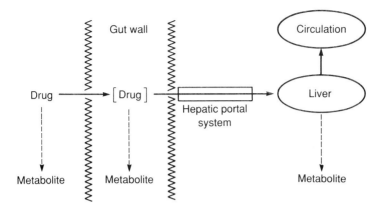

Fig. 4.5 Barriers to oral absorption.

E. Implication of Pharmacokinetics for Oral Administration

The oral route, preferably 'once a day', is the most favoured clinical option in terms of patient compliance. It is also the route where most pharmacokinetic/metabolic barriers exist (see Fig. 4.5), presumably because of the evolutionary need to avoid unwanted, or toxic, substances in diet entering the systemic circulation.

Once a drug is in the GI tract it needs to have appropriate physicochemical properties to dissolve and be absorbed (discussed above). However, as this is happening, it may be eliminated by enzymic or chemical attack in the gut contents (peptidic drugs are particularly prone to attack). After absorption it passes through the gut wall where it must 'run the gauntlet' of a rich variety of hydrolytic/oxidative and conjugative enzymes. It will then enter the hepatic portal system which carries blood that perfuses gut straight to the liver. In the liver a drug encounters an even wider variety of highly active metabolizing enzymes. Even at this stage, avoiding metabolism does not guarantee passage into the systemic circulation since an intact drug may be extracted into the bile and excreted into the gut contents to recycle. Once a drug has entered the systemic circulation then it needs to gain access to its site of action before it is eliminated. This *exclusive* passage of the drug through three metabolizing systems (i.e. the liver, gut wall and contents) before entering the circulation is peculiar to oral administration and is a frequent cause of failure. The term 'first pass effect' is used to describe this problem of drug loss between its site of administration and the systemic circulation (Johnson, 1980).

III. What to do in the Event of Failure?

Understanding the principles of drug access is extremely important because it provides the medicinal chemist with a framework of ideas to remedy failure. When a drug access failure is observed, the potential cause or causes must be identified. Options that are available for a remedy may be taken from three areas:

(i)   Pharmaceutical assistance.
(ii) Analogue synthesis.
(iii) Prodrug synthesis.

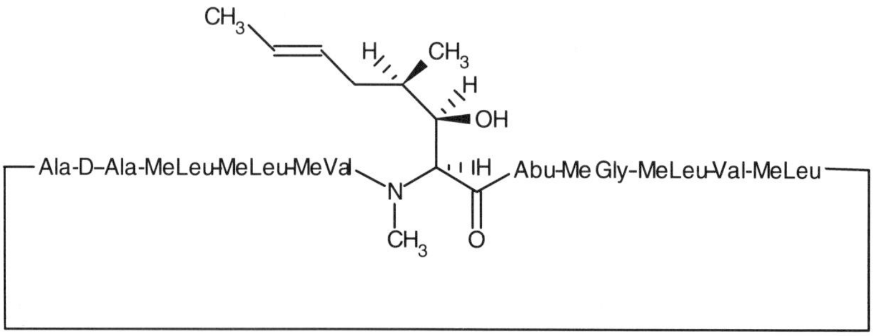

Fig. 4.6 Cyclosporin, a poorly soluble compound that requires formulation.

A. Pharmaceutical Assistance

Pharmaceutical assistance may be provided by formulation or alternative routes of delivery. Dissolution rate can be increased by formulation in a solubilizing vehicle or by finely dividing drug particles by milling. Cyclosporin (Fig. 4.6) is an example of a drug which requires a solubilizing vehicle to attain good bioavailability [40–60% in man (Davies, 1989)].

Slower dissolution rates may also be achieved using 'slow release' formulations, a strategy often employed for increasing the duration of action of a drug or reducing peak plasma levels immediately after dosing.

Formulation as an enterically coated tablet avoids chemical instability to acid as it dissolves only after passage through the acidic contents of the stomach. The two areas where formulation is of little value are increasing the absorption of polar drugs and protecting drugs against clearance after absorption.

A long list of alternative routes to oral administration exists which offer a varying amount of acceptibility and applicability, depending on the disease area. These routes include injection of solutions or polymer depots or by absorption from other sites, e.g. the mouth either under the tongue (sublingual) or through the cheek (buccal), the nose (intranasal), the rectum, the lungs (by inhalation), the skin (transdermal) and even the eye (transocular). All share the important advantage over oral administration of reduced first pass metabolism; however, they do frequently raise major issues of dose size/ presentation and clinical acceptability.

In order to achieve systemic exposure after administration by these routes, the usual constraints on lipophilicity and solubility need to be considered. Good absorption by these alternative routes may be as, or even more, difficult to achieve them for oral administration. Transdermal absorption, for example, is particularly difficult as drugs must penetrate through layers of dead cells on the skin's surface before reaching blood in perfused tissue. Reduced surface area for absorption can also cause problems, for example rectal absorption (deBlaey and Polderman, 1980) occurs through an area of $0.05\,m^2$ which compared with $70\,m^2$ for the small intestine!

An alternative route of administration or formulation may provide a convenient solution to a number of problems, but the technique most frequently applied is the analogue approach.

Fig. 4.7   The prodrug principle.

B.   Analogue Synthesis

Synthesis of analogues with superior drug access properties is the reflex response of the medicinal chemist to failure in this area. This approach has great potential since it may be applied to any problem of drug access. A whole range of properties (e.g. solubility, lipophilicity, metabolic stability of pharmacokinetics) may become a target for systematic exploration as part of the drug discovery synthetic programme.

Even if a calculated decision is taken not to explore such properties they may be optimized 'inadvertently' from *in vivo* data, since a better translation of *in vitro* potency/ duration to an *in vivo* model will result from improvements in drug access. A good translation of potency may result in that analogue becoming a lead for further structural modification, or for clinical evaluation. Reducing the 'inadvertent' nature of this process by due consideration of drug access properties increases the probability of optimizing the access profile of a series.

The big limitation of this approach is its resource intensiveness and the challenge of reconciling the desired modification (e.g. reducing lipophilicity) with good biological potency. Satisfying structure–activity relationships, whilst optimizing physicochemical and pharmacokinetic parameters, may test the structural ingenuity of the chemist to the limit. In the extreme case, where potency and drug access properties cannot be reconciled, then a prodrug approach may provide the best answer.

C.   Prodrug Synthesis

A prodrug is a *structural derivative* of a drug which must be chemically transformed within the body to exert its pharmacological or therapeutic action. The modifying unit of structure is attached covalently to the drug and known as a promoiety (Fig. 4.7).

Designing a structure so that it will undergo this transformation at an appropriate place and rate creates an extra level of complexity over a *normal* drug. This additional complexity arises from issues such as production costs, toxicity, chemical instability and predictability of performance across species. Interspecies variability is a particular problem, and a classic example is provided by simple aliphatic esters of penicillin (Kirchner *et al.*, 1949). These are prodrugs in mouse or rat but fail to be transformed to the parent drug in rabbit, dog and *man* in particular!

Prodrugs may be used to deal with pharmaceutical or pharmacokinetic issues. Pharmaceutical issues include unpalatability, formulating an intravenous (i.v.) dose, pain on injection or gastrointestinal irritation. Problems in the pharmacokinetic area may be subdivided into a further four different aspects:

(i)   Improving tissue penetration by altering lipophilicity or solubility which may result in better absorption or access to the site of action.
(ii)   Reducing presystemic metabolism to improve both the magnitude and consistency of bioavailability.

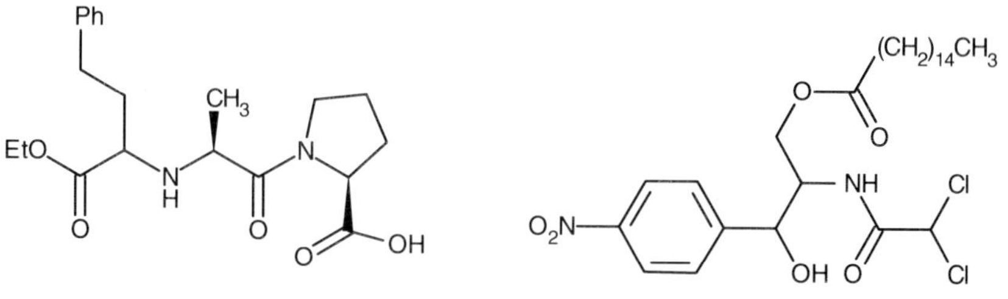

Enalapril   Chloramphenicol palmitate

Fig. 4.8   Aliphatic esters to alter absorption or aqueous solubility.

(iii) Obtaining selective transformation to produce a drug in the target tissue and thereby avoid or reduce the systemic effects of the drug.
(iv) Altering the rate of onset or duration of action of a drug.

The remainder of this chapter is devoted to examples of prodrugs that represent attempts to deal with these pharmaceutical and pharmacokinetic issues. The list of examples is certainly not comprehensive but is intended to illustrate the chemical and mechanistic principles which have been applied to this area.

IV.   Design of Prodrugs

A.   Prodrugs Derived from Carboxylic Acids on either the Drug or Promoiety

A large proportion of all prodrugs fall into this category. Its popularity arises as the body has a great variety of enzymes capable of hydrolysing esters which are distributed throughout most tissues. Consequently, a broad range of esters are hydrolysed enzymatically, at different rates and across a wide spectrum of tissues. This provides plenty of scope to achieve successful prodrug design. Masking carboxylic acids or hydroxy groups as esters is also popular as it increases lipophilicity.

a.   Simple Aliphatic and Aryl Ester Prodrugs

Simple aliphatic and aryl esters are particularly attractive as they are cheap to prepare, chemically stable and their hydrolysis products are usually toxicologically bland.

Enalapril (Fig. 4.8) is the ethyl ester prodrug of a highly polar angiotensin-converting enzyme inhibitor (enalaprilat) used to control hypertension (see Chapter 9). Owing to its extremely polar nature, enalaprilat is very poorly absorbed (i.e. $< 12\%$) whereas the more lipophilic prodrug, enalapril (log P 0.16) is between 50 and 75% absorbed (Ulm et al., 1982; Swanson et al., 1984). However, it has been suggested (Amidon and Friedman, 1989b) that the absorption of enalapril is mediated by a peptide transport system and not by passive diffusion.

In contrast, chloramphenicol palmitate (Fig. 4.8) (Glazko et al., 1952) is extremely lipophilic, which results in a very low aqueous solubility. It is through this low aqueous solubility that the extremely bitter taste of the antibiotic chloramphenicol is avoided. As expected, this highly insoluble prodrug is poorly absorbed as a result of its low dissolution rate. In fact, three of the four possible solid forms (polymorphs) of the prodrug fail to give plasma levels of drug because of poor dissolution. The fourth succeeds

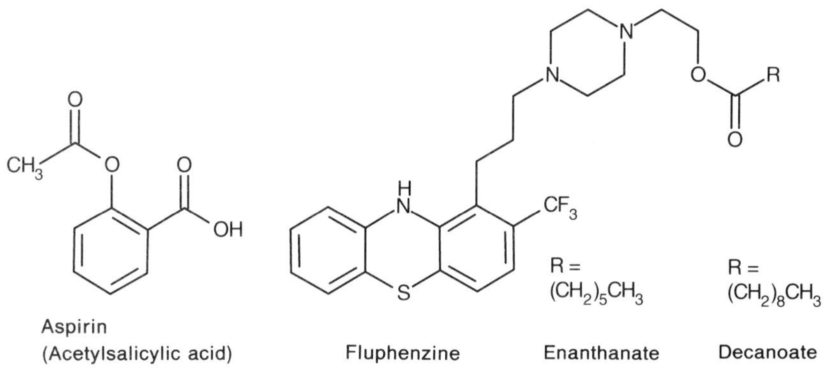

Fig. 4.9   Reducing local irritation or prolonging duration of action using simple esters.

because it is a substrate for an enzyme [pancreatic lipase (Andersgaard *et al.*, 1974)] which can convert it from the solid form into chloramphenicol, which is water soluble and absorbed.

Gastric irritation is reduced by one of the oldest known prodrugs, aspirin. Acetyl-salicylic acid (aspirin) (Fig. 4.9) is quantitatively deacetylated by the action of esterases after absorption to release the drug salicylic acid. This avoids direct contact of the gut wall lining with salicylic acid which is an irritant.

Lipophilic esters of the antipsychotic drug, fluphenazine, such as the enanthanate and decanoate (Fig. 4.9), may be used to prolong its duration of action when given by intramuscular injection. Interestingly, the rate-limiting step that results in prolonged duration of action is diffusion from the site of injection into the systemic blood supply (Dreyfus, 1976). Intravenous administration results in a similar plasma profile to the parent drug, illustrating that the route of administration as well as physical properties should be considered in prodrug design.

Usually, prodrugs are low molecular weight (< 600) therapeutic agents; however their application to high molecular weight enzyme complexes can be clinically useful. Anistre-plase (or anisoylated plasminogen streptokinase activator complex, APSAC) is an anisyl ester of the active-site serine of a protein complex used to dissolve blood clots (Smith *et al.*, 1981). After intravenous administration, this prodrug rapidly adheres to blood clots, before slower ester hydrolysis restores the enzymic activity which dissolves the clot. This reduces peripheral haemorrhage compared with the parent drug and allows administration as an intravenous bolus (i.e. one injection) rather than an infusion.

b.   Amides as Prodrugs

The slow rate of cleavage of amides *in vivo* has limited their application in prodrug design. An exception to slow cleavage is provided by α-amino acids/peptides, and attempts to apply these to clinically effective prodrugs have been made. An elaborate system relying on hydrolysis followed by cyclization has been reported by Thomas (1986). LDZ is a prodrug of diazepam in which the lysine promoiety is hydrolysed by a specific class of enzymes (the aminopeptidases). A chemically unstable amine is generated which spontaneously cyclizes at physiological pH to give diazepam (Fig. 4.10).

Studies in man revealed that this helps to avoid a side effect of drowsiness by reducing peak plasma levels but does not improve bioavailability. Acyclic prodrugs of cyclic drugs

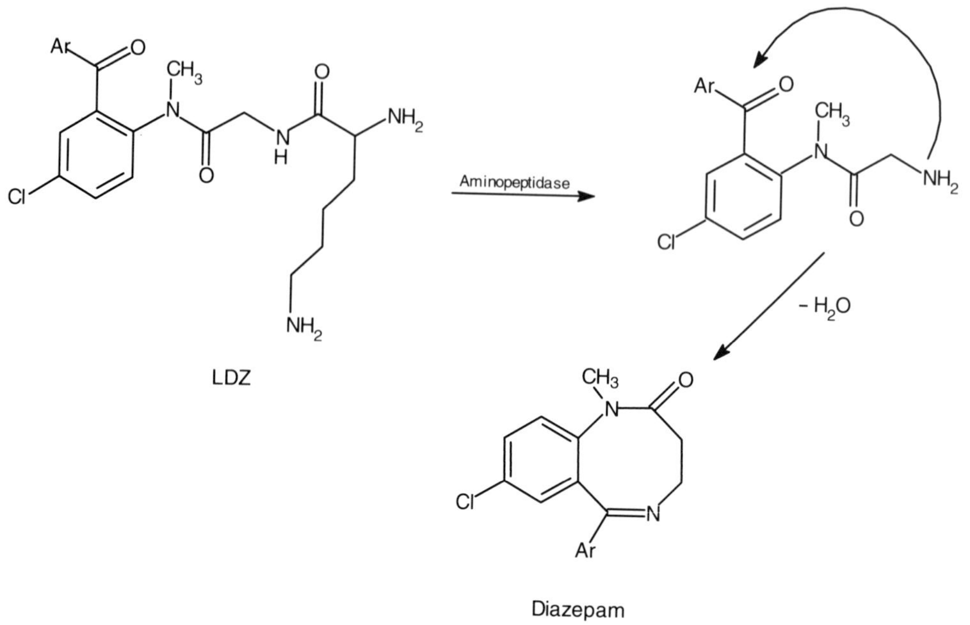

Fig. 4.10   An amide hydrolysis/cyclization cascade to diazepam.

have been used for a variety of other types of compound (Bundgaard, 1985a), e.g. barbituric acids, hydantoins, 2,4-oxazolidinediones, imides, lactones and cyclic quaternary ammonium compounds.

c.   *Use of Ester-derived Prodrugs to Increase Polarity for Enhancing Aqueous Solubility*

Poor aqueous solubility may limit oral absorption but is particularly a problem for intravenous administration, where high concentrations are required.

Carboxylic acid esters have been used to increase polarity by incorporating ionized functionality into the promoiety. Alcohols have often been derivatized as succinates or as α-amino esters, whereas β-amino alcohols (e.g. dimethylaminoethyl) have been used to derivatize acids. Prodrug esters of α-amino acids are particularly attractive as the promoiety can be selected to release an endogenous material and are readily hydrolysed by many peptidases, e.g. α-aminopeptidases. The hydrolysis of succinate and dialkylaminoethyl esters is noteworthy as it may proceed chemically under physiological conditions as a result of neighbouring group participation (Fig. 4.11). Two examples of prodrugs from this class are chloramphenicol succinate and lysine oestrone ester (Fig. 4.11).

Chloramphenicol succinate (Glazko et al., 1957) is used for intravenous administration. Its high solubility is in complete contrast with the palmitate described earlier, demonstrating the extreme changes in physical properties that are possible through use of prodrugs.

Lysine oestrone ester was prepared to improve intestinal absorption (Amidon et al., 1980) and increases the absorption rate relative to oestrone by up to five orders of magnitude!

Esters derived from inorganic acids, such as sulphuric and phosphoric acid have been

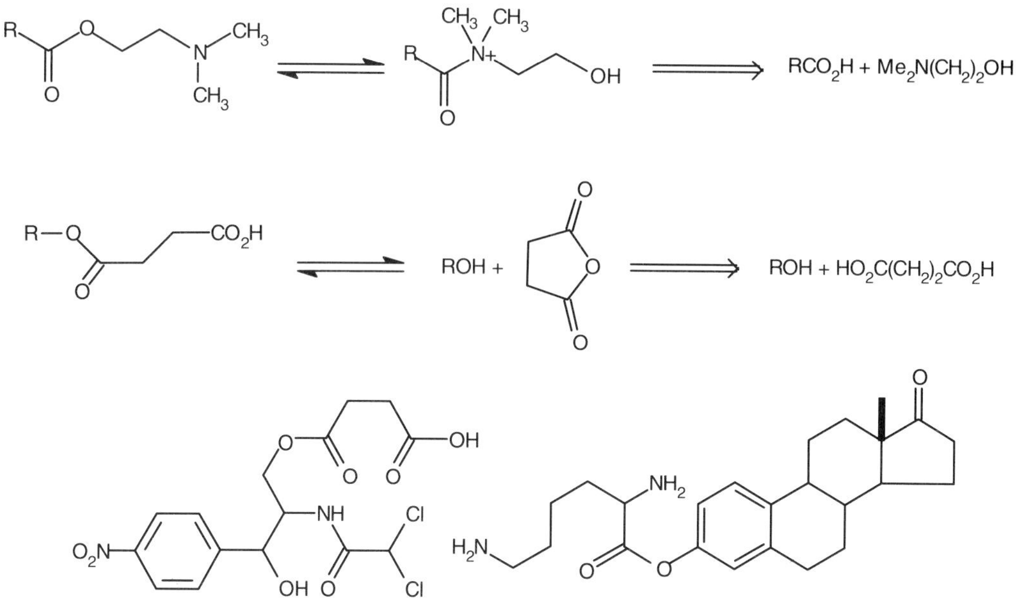

Chloramphenicol succinate

Lysine oestrone ester

Fig. 4.11 Neighbouring group participation within, and examples of, prodrugs to enhance aqueous solubility.

proposed to increase aqueous solubility. The use of sulphates has not been very successful owing to very poor hydrolysis *in vivo* (Bundgaard, 1985b). Phosphates, however, are labile as a result of hydrolysis by a group of enzymes known as phosphatases. An example is clindamycin phosphate (Fig. 4.12) (Edmondson, 1973; Gray *et al.*, 1974), which avoids the pain resulting from intramuscular injection of the parent drug. The prodrug has an improved aqueous solubility of > 150 mg/ml from 3 mg/ml for clindamycin and a 10 min half-life for hydrolysis *in vivo*.

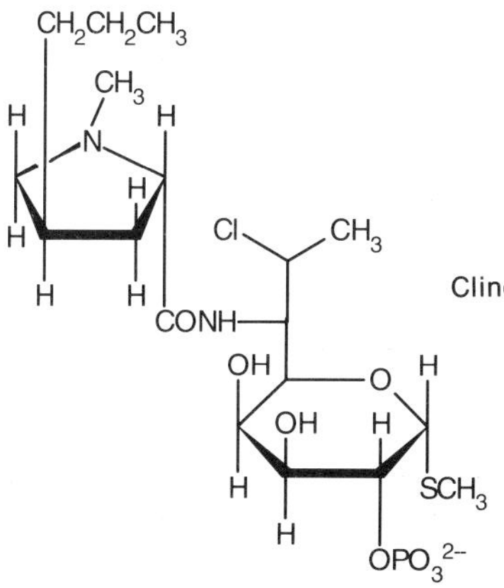

Clindamycin phosphate

Fig. 4.12 Phosphate as a solubilizing group.

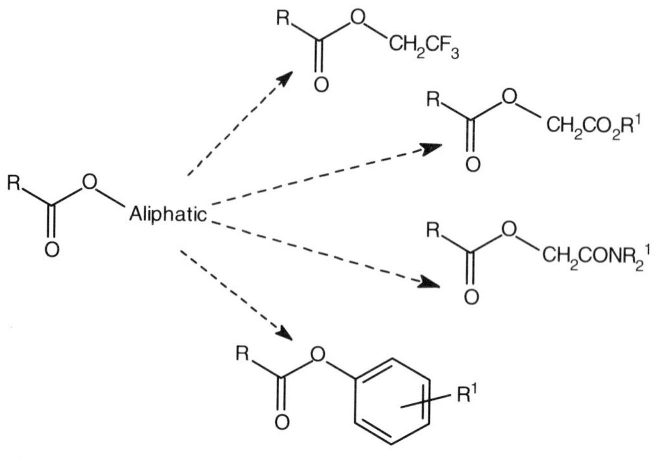

Fig. 4.13   Esters with enhanced lability.

d.   Modifications to Carboxyl Ester Prodrugs to Improve Hydrolysis Properties
The rate of prodrug hydrolysis by esterases is usually dependent on the structural
features on both sides of the ester linkage. In favourable examples, even simple aliphatic
esters are well hydrolysed *in vivo*.

Although structural features on the promoiety are easy to alter, difficulties arise when
the drug's structure makes it a poor substrate. One class of drugs where this has been
a particular problem is the β-lactam series of antibiotics (Ferres, 1980). These drugs are
polar carboxylic acids, and are poorly absorbed. Early attempts to 'prodrug' penicillins
relied on simple aliphatic esters to increase lipophilicity, but, as discussed earlier,
hydrolysis was very poor in man. Consequently, modifications to the alcohol moiety
were sought to improve hydrolysis rate. Attempts at improving hydrolysis rates in other
drug series have relied on increasing inductive effects by α-substituted alcohols or aryl
esters, e.g. Fig. 4.13.

Whether greater lability *in vivo* is a consequence of greater chemical lability or better
recognition by the hydrolysing enzymes is difficult to determine. Aryl esters have been
successfully applied in a β-lactam series for side chain carboxyls, e.g. carindacillin (Fig.
4.14); however, they are still not labile enough for the more hindered thiazoline carboxyl.

A solution to 'prodrugging' the 'esterase-resistant' thiazoline carboxyl was found by
Jansen and Russell (1965). They prepared diester derivatives of gem diols in which one

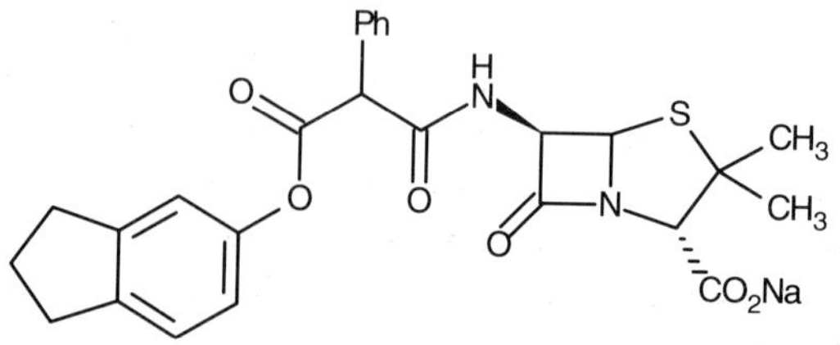

Fig. 4.14   An aryl ester of a β-lactam antibiotic.

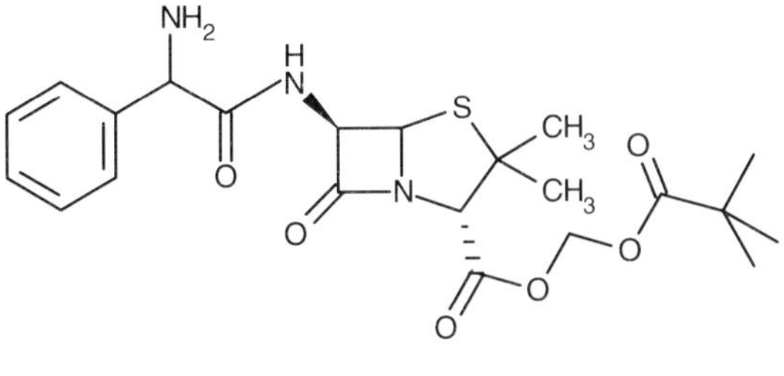

Pivampicillin

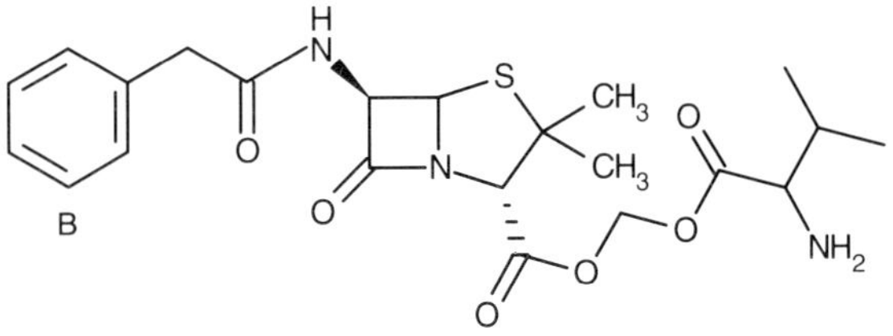

B

Fig. 4.15 Acyloxyalkyl esters as activated prodrugs of β-lactam antibiotics.

hydroxyl is acylated by a β-lactam and the other by any acid of choice. These 1-acyloxy-lalkyl esters offer great advantage over other promoieties, since productive hydrolysis is possible at a distal site. Hydrolysis gives a hemiacetal ester which then spontaneously loses aldehyde to reveal the other carboxylic acid. Pivampicillin is an example of a β-lactam prodrug which works via this strategy showing 2–3-fold improvement in absorption over the parent drug, pivampicillin (Ferres, 1980).

Application of 1-acyloxyalkyl esters to penicillin G was less successful until it was realized that prodrugs were failing as a result of poor aqueous solubility. Increasing the solubility by using the α-amino-acid derived prodrug B improves absorption to 40% from 14% for the parent drug (Fig. 4.15) (Ferres, 1980).

Generally, good esterase hydrolysis is observed for 1-acyloxyalkyl prodrugs of widely differing structure. Hence, a wide variety of acids/aldehydes may be released allowing broad scope to modify physical properties and hydrolysis rates. Also, owing to the distal nature of the hydrolysis, one of the carboxylic esters can be replaced with a whole range of functional groups (Bodor et al., 1983; Bundgaard, 1985c; Alexander et al., 1991) (e.g. Fig. 4.16).

Therefore, an approach originally designed to improve hydrolytic lability has yielded a whole family of prodrugs capable of derivatizing many functional groups.

A disadvantage of this approach is that prodrugs may release potentially toxic aldehydes, e.g. formaldehyde. The issue of formaldehyde toxicity is one of quantity since it is usually present in vivo as a result of demethylation of dietary components or, sometimes, drugs. Furthermore, acute-use drugs such as antibiotics present less concern than chronic therapies, which result in exposure to elevated levels for much greater time periods. Formaldehyde release may be avoided by using higher aldehydes, e.g. acetalde-hyde, but this introduces a new chiral centre. Encouragement from the regulatory authorities to develop single enantiomers only is making this option less desirable. Replacing aldehydes with symmetrical ketones is very difficult, since the intermediate 1-(acyloxyalkyl)-tertiary halides are chemically unstable (Bigler and Neuenschwander,

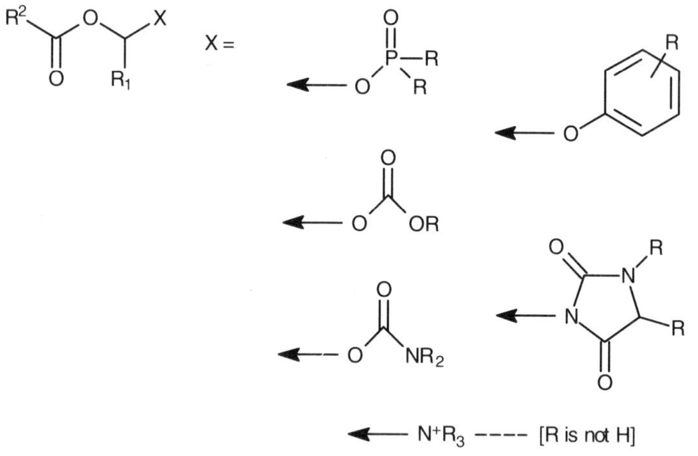

Fig. 4.16 Acyloxyalkyl derivatives of functional groups other than carboxylic acids.

1978). However, chirality complications and formaldehyde release may be avoided by arranging hydrolysis to occur through substituted vinyl esters (Fig. 4.17) (Sakamoto *et al.*, 1984).

B. Activation of Prodrugs by Oxidation or Reduction

This method of prodrug transformation is not as common as hydrolysis, presumably because substrate requirements are more demanding. However, like esterases, oxidative

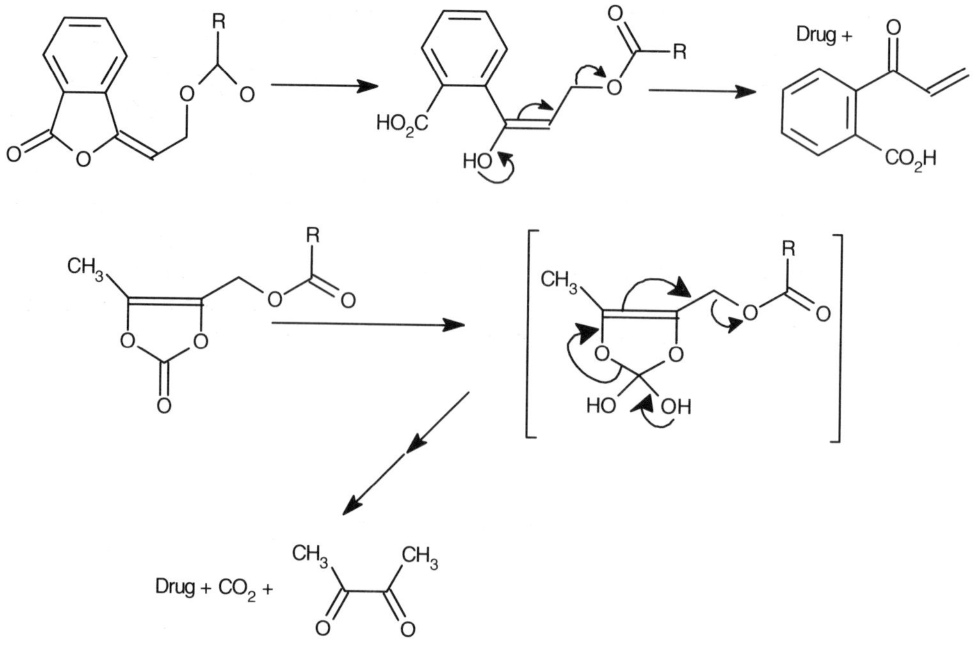

Fig. 4.17 Distal hydrolysis of achiral promoieties.

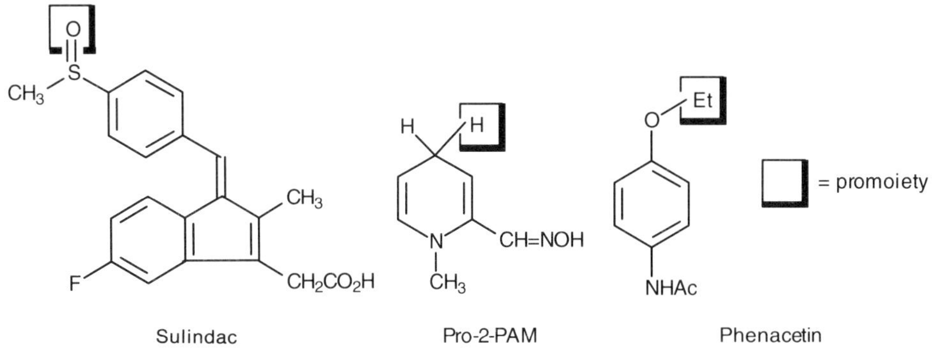

Fig. 4.18 Prodrugs activated by oxidation or reduction.

and reductive enzymes occur in a wide variety of tissues. Dealkylation of aryl ethers, amines, etc. also falls into this category, since the initial step is that of oxidation. Potentially the promoieties for this approach may be much smaller (e.g. H) than those for hydrolysis. Sulindac, proPAM and phenacetin (Fig. 4.18) are three examples activated by reduction, oxidation and dealkylation respectively (Duggen, 1981; Shek *et al.*, 1976; Pang and Gillette, 1978).

The sulphoxide of sulindac is reduced *in vivo*, after oral absorption to yield the anti-inflammatory drug, sulindac sulphide. The prodrug reduces gastrointestinal toxicity and increases aqueous solubility by approximately 100-fold.

ProPAM is the dihydropyridine form of the highly polar quaternary 'pyridinium' cholinesterase reactivator 2-PAM (pralidoxime). It improves brain concentrations of the parent drug as it passes more readily through the blood–brain barrier. Site-specific drug delivery (see later) was the intended objective for this prodrug, but it falls short of truly reaching this demanding target as most of the dose is oxidized in other tissues.

Phenacetin is the *O*-ethyl derivative of the drug paracetamol and highlights some potential problems when a prodrug campaign is embarked upon. Removal of the *O*-ethyl promoiety occurs in the liver by oxidation as illustrated in Fig. 4.19.

Liver clearance of paracetamol from the systemic circulation occurs through phenolic conjugation (e.g. as a sulphate derivative). Therefore, one might expect protection of the phenol to reduce this liver clearance and, hence, reduce first pass effects after an oral dose. Studies in isolated perfused livers (Pang and Gillette, 1978) indicate the opposite effect! Conjugation of paracetamol by the liver is *more* extensive when perfused as the prodrug rather than the drug. This alarming observation has been referred to as sequen-

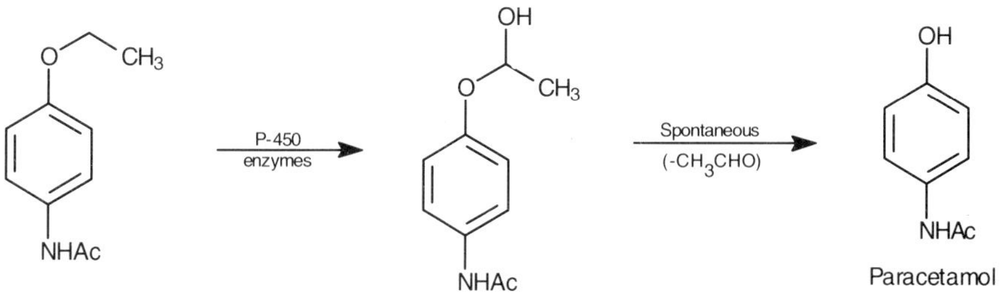

Fig. 4.19 Dealkylation sequence for phenacetin.

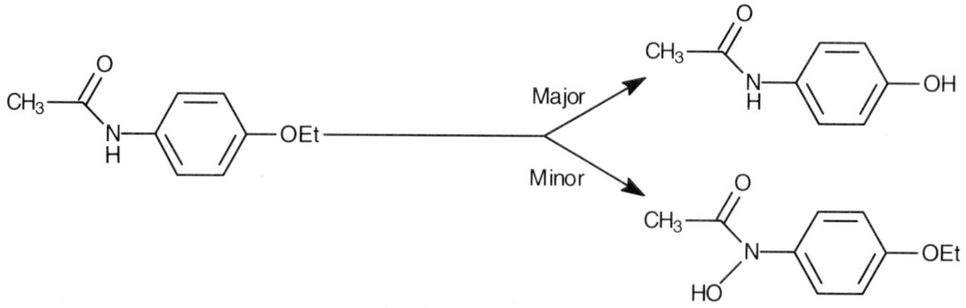

Fig. 4.20   Increased toxicological risk through metabolism.

tial metabolism, and its implication for masking metabolic vulnerability is that prodrug cleavage must occur after and not during/before perfusion through the clearance organ.

Although the major metabolic pathway for phenacetin is 'de-ethylation' one of the minor routes is N-hydroxylation to the hydroxamic acid (Hinson and Mitchell, 1976), a transformation known to increase the risk of toxicological activity (Fig. 4.20).

Paracetamol is considerably more resistant to N-hydroxylation and therefore avoids this toxicological concern. Great care needs to be exercised to ensure that a prodrug does not introduce additional toxicological problems, through either the fragments released or formation of toxic metabolites (Gorrod, 1980).

C.   Mannich Bases as Prodrugs

The hydrolysis of Mannich bases [Fig. 4.21 (1)] proceeds through the same 1-hydroxyalkyl intermediate observed for hydrolysis of 1-acyloxyalkyl esters [Fig. 4.21 (2)] and oxidative dealkylation [Fig. 4.21 (3)]. Mannich bases differ in their chemically rather than enzymically mediated fragmentation.

Amines and NH acidic compounds (e.g. amines, amides, imides, carbamates, hydantoins and ureas) have been 'prodrugged' by this approach generally with the aim of increasing aqueous solubility and dissolution rate. Hydrolysis rates are dependent on the pK_a of the drug derivatized and on steric hindrance. Since hydrolysis is chemical, the rate can be quantitatively predicted (Bundgaard, 1985d) from a knowledge of these parameters. Hetacillin and hexamine (Fig. 4.22) provide examples of prodrugs where the Mannich base principle has been used to solve problems other than solubility.

Hexamine chemically degrades to formaldehyde, a contact antibacterial, in aqueous solution of pH less than 5. It has been used to treat urinary tract infections. The prodrug

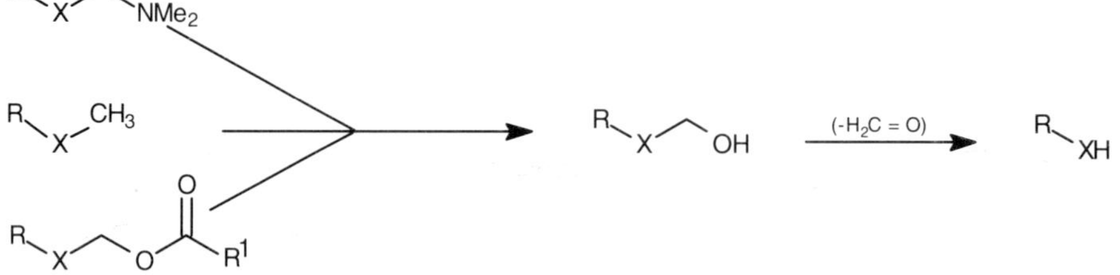

Fig. 4.21   Similarity between cleavage of Mannich bases and dealkylation or hydrolysis of acyloxyalkyl prodrugs.

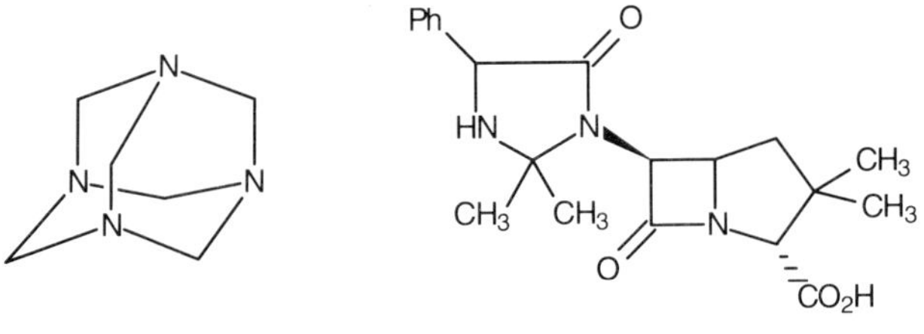

Hexamine   Hetacillin

Fig. 4.22   Prodrugs derived from Mannich bases to solve problems other than solubility.

is eliminated into the acidic infected urine where it release the potentially toxic for-maldehyde directly at its site of action. Oral administration of hexamine requires formulation as an enterically coated tablet to prevent dissolution and release of for-maldehyde in the acidic contents of the stomach. This provides an example where both formulation and the prodrug approach are required for clinical success.

Hetacillin, a prodrug of ampicillin, is used to improve the chemical stability of concentrated aqueous solutions (Schwartz and Hayton, 1972). Ampicillin solutions decompose as a result of intermolecular attack of the side chain amino group on the β-lactam ring. In hetacillin, this amino group is protected by both steric hindrance and reduced basicity arising from incorporation of the amine in a 4-imidazolidinone ring. On dilution or after administration, hetacillin decomposes with a half-life of approximately 11–18 min (Fig. 4.23).

D.   Site-specific Delivery using the Prodrug Approach

Localized release of drug to its site of action after systemic administration of a prodrug is a conceptually exciting idea. Potentially this can offer convenient and effective therapy whilst minimizing toxicity or unwanted systemic effects. Regrettably, its conceptual potential has not as yet been realized in clinical success. The potential of this approach has been reviewed by Stella and Himmelstein (1980) who concluded that at least three aspects of performance need to be optimized to succeed:

(i)   Transport to and uptake by the site of action should be rapid and essentially perfusion-rate-limited.
(ii) Selective cleavage of the promoiety must occur at the active site and not at other often more highly perfused sites such as the liver, kidney, etc.
(iii) Once a drug is released at its site of action, it must reside there for a suitable period of time and not rapidly distribute to other tissues.

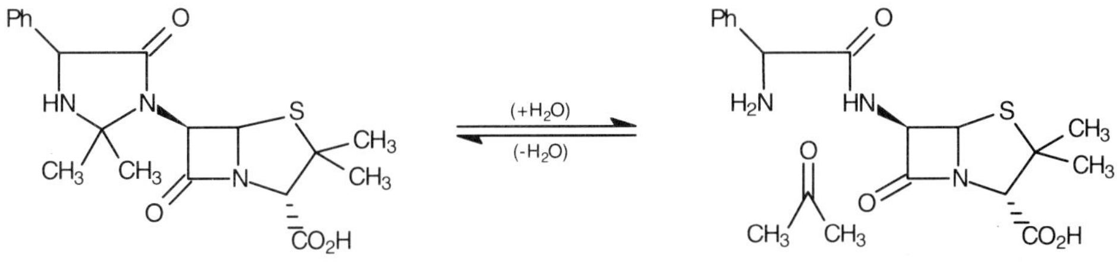

Fig. 4.23   The equilibrium of hetacillin with ampicillin in aqueous solution.

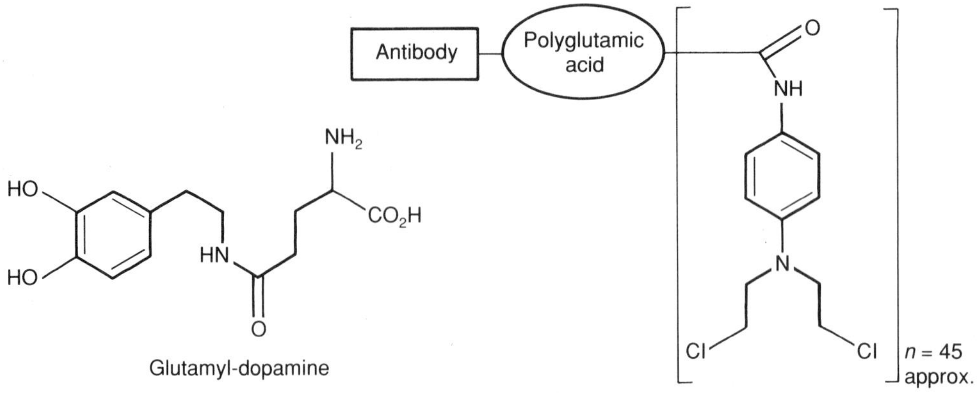

Fig. 4.24   Use of glutamic acid derivatives in site-specific delivery.

Selective cleavage at the target site is fundamental to this approach, and may be undermined by differences in perfusion rate, i.e. prodrug cleavage may be slower in non-target tissues, but they may still act as the main site of activation as the prodrug is exposed to them for longer because of higher perfusion. Following *site-specific release*, a drug needs to stay in that area. A majority of conventional drugs rapidly distribute between tissues, so even with the perfect promoiety, site-specific delivery will fail.

Despite the difficulties of achieving 'total' site-specific delivery, many prodrugs do succeed in reducing exposure of certain tissues to drug, i.e. they are site biased rather than specific. Aspirin, hexamine and APSAC all provide examples and have already been discussed, but I would like to supplement these with two examples where site-specific delivery was a rational objective in their design.

Glutamyl-dopamine is a kidney-targeted prodrug of dopamine (Kyncl *et al.*, 1979). This drug has a vasodilating action on the kidney and is useful in the treatment of shock, but it also has many other systemic effects. Hence, site-specific delivery is attractive. The enzyme glutamyltranspeptidase was selected for cleavage of the promoiety since it is known to occur at high levels in the kidney and effectively defines the promoiety used (Fig. 4.24).

A second approach (Rowland *et al.*, 1975) also uses glutamic acid but as a polymer to derivatize an anticancer drug *p*-phenylenediamine mustard (PDM).

The polyglutamic acid–PDM macromolecule is linked to an antibody that recognizes and adheres to tumour cells. This macromolecule prodrug does show marked reductions in toxicity compared with PDM; however, it has not been proven whether the drug is 'released' from its carrier to exert its action.

The area of macromolecular drug targeting in cancer therapy has been reviewed (Sezaki and Hashida, 1984).

V.   Summary

As the complexity of lead structures increases, so does the need for adequate consideration of its consequences for drug delivery. A balance between aqueous solubility/lipophilicity/metabolic stability and pharmacokinetics should be sought from the earliest stage in the drug discovery process, to ensure that a series may ultimately yield success. If the properties of a series cannot be manipulated sufficiently whilst retaining activity, then formulation and prodrug approaches should be considered. Formulation is useful in protecting against acid instability and altering dissolution rates, but can do little to assist problems of low lipophilicity or metabolic clearance subsequent to absorption. Prodrugs can improve lipophilicity and suppress metabolic vulnerability, as well as helping with pharmaceutical problems such as taste, pain on injection or aqueous solubility. The prodrug concept may be applied to reducing toxicity or systemic side effects through the use of site-specific delivery if a set of rigorous conditions can be fulfilled. Finally, each approach should not be considered in isolation since a combination may provide the most direct way to overcome a series of problems.

Suggested Reading

Bundgaard, H. (1985). *In* 'Design of Prodrugs' (H. Bundgaard, ed.), p. 1–92, Elsevier Science Publishers B.V., Amsterdam.

Ferres, H. (1980). *Chem. Ind.*, 435.

Johnson, P. (1980). *Chem. Ind.*, 433.

Rekker, R.F. (1977). *In* 'Pharmacochemistry Library, Vol. 1, The Hydrophobic Fragmental Constant' (W.Th. Nauta and R.F. Rekker, eds), Elsevier Scientific Publishing Company, Amsterdam.

Stella, V.J., Charman, W.N.A. and Naringrekar, V.H. (1985). *Drugs* **29**, 445.

Yalkowsky, S.H. and Morozowich, W. (1980). *In* 'Drug Design 1X' (E.J. Ariëns, ed.) pp. 121–185, Academic Press, London.

References

Alexander, J., Fromtling, R.A., Bland, J.A., Pelak, B.A. and Gilfillan, E.C. (1991). *J. Med. Chem.* **34**, 78.

Amidon, G.L. and Friedman, D.I. (1989a). *J. Pharm. Sci.* **78**, 995.

Amidon, G.L. and Friedman, D.I. (1989b). *Pharm. Res.* **6**, 1043.

Amidon, G.L., Leesman, G.D. and Elliot, R.L. (1980). *J. Pharm. Sci.* **69**, 1363.

Andersgaard, H., Finholt, P., Gjermundsen, R. and Hoyland, T. (1974). *Acta Pharm. Suecica* **11**, 239.

Bigler, P. and Neuenschwander, M. (1978). *Helv. Chim. Acta* **61**, 2165.

Bodor, N., Sloan, K.B., Kaminski, J.J., Shih, C. and Pogany, S. (1983). *J. Org. Chem.* **48**, 5280.

Bundgaard, H. (ed.) (1985a). *In* 'Design of Prodrugs' p. 51, Elsevier Science Publishers B.V., Amsterdam.

Bundgaard, H. (ed.) (1985b). *In* 'Design of Prodrugs', p. 9, Elsevier Science Publishers B.V., Amsterdam.

Bundgaard, H. (ed.) (1985c). *In* 'Design of Prodrugs', p. 21, Elsevier Science Publishers B.V., Amsterdam.

Bundgaard, H. (ed.) (1985d). *In* 'Design of Prodrugs', p. 12, Elsevier Science Publishers B.V., Amsterdam.

Davies, S.S. (1989). *In* 'Therapeutic Peptides and Proteins: Formulation, Delivery, Targeting' (D. Marshak and D. Lin, eds), p. 41, Cold Spring Harbor Laboratory, Cold Spring Harbor, NY.

deBlaey, C.J. and Polderman, J. *In* 'Drug Design, Vol IX' (E.J. Ariëns, ed.), p. 243, Academic Press, London.

Dreyfuss, J., Shaw, J.M. and Ross Jr, J.J. (1976). *J. Pharm. Sci.* **63**, 1310.

Duggan, D.E. (1981). *Drug Metab. Rev.* **12**, 325.

Dunn III, W.J., Koehler, M.G. and Grigoras, S. (1987). *J. Med. Chem.* **30**, 1121.

Edmondson, H.T. (1973). *Ann. Surg.* **178**, 637.

Ferres, H. (1980). *Chem. Ind.*, 435.

Glazko, A.J., Edgerton, W.H., Dill, W.A. and Lenz, W.R. (1952). *Antibiot. Chemother.* **2**, 234.

Glazko, A.J., Carnes, H.E., Kazenko, A., Wolf, L.M. and Reutner, T.P. (1957). *Antibiot. Annu.* 792.

Gorrod, J.W. (1980). *Chem. Ind.*, 457.

Gray, J.E., Weaver, R.N., Moran, J. and Feenstra, E.S. (1974). *Toxicol. Appl. Pharmacol.* **27**, 308.

Hinson, J.A. and Mitchell, J.R. (1976). *Drug Metab. Dispos.* **4**, 430.

Ho, N.F.H., Park, J.Y., Morozowich, W. and Higuchi, W.I. (1977). *In* 'Design of Biopharmaceutical Properties through Prodrugs and Analogues' (E.B. Roche, ed.), A.Ph.A., Washington.

Jansen, A.B.A. and Russell, T.J. (1965). *J. Chem. Soc.*, 2127.

Johnson, P. (1980). *Chem. Ind.*, 443.

Kirchner, F.K., McCormick, J.R., Chester, J., Cavallito, C.J. and Miller, L.C. (1949). *J. Org. Chem.* **14**, 388.

Kyncl, J.J., Minard, R.N. and Jones, P.H. (1979). *In* 'Peripheral Dopaminergic Receptors' (Impsand and Schwartz, eds), Pergamon Press, New York.

Pang, K.S. and Gillette, J.R. (1978). *J. Pharm. Exp. Ther.* **207**, 178.

Preston Mason, R., Campbell, S.F., Wang, S.D. and Herbette, L.G. (1989). *Mol. Pharmacol.* **36**, 634.

Rekker, R.F. (1977). *In* 'Pharmacochemistry Library, Vol. 1, The Hydrophobic Fragmental Constant' (W.Th. Nauta and R.F. Rekker, eds), Elsevier Scientific Publishing Company, Amsterdam.

Rowland, G.F., O'Neill, G.J. and Davies, D.A.L. (1975). *Nature* **255**, 487.

Sakamoto, F., Ikeda, S. and Tsukamoto, G. (1984). *Chem. Pharm. Bull.* **32**, 2241.

Schwartz, M.A. and Heyton, W.L. (1972). *J. Pharm. Sci.* **61**, 906.

Sezaki, H. and Hashida, M. (1984). *In* 'CRC Critical Reviews in Therapeutic Drug Carrier Systems' (S.D. Bruck, ed.), p. 1, CRC Press.

Shek, E., Higuchi, T. and Bodor, N. (1976). *J. Med. Chem.* **19**, 113.

Smith, R.A.G., Dupe, R.J., English, P.D. and Green, J. (1981). *Nature* **290**, 505.

Stella, V.J. and Himmelstein, K.J. (1980). *J. Med. Chem.* **23**, 1275.

Swanson, B.L., Vlasses, P.H., Ferguson, R.K., Berquist, P.A., Till, A.E., Irvin, J.D. and Harris, K. (1984). *J. Pharm. Sci.* **73**, 1655.

Thomas, W.A. (1986). *Biochem. Soc. Trans.* **14**, 383.

Ulm, E.H., Hichens, M., Gomez, H.J., Till, A.E. and Hand, E. (1982). *Br. J. Pharmac.* **14**, 357.

Yalkowsky, S.H. and Morozowich, W. (1980). *In* 'Drug Design IX' (E.J. Ariëns, ed.), p. 123, Academic Press, London.

Young, R.C., Mitchell, R.C., Brown, T.H., Ganellin, C.R., Griffiths, R., Jones, M., Rana, K.K., Saunders, D., Smith, I.R., Sore, N.E. and Wilks, T.J. (1988). *J. Med. Chem.* **31**, 656.

–5–

QSAR and the Role of Computers in Drug Design

E.G. MALISKI
Glaxo Research Institute,
Research Triangle Park,
NC 27709, U.S.A.

J. BRADSHAW
Glaxo Group Research Ltd,
Park Road,
Ware,
Herts SG12 0DP, U.K.

Dr Ed Maliski is Manager, Research Computing in the Glaxo Research Institute, Glaxo Inc. He was recruited by Glaxo after spending 10 years at Sterling Drug in Rensselaer, NY, following his degree studies at the State University of New York. At Sterling he was, at first, a bench chemist before starting the first computational group. His research interests include graph theory as applied to chemistry and non-linear dynamics, and the applications of fractals and chaos to chemical and biological phenomena.

Dr John Bradshaw is currently Research Associate in Computational Chemistry within Glaxo Group Research Ltd. He joined Allen and Hanbury's Ltd at Ware in 1971 as a Senior Chemist, working in synthetic chemistry, after completing his University education at the University of Manchester. In 1976 he set up the Chemometrics group which formed the nucleus of what is now Computational Chemistry. His research interests include the application of mathematical and statistical methodology to problems of drug design.

MEDICINAL CHEMISTRY 2nd Edition
ISBN 0-12-274120-X

I. Introduction

The purpose of this chapter is to examine the role of the computer in the design and discovery of drugs. Computers augment the scientific process in drug discovery by assisting the chemist with collecting, storing, manipulating, analysing and viewing data. Further, computers provide a link to theoretical chemistry and graphic modelling, providing calculated estimates of molecular properties, models of molecules, models of biological sites and sometimes even models of drug–receptor interactions. This chapter will focus on the use of computers in one type of data analysis: quantitative structure–activity relationships (QSAR). The first part of the chapter will examine the role of QSAR in the discovery process.

It is important to note that what current computer systems do *not* do well is interpret data, or interpret the results of a data analysis. Only the scientist can decide whether the results obtained are *valid* or *meaningful*. This point will be the focus of the rest of the chapter.

II. QSAR in the Drug Discovery Process

The development of a QSAR acts as a catalyst in the hypothesis–testing–analysis cycle of the normal scientific process as it is applied to drug discovery and drug design (Fig. 5.1).

The process of drug design begins with the identification of a compound with interesting biological activity. An hypothesis is conceived suggesting which chemical features are related to the bioactivity. The hypothesis is evaluated by synthesizing and testing compounds in an appropriate biological assay. On the basis of the analysis of these biological results, the chemist can modify the original hypothesis and then repeat the process if necessary. Part of the analysis is intended to ascertain a structure–activity

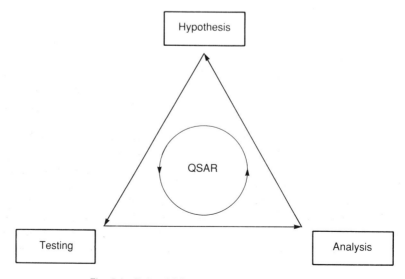

Fig. 5.1   Role of QSAR in the scientific process.

TABLE 5.1
4-Substituted Analogues of AH19065

(1)

Registry number	R	ED_{50} (mg/kg)	DR_{10} (μg/ml)	π	$\mathscr{F}$	$\mathscr{R}$	MR
AH19065	H-	0.18	0.14	0.00	0.00	0.00	1.03
AH20403	CH_3-	0.25	0.24	0.56	0.04	−0.13	5.65
AH21061	$CH_3CH_2OC(=O)$-	2.70	> 10	0.51	0.33	0.15	17.47
AH21514	$(CH_3)2CH$-	0.33	0.20	1.53	−0.05	−0.10	14.96
AH21561	$HOCH_2$-	1.72	1.00	−1.03	0.00	0.00	7.19
AH21830	Br-	0.29	0.46	0.86	0.44	−0.17	8.80
AH21936	CH_3OCH_2-	0.21	0.13	−0.78	0.01	0.02	12.06

relationship, or SAR, that is, to identify whether there is a pattern within the *changes* of molecular features which may explain the *changes* in biological activity. Table 5.1 shows a series of 4-substituted ranitidine analogues (**1**) that were tested for histamine H_2-antagonist activity. The 4-position was selected for modification as it was felt it would influence the binding of the key basic dimethylaminomethyl function to its receptor (Judd, 1980).

The compounds, which were chosen so as to maximize the diversity of physico-chemical properties consistent with synthetic accessibility (see below), are shown in Table 5.1. In this case the atria results are reported as DR_{10}, the dose of compound necessary to shift the histamine dose–response curve 10-fold to the right. The ED_{50} (the dose giving half-maximal response) is for the perfused rat stomach preparation. Both of these tests are described in Chapter 12.

Physicochemical data were taken from the corresponding benzenoid system (Leo *et al.*, 1971), a practice we would no longer recommend (Bradshaw and Taylor, 1989). The relative lipophilicity of the substituent is measured by π, the inductive electronic effect by $\mathscr{F}$ and the mesomeric effect by $\mathscr{R}$. The overall bulk of the system is measured by the molar refractivity MR (further explanation of this terminology is given at the end of this chapter).

There is no simple relationship between the biological activity and any of the parameters in Table 5.1 taken singly, and the use of two or more parameters is invalid (Topliss and Costello, 1972) for such a small number of compounds. However the directional steric parameters of Verloop (Verloop *et al.*, 1976) did provide a possible correlation. Using the minimum van der Waals radius of the substituent, B_1, as the independent variable, a simple linear regression was developed for both pED_{50} and pDR_{10}. The ester, AH21061, and hydroxymethyl derivative, AH21561 were outliers (i.e. anomalies), using both *in vitro* and *in vivo* data. Verloop has shown for substituents such as these that the

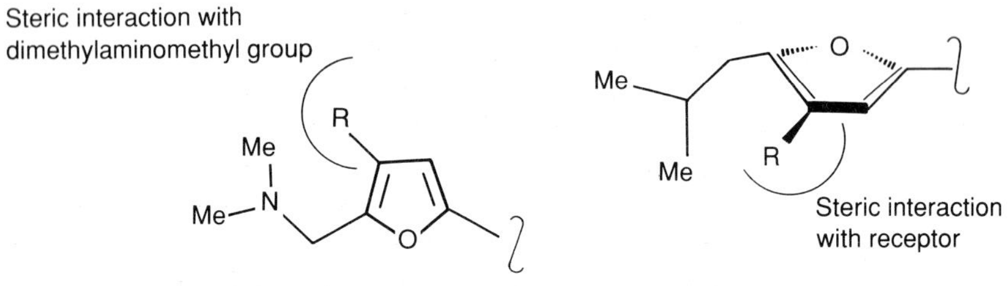

Steric interaction with
dimethylaminomethyl group

Fig. 2 Alternative steric interactions of the 4-substituent in ranitidine.

radius value for the radius 'opposite' to the B_1 value may need to be used. When this is done these points are no longer outliers and we get the equations

$$pED_{50} = 0.646(\pm 0.488) - 0.366(\pm 0.188)B_1 \qquad (1)$$

$$n = 7, r = 0.913, s = 0.198$$

$$pDR_{10} = 1.211(\pm 0.457) - 0.624(\pm 0.192)B_1 \qquad (2)$$

$$n = 7, r = 0.966, s = 0.203$$

In all the equations in this chapter, n is the number of compounds, r is the correlation coefficient and s is the standard error of the fitted lines. Values in parentheses are the 95% confidence limits for the independent parameters.

These equations were consistent with the idea that the group R (Table 5.1) was forcing the Mannich base into an unfavourable conformation for activity. One option was that the substituents were orienting themselves so that their minimum radius was towards the receptor. Whilst this could be true for the ones where B_1 was used, it would be unlikely that the ester in AH21061 was both out of conjugation *and* presenting its maximum radius to the receptor. The hydroxymethyl derivative AH21561 would have to be accommodated by assuming hydrogen-bonding to the receptor.

In this example the hypothesis was that the 4-substituent would influence the binding of the Mannich base in ranitidine (AH19065) (Fig. 5.2). The QSAR uses the same approach except it expresses the relationship between structure and activity in the natural language of science, i.e. in terms of a mathematical model. It is important to note that QSAR does *assume* that both the observed activity and the physicochemical properties of the molecule can somehow be quantified. The data in Table 5.1 combined with an appropriate chemical descriptor for the size of the 4-substituent produce the quantitative models given by Eqs (1) and (2). Notice the relationship between the hypothesis suggested earlier and this equation. It follows then that one purpose of QSAR is the development and the interpretation of mathematical equations useful for providing insight into the importance of certain chemical properties of drugs.

III. Applications Related to Biological Processes

QSAR can be used to 'design-in' biological properties such as potency, efficacy, selectivity and bioavailability. Most successful drugs have at least these four characteristics. A detailed explanation of these terms can be found in Kenakin's (1987) book. This section will briefly review these terms and examine opportunities for QSAR to address each of them.

Potency and efficacy, especially during the discovery process, generally refer to building in the desired biological effects. Potency refers to the amount of compound needed to produce a given effect whilst efficacy, by Kenakin's definition, refers to the maximum level of effect a drug can produce. These are subtle but important differences. For the moment, consider this trivial example from medicine. Given a choice, most general practitioners would probably recommend 500 mg of acetylsalicylic acid (aspirin) instead of 500 mg of acetic acid (vinegar) for the relief of minor headache pain (analgesia). Although such an example may seem obvious and trivial, it does dramatize two very important points in medicinal research. First, all chemicals do not produce the same biological effects, in fact, some produce no apparent effect at all. Secondly, although this difference in analgesic effect seems obvious from experience, the chemical reason or reasons for the difference is not necessarily so obvious. Put another way, this means that there must be *some differences* in the chemical properties and/or structure of aspirin and vinegar which explains the *different* observed analgesic effects (activity) produced and further there must be *something specific* about the structure of aspirin which produces an analgesic effect.

An example of how QSAR was used to assist in increasing the potency during the design process of compounds of this type is described by Unger (1984). In a series of benzoylpyrrolopyrroles (**2**) the 4-substituent was varied and the resulting QSAR equation suggested that the vinyl substituent should be prepared. The company called Syntex took this compound into advanced pharmacology and into the clinic in order to be tested in man (Unger, 1984).

(2)

$$\log A = -4.45 - 0.73\text{HA} + 6.5B_3 - 1.55(B_3)^2 \tag{3}$$

$$n = 12, \ s = 0.550, \ r = 0.832$$

In Eq. (3), A is the analgesic dose relative to aspirin, HA indicates whether the 4-substituent is a hydrogen bond acceptor and B_3 is one of the Verloop size parameters described earlier.

There is a need for most drugs to be selective about which biological site or sites they influence. Selectivity considerations are necessary during the design process to assure that undesirable side effects are minimized or eliminated. The process of designing in selectivity is called 'drug targeting'. This can be exemplified by further consideration of aspirin. Aspirin is a very non-toxic drug which has often been prescribed to provide relief from headache pain, to reduce the pain and inflammation associated with arthritis or to act as an anticoagulant for those prone to heart attack or stroke. One additional biological effect of aspirin is that it appears, in certain instances, to lead to the undesirable side effect of producing stomach ulcers. The goal of drug targeting or drug selectivity is to minimize such side effects while still providing the desired biological effect. This often means adjusting the shape and physicochemical properties of drug molecules to take advantage of the subtle differences in the environments of biological sites throughout the body. Considerable time and effort is still being spent to find alternatives to aspirin which have its potency and safety but without the ulcerogenic properties.

A simple example of using QSAR to give the required selectivity is shown with the antihypertensive (blood-pressure lowering) agents (3) related to labetalol (Trandate®) a compound with activity as an antagonist at α- and β_1-receptors. We showed that the α-effects and β-effects were related to different manifestations of the electronic nature of the aromatic ring. The α-effects were related to the mesomeric effect measured by σ_R and the β-effects to the inductive effect measured by σ_1 (Bradshaw, 1983).

R³⧵

R⁴ — ⟨ring⟩ — CH(OH)CH₂NHCH(Me)CH₂CH₂Ph

(3)

where $R^3 = CONH_2$ or SO_2NH_2
and $R^4 = H$, OH, O-alkyl, halogen, NH_2, NH-alkyl, N(alkyl)$_2$

H₂NOC⧵

⟨thiophene ring, S⟩ — CH(OH)CH₂NHCH(Me)CH₂CH₂Ph

(4)

For α-adrenoreceptor antagonism:

$$pDR_{10} = -2.65(\pm 0.20) - 2.00(\pm 0.45)\sigma_R \qquad (4)$$

$$n = 8, r = 0.876, s = 0.237$$

For β-adrenoreceptor antagonism:

$$pDR_{10} = -0.09(\pm 0.08) + 1.76(\pm 0.34)\sigma_1 \qquad (5)$$

$$n = 10, r = 0.866, s = 0.150$$

Although the quality of the QSAR equation (4) and (5) was not good, it did allow the suggestion to be made that the corresponding thiophene (4) would be a good compound *if the model were correct*. In the event, that proved to be the case and the compound was a potent α- and β-adrenoceptor antagonist.

Drug delivery refers to the process of ensuring that sufficient drug actually reaches the appropriate site within the body to produce the desired biological effect. There are many reasons why some chemicals make poor drugs. These include problems associated with properties such as solubility, transportability, stability and metabolic susceptibility. Two prominent reasons why these problems occur are as follows. First, the body's own defence mechanisms tend to intercept the drugs before they reach the desired biological site; secondly, the path between where a drug can be introduced into the body and the desired biological site is often associated with a complex series of varying polar (aqueous) and non-polar (organic) environments. Structural modifications of a drug molecule must be considered to overcome the problems of what is termed 'bioavailability'.

The use of QSAR in the optimization of bioavailability can be exemplified by work on a series of cephalosporins of the type (5).

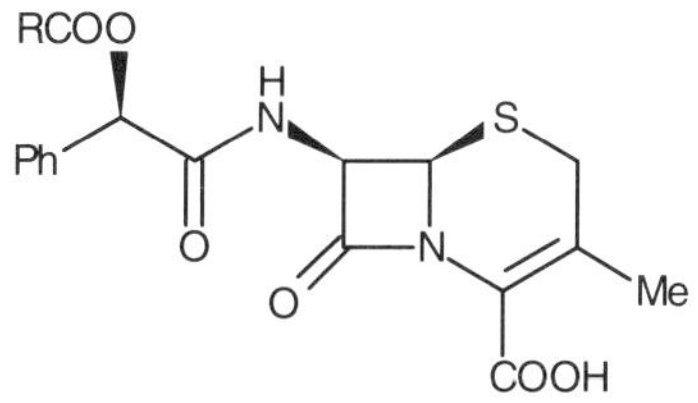

(5)

Workers at Takeda (Yoshimura *et al.*, 1985) showed that the relative bioavailability, BA, of these compounds could be related to the logarithm of the octanol/water partition coefficient, log P, by Eq. (6):

$$\log \text{BA} = 0.763 \log P - 0.226 \log P^2 + 0.879 \qquad (6)$$

$$n = 8, r = 0.946, s = 0.117$$

Lipophilicity was found to be a critical factor for oral absorption. It was established that ester prodrugs of the parental cephalosporin having a 7-mandeloylamino group (**5a**) instead of the more usual 7-phenylglycylamino group (**5b**) were absorbed effectively through the gastrointestinal tract if log P values were in the range 1.23–2.14.

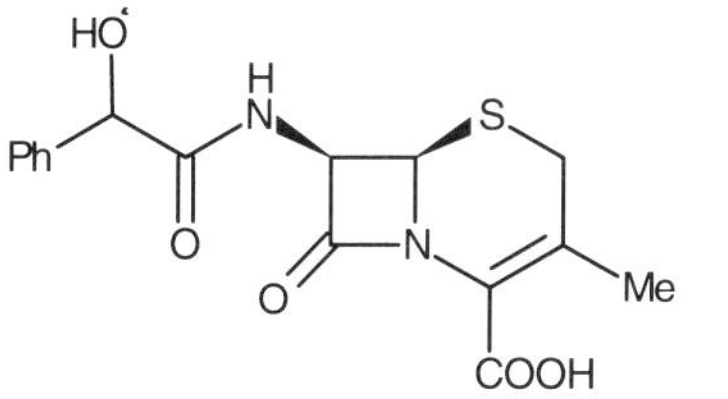

(5a) 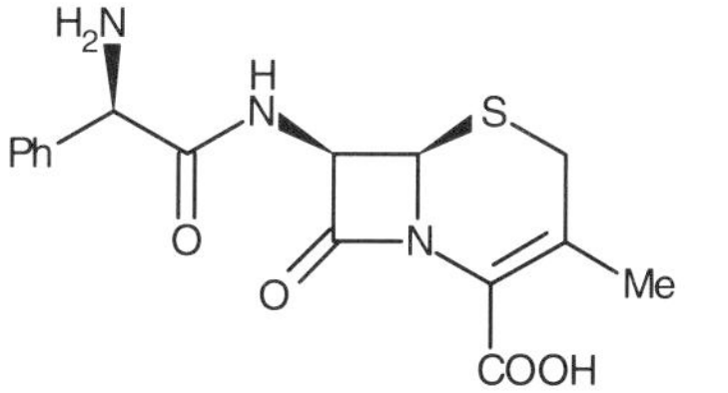   (5b)

IV. Applications Related to the Drug Discovery Process

There is another way in which QSAR and computers can assist the scientist and that is by increasing the *efficiency* of the discovery process. Chemists must deal with the reality of limitations of available time and resources. As of this writing, *Chemical Abstracts Service* has catalogued in excess of 11.4 million structures from chemical literature with about 9000–12 000 new compounds added *each week*. It is unlikely that all these compounds could be synthesized and tested for every known biological activity even with the advances made in high-volume biological testing. Generally the chemist will select a more manageable target such as a class of compounds. However, even in this more limited case the number of potential compounds grows quickly. Consider the example of the discovery of ketoconazole (**6**) as described by Heeres (1985).

Ketoconazole analogues produced by making changes to groups R^1, R^2 and R^3 are shown in Table 5.2. The number of compounds that theoretically could be considered, using these changes, can be calculated by multiplying together the number of variations at each site. In this example it becomes: 2 *cis/trans* possibilities × 21 phenyl ethers (R^1) × 2 heterocyclics (R^2) × 16 phenyl ketals (R^3) = 1344 unique structures. It is interesting to note that a more effective compound, terconazole (**7**) would not have been one of the 1344 structures mentioned.

TABLE 5.2

Ketoconazole Analogues Available by Making
Indicated Changes to the Groups R^1, R^2 and R^3

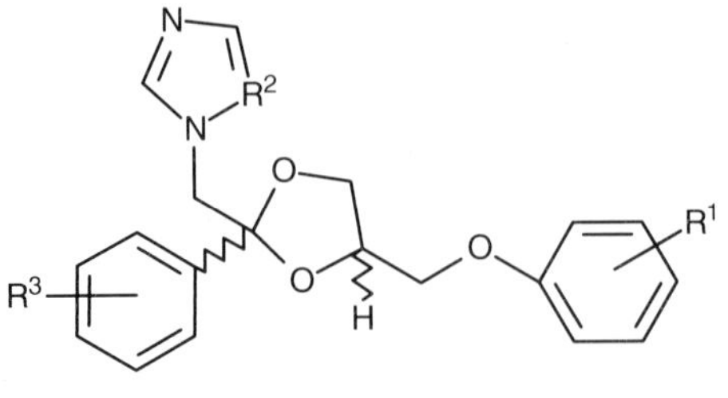

(6)

R^1	R^2	R^3
H	CH	H
2Cl	N	2-Cl
3-Cl		3-Cl
4-Cl		4-Cl
4-F		3-Br
2-Br		4-Br
4-Br		3-Me
2,4-di Cl		4-Me
3,4-di Cl		4-MeO
2-Ph		2-Cl, 4-MeO
4-Ph		2-Cl, 4-Me
4-OMe		2-Cl, 4-F
4-CN		2-Cl, 4-Br
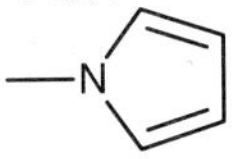		2,4-di Br
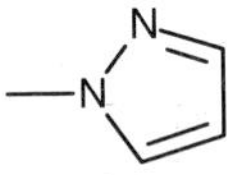		2-Br, 4-Cl
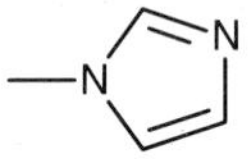		2,4-diCl
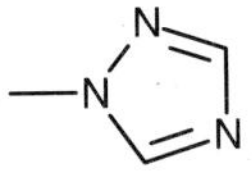		2,3,4-triCl
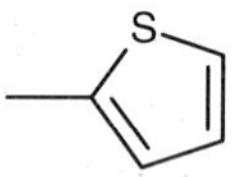		
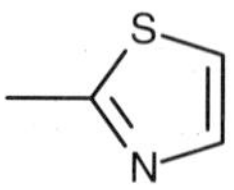		
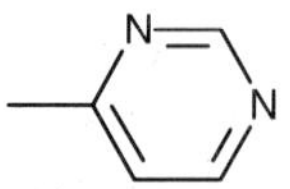		
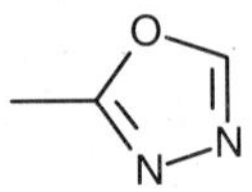		

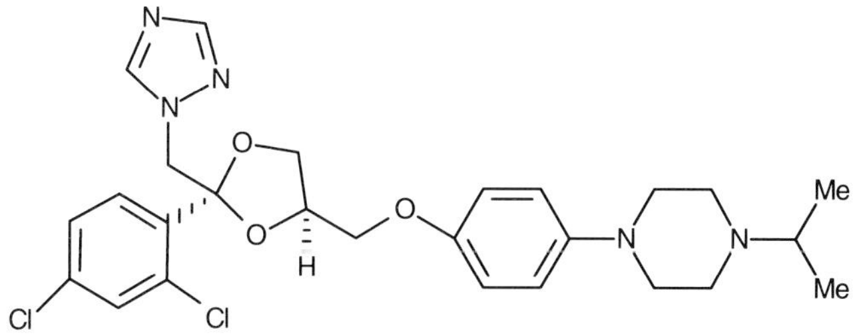

(7)

Another example of the combinatorial problem can be demonstrated by simple substitution of 20 rather common aromatic substituents on as few as three sites on indomethacin (8). This leads to 20^3 or a staggering 8000 unique structures.

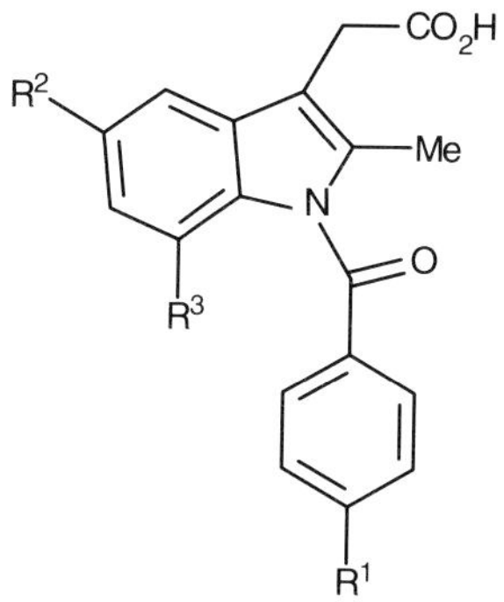

(8)

Where R^1, R^2 or R^3 = H, Cl, F, Br Me Et, i-Pr, t-Bu, OMe, OEt, CN, OH, NO_2, NH_2, CH_2OH, CH_2OMe, SH, SMe, CF_3, CH_2Ph

The traditional way to solve this combinatorial problem is to optimize one site at a time holding the other sites constant. In the previous indomethacin example that would mean finding the 'best' R^1 substituent. Then holding R^1 constant find the 'best' R^2. Finally, holding R^1 and R^2 constant find the best R^3. In this way only 20 + 20 + 20 or 60 compounds need be made which is an improvement over the 8000 suggested earlier. In statistical terms this approach is refered to as an 'L' design. This seems like a good approach and frankly most drugs on today's market were discovered in this manner, and, before the availability of computers, this was the only realistic alternative. However, 'L' designs make at least one assumption, namely, that the substituent changes are acting *independently*. That is changes in one part of the molecule have little or no effect elsewhere in the molecule and therefore the relationship between the substituents is additive. It turns out that such an assumption is not necessarily correct. Table 5.3. shows that, even though the *para*-substituent is not adjacent to the phenolic group, it has considerable effect on the pK_a of the molecule.

This extreme case of the interdependence of substituents is chosen because it is so obvious and as such predictable. Often these interactions are more subtle and therefore unexpected but the effects can be quite dramatic. This is demonstrated by the seemingly small change caused by the introduction of a methyl substituent into ranitidine as given in Fig. 5.3. The reader will recollect from Table 5.1 that replacement of the hydrogen atom in the 4-position by a methyl group had only a small effect on activity.

TABLE 5.3
pK$_a$ Values of Some Substituted Phenols

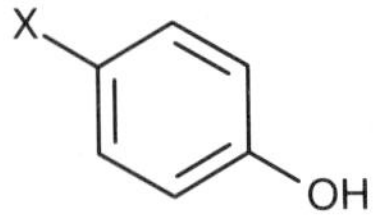

Substituted (X)	pK$_a$
H	10.02
Cl	9.38
NO$_2$	7.15

Several approaches have been suggested by QSAR practitioners which address the combinatorial problem whilst avoiding the linear assumption. Discussion will be limited here to only three approaches which offer diverse solutions to this problem.

One of the earliest approaches introduced has become known as the Free–Wilson approach (Free and Wilson, 1964) named after its authors. This approach, sometimes referred to as the *de novo* approach, is actually an extension of the traditional method described earlier. Like the traditional approach the assumption is made that the substituent effects are additive. The reader is referred to the original paper for a more detailed description of the technique. The advantages offered by the Free–Wilson approach include the ability to change the substituent at several sites (multiple substitution) in an organized and controlled manner and the ability to deal with the changes in the chemical properties as the substituents change without actually identifying what those changes are. What this means is that in the indomethacin example the 60 compounds required on the basis of traditional additivity can be reduced to 40 compounds with the same amount of information. The disadvantages of this approach are its empirical nature and the limitations imposed by the additivity assumption. The empirical nature refers to the fact that the approach can be helpful in identifying what combination or combinations of the substituents examined are most likely to be active but it provides little direct physicochemical insight into why these are good combinations. The result is that it is sometimes difficult to extend the analysis to substituents

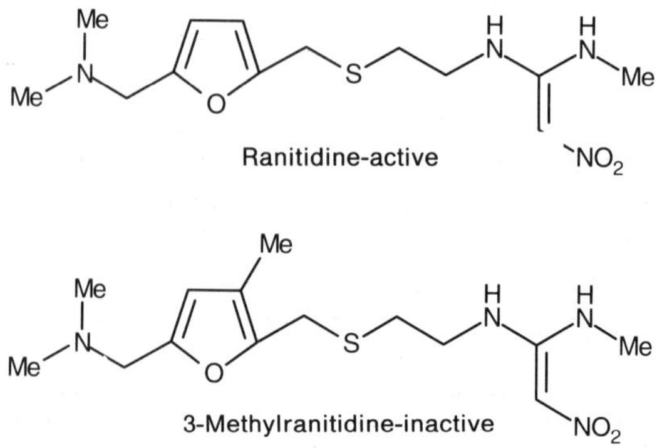

Fig. 5.3   Effect of substituting in the 3-position of ranitidine.

other than those tested. However, it still remains a useful technique in cases where additivity is a reasonable assumption and where only a limited number of substituent changes are chemically allowed or available.

More recently Austel (1982) introduced an approach based on a statistical technique for selecting a chemically diverse subset of compounds from a larger set of compounds. This statistical technique is called fractional factorial design. The extension of fractional factorial design to include multiple site substitution is described in a Ph.D. thesis by Hellberg (1986). The general concept begins by selecting multiple sites to be modified on a lead molecule. Then a set of substituents to be used at each site is chosen from a list of 28 common substituents. The selection of substituents is based on synthetic feasibility and chemical diversity. Substituents which are difficult or impossible to introduce at a particular site, as determined by the chemist on the basis of his or her experience, are eliminated from consideration. The chemical diversity is initially introduced by considering four general descriptors of each substituent; size, lipophilicity,* electronic factors and the proton donor/acceptor nature of the substituent. Finally, fractional factorial design strategy is applied and a statistically representative subset is identified. Without going into detail, this means that the indomethacin example can be reduced to a subset of 16 compounds. Again the reader is referred to the original work for more detailed explanations. This approach has many advantages: it is easy to do, it requires only a small number of compounds to be made and it does not assume additivity. Most importantly, any QSAR equation generated will have chemical meaning. For example, an equation which indicates that large substituents are not well tolerated at the R^3 position in the indomethacin example provides an insight useful to the chemist in future compound design. The disadvantages of this approach are that the chemist is more or less restricted to the 28 substituents listed and that some of the variables such as the lipophilic- and electronic-influencing variables can be greatly distorted by their chemical environment even to the point of 'reversing' their character. For example, the introduction of a 2-amino substituent into pyridine produces hardly any change in the lipophilicity of the compound as measured by the octanol/water partition coefficient. Normally, one would expect an amino group to markedly decrease the lipophilicity, as is seen in going from benzene to aniline (Fig. 5.4).

This latter disadvantage represents special circumstances and can often be avoided by applying a little chemical common sense.

The last approach to be discussed is called whole molecule design (Maliski and Bradshaw, 1992). This approach is based on the assumption that researchers usually have several ideas for substituent changes and several sites on the molecule to consider. Further, these changes may not be accommodated by the usual parameterized substituent approaches. For example, in the indomethacin design problem, incorporating changes of the ring from an indole to a benzofuran or a benzothiophene would be difficult to do properly using a substituent-based approach. Whole molecule design easily accommodates such changes. In this approach the chemist defines all of the molecular changes of his or her hypotheses as a set of molecular fragments, and all the possible combinations of these fragments are computer-generated. Next, the chemist can add any

---

*Lipophilicity, or hydrophobicity, usually refers to the tendency of a drug to distribute or partition amongst lipid and aqueous layers. This is usually expressed quantitatively as log of the partition coefficient (log P) where log P = log (conc. in lipid layer) − log (conc. in aqueous layer).

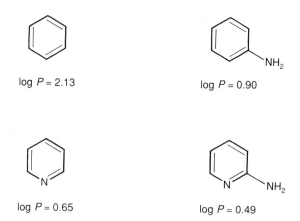

log P = 2.13   log P = 0.90

log P = 0.65   log P = 0.49

Fig. 5.4   Effect on lipophilicity following the introduction of an amino group into benzene and pyridine.

rules based on experience and intuition such as: 'eliminate any compound with more than one nitro group' or 'keep only those compounds where the estimated log P is between 2.0 and 2.5'. There are no limits on the number of rules that can be incorporated. After the rules are applied, the remaining compounds are parameterized on the basis of the size of the remaining set of compounds, the nature of the particular project, experience, and hardware/software considerations. A representative subset is then selected using clustering techniques (grouping compounds by similarity) or experimental design. The advantages of whole molecule design are its ability to accommodate any synthetic modifications suggested by the chemist and its ability to incorporate intuition into the design process. A limitation of this approach is that by concentrating on the whole molecule the 'detail' of the parts may be overlooked. For example, the dipole moment might be considered as a whole molecule descriptor; however, both methane and carbon tetrachloride have similar dipole moments but the electronic nature of the carbon atom differs markedly between the two compounds.

V.   Interpretation

A.   The Nature of Biological Data

The application of computer techniques to the discovery process is analogous to doing synthesis. Not all synthetic chemistry is Diels–Alder chemistry and not every compound can be made by a Michael addition reaction. Just as one must choose a synthetic route based on the desired chemical target and on the available starting materials, one must choose a mathematical technique based on the nature of the biological data and the availability of chemical descriptors. The next section will discuss some of the important considerations in evaluating biological data, selecting chemical descriptors and interpreting mathematical models. Carrying this analogy to synthetic chemistry a little further raises another important consideration: in synthetic work it is not at all uncommon to find that the proposed synthetic route did not work as planned and modification or redirection of effort is needed. The same holds true for mathematical models. Assumptions change as more compounds are made and tested. New insight may dictate new models. This should not be seen as an indictment of the models but as a logical consequence of the research process.

As is noted in the first chapters of this book, actual biological systems of interest to the pharmaceutical industry, ultimately the human body, are complex dynamic chemical environments subject to variability by their very nature. Such variability must be considered when interpreting any biological data. This is especially true for the selection of the mathematical modelling tool and in the interpretation of the resulting equation.

Further, almost all primary biological testing is conducted in models of the disease state rather than in the disease state itself. For example, potential antibacterial drugs are usually tested first against the isolated bacteria in a Petri dish, then advanced to animal models, and, if effective, to man. The assumption here is that any compound that is ineffective in inhibiting bacteria in a Petri dish will also be ineffective in man. The fact that this type of testing produces many false positives, that is compounds that are effective in the Petri dish but not in man, is accepted. Therefore the pre-screening in the Petri dish is assumed to be a *necessary but not sufficient* model of the disease state. Similarily, no QSAR model can be more accurate than this assumption allows.

Wold *et al.* (1984) suggest that an alternative for improving the information content of the biology is to measure the desired biological effect in several different biological tests and in several different biological systems. Their argument is based on the assumption that no single biological test is an exact predictor of the ultimate biological event in man, therefore each test may be a partial descriptor, implying that a battery of tests considered simultaneously may be more indicative of the desired state in man. This approach can be referred to as developing a biological profile. The statistical methodology used for evaluating many biological responses at the same time is called 'multivariate analysis'. There is a certain parallelism to the 'L' design of a chemical series mentioned earlier. In this case in 'optimizing' a single biological test before proceeding to the next test one runs the risk of falling victim to the prejudice of the particular test. If the tests are optimized in succession, the prejudice of the first test will be added to the second, and the first and second to the third, and so forth. What multivariant techniques provide is a method of analysing the biological data for commonality and differences while classifying the chemical compounds into groups or classes and sometimes even identifying chemical properties which may be related to those classifications.

Finally, the nature of the data must be reviewed before development or interpretation of a QSAR model. All data are not created equal. When the difference between the reported data is greater than the error in the experimental measurement, the data are considered continuous. For example, if the data are 3.0, 4.0, 7.0, 9.0, 15.0 and the error in measurement is 0.2, the data may be considered continuous. If on the other hand these same data have an associated error of say 2.0, then 3.0 is not different from 4.0, and 7.0 is not different from 9.0. Such data then fall into three groups or categories; (3.0, 4.0), (7.0, 9.0) and (15.0). This data type is usually referred to as ordinal or interval data or often simply categorical data. A special type of categorical data which has no numeric value associated with it is nominal data. An example of this is the classification of compounds as active or inactive. Each data type represents legitimate data and each data type has several methods available for appropriate analysis. One of the most common problems in QSAR is that data are too often reported in such a way as to appear as if they are continuous when in reality they are categorical.

Another common problem associated with biological data is that, for convenience of testing, biological data are often measured in units of mass/unit volume but for QSAR the required units are moles/unit volume. This, of course, is an easy conversion to do but also an easy one to forget.

B.   The Nature of Chemical Descriptors

The ways that have been used to describe chemical structures as a number or a set of numbers are too numerous to mention even briefly, so the next section will concentrate on three approaches used to 'quantify' a chemical compound.

The first quantitative chemical descriptors reported in the literature are related to measured or calculated physicochemical properties of a compound. Simple examples include molecular weight, melting/boiling point and instrument measurements such as nuclear magnetic resonance (NMR) shifts. Although such descriptors are meaningful to most chemists, many of these descriptors do require that the compound must be made in order for the descriptor to be measured. This requirement for the compound to already exist precludes using any resulting QSAR equation to predict the biological activity of hypothetical compounds. Such predictions are often an important contribution of QSAR to the chemist in his or her work. Other physicochemical parameters have been constructed from measured values in such a way as to be considered generalizable. One of the most well known of these are the π values used to estimate the hydrophobic nature of substituents on an aromatic ring. For example, the π value for a *para*-substituted chlorine atom is estimated from the differences in the logarithms of the octanol/water partition coefficients of *p*-chlorobenzoic acid and of benzoic acid:

$$\pi p\text{-Cl} = \log P_{(p\text{-ClPhCOOH})} - \log P_{(PhCOOH)}$$
$$= 2.65 - 1.87$$
$$= 0.78$$

The assumption is that the contribution of the *p*-chloro substituent will remain more or less constant for most aromatic systems, an application of the linear free energy (LFER) approach. The limitations of this linear additive assumption have been discussed in an earlier section. The reader will note that this concept is entirely analogous to the Hammett σ parameters derived from the pK_a values of substituted benzoic acids.

Substituent descriptors based on physical measurements of molecular properties remain a very popular source of parameters for modelling a chemical entity. There are many such descriptors detailed elsewhere (Martin, 1978; Plummer, 1990).

A second source of number descriptors for structures is computational chemistry. For instance, most large molecular modelling packages will provide routines for the calculation of some estimate of molecular volume and molecular surface area. Quantum mechanics calculations are often used to provide descriptors related to the electronic nature of molecules. These descriptors include estimates of point charges on particular atoms, of dipoles and of polarizability. Further, both quantum and molecular mechanics calculations have been used to estimate the energies of molecules in a given conformation or set of conformations. These energy calculations or the difference in energy between a selected conformation and a local minimum energy conformation are used to simulate the strain on a molecule to fit it into some predetermined orientation. A limitation often associated with quantum calculations is that the high level calculations which are usually recommended require either a large computer or a fair amount of time on smaller

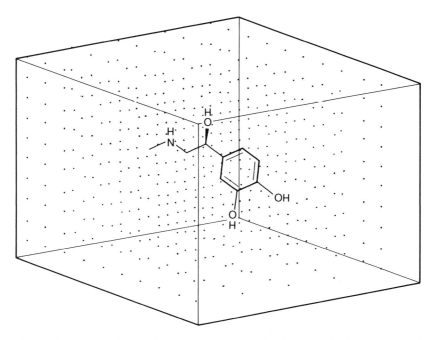

Fig. 5.5   Pictorial representation of the fields around a molecule using the 'grid of points in a box' approach.

machines. This often restricts the number of compounds that can be considered in the study.

One of the more recent contributions from computational chemistry is a method called comparative molecular field analysis or CoMFA (Cramer *et al.*, 1988). In this method the steric and electrostatic fields are represented as grid points within a 'box' constructed in such a way as to enclose completely the molecules of interest (Fig. 5.5). Each member of this set of molecules is oriented by aligning it with an arbitrary 'standard' molecule. It is then placed into the 'box' and its individual fields calculated. The resulting set of grid points becomes the chemical descriptors whilst the biological results become the left hand side of the QSAR equation. As these grids usually contain thousands of points, special mathematical techniques such as partial least squares are used in the analysis. One of the basic limitations of all of these methods is the dependence on the choice of the three-dimensional conformation of a molecule to be used in the calculation. Since most molecules of interest to medicinal chemists have at least some degree of flexibility, chosing a 'meaningful' conformation can be difficult.

A third means of creating numerical chemical parameters is from mathematical theory which, in this discussion, will be exemplified by applied graph theory. The basic premise is that a chemical structure can be thought of as a series of nodes (atoms) and edges (bonds). Since different molecules each have a unique arrangement of atoms and bonds, connectivity patterns called 'paths' can be calculated and combined to form a molecular connectivity index or as it is more commonly referred to, a topological index. Literally hundreds of such indices have been proposed each varying to some degree by the values used to represent the atoms and the bonds and in the way the path is calculated. Kier

(1986) provides an excellent overview of the topic and some relevant applications to chemistry. The main appeal to this approach is that such indices can *always* be calculated for any molecule that can be drawn. There are no requirements to conduct any actual physical measurements and they are conformationally independent. Additionally, since such calculations require only simple mathematics, involving exact counting rather than machine- or theory-limited measurement, they can easily and accurately be applied to large sets of molecules numbering several million structures. The major limitations of these parameters is that they are often hard to relate back to the physical world. Thus QSAR models formed from these indices and any resulting predictions of biological activity of untested molecules sometimes must be accepted on faith.

C.   Interpreting the QSAR Model

Interpretation of any QSAR model should begin with asking two fundamental questions:

1. Is the equation mathematically sound?
2. Is the model chemically reasonable?

Mathematical soundness refers to the validity of the statistical technique as applied [see also Wold (1991)]. Chemical reasonableness refers to the relevance of the chemical descriptors to the problem. Although these questions may seen obvious, it becomes easy to overlook them during the course of the drug discovery process.

  The best way to ensure that a model is mathematically valid is to consult with an independent statistician. Every statistical technique has its strengths and limitations. As stated earlier, usually it is the amount, nature and type of data, both chemical and biological, that determine the appropriateness of a technique for a particular problem. It is worth mentioning here that there is increasing evidence that Nature can best be described with non-linear models. Although this may be the case, non-linear models require more care to develop and are less obvious to interpret. For this reason some QSAR practitioners have traditionally preferred and will continue to prefer linear models. Caution, however, must be exercised when using linear models for prediction. Since linear models are approximations for a small part of a non-linear surface, extending the model to predict values beyond those used to form the model can be misleading.

  Suppose the true model is represented by the heavy line in Fig. 5.6. If, however, the only data we have are for compounds with log P in the range -1.0 to 2.0 (from points A to B), these data can be fitted to a statistically significant parabola, i.e. a linear combination of log P and log P^2.

$$-\log (\% \text{ inhibition}) = 0.4 \log P^2 + 0.7 \log P - 1.1$$

This is shown as the dotted line in Fig. 5.6.

  This partial representation of the model will do an adequate job at predicting point C (log $P = 1.8$) but will fail to predict any values greater than 2.0 such as point D or activities of compounds with log P below -1.0. However, provided that care is taken and false expectations are not set for extrapolated values, linear models do still provide valuable insight during the experimental process, as has been shown in the examples earlier in this chapter.

  One of the most common traps of QSAR is to become so involved and excited about the model and the fit of the data to the model that the underlying science gets forgotten

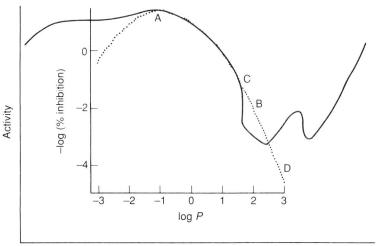

Real world independent variable

Fig. 5.6   Relationship between a linear model and a section through a non-linear hypersurface.

or ignored. It is always best to review the model just before interpretation to ensure that the model makes sense. The importance of considering the experimental error in determining the type of biological data has been previously discussed. Chemical data too can be misleading. Hopefully the error in the chemical descriptor was considered before analysis. Even so there are still two important considerations to review.

1. Can a logical potential mechanism of drug action be developed which would fit the QSAR model? If not, the model may be spurious.
2. Are the chemical descriptors that are most influential in the model varying enough to warrant such significance? For example a model which relies heavily on log P when the data for log P varies only between 1.2 and 1.5 (experimental error 0.2) should be viewed with great scepticism (Martin and Panas, 1979).

VI.   Summary

What we have attempted to do in this chapter is show both the value and limitations of the use of computers in medicinal chemistry. This has been illustrated by applications of QSAR.

The process of drug design is ill understood. A commercial drug does not bear the same relationship to its design process that an aeroplane or automobile does. Most of this book illustrates the need for lateral thinking and taking advantage of chance occurrences. The current generation of computers possesses neither of these attributes.

What computers *are* allowing us to do is *quantify* hypotheses such that a practised medicinal chemist can suggest compounds to test out his/her ideas.

The expansion of databases containing chemical information was remarked on earlier in this chapter; what is now being developed is software to re-create the 'browsability' of the original printed text and also to minimize the prejudices of the abstracter. It is important to remember that information, ideas, hypotheses, etc. gain no value by being passed through a computer, and manipulation of data does not turn bad science into good science.

The medicinal chemist now has unprecedented access to information world-wide and the ability to collaborate with other scientists separated by many thousands of miles. For example, the two authors of this chapter worked on the same electronic copy even though they were separated by 3000 miles and five time zones.

Medicinal and organic chemists have in general been slow to move into the use of computers in their work, a fact reflected in the low level of computing in most under-graduate courses. What we would like to suggest is that computers *can* be useful in medicinal chemistry, and used appropriately, they can assist the chemist in 'doing good science'.

VII. Further Explanation of Terminology

The following terms, used within this chapter, may be unfamiliar to some readers: π, σ, σ_I, σ_R, $\mathscr{F}$, $\mathscr{R}$, L, B_1–B_4.

π is a measure of the change in lipophilicity of a compound (i.e. its altered preference for a non-aqueous environment), caused by the introduction of a substituent X. Formally it is defined as the difference in the logarithm of the octanol/water partition coefficient of the parent compound, $\log P_{R-H}$, and that of the compound itself, $\log P_{R-X}$.

$$\pi_X = \log P_{R-X} - \log P_{R-H}$$

Positive values of π_X indicate that the substituted compound is more lipophilic than the parent. Also as π_X is the difference between two equilibrium constants, it is proportional to a change in Gibbs' free energy. This change in energy, $\Delta\Delta G$, can be quantified as

$$-\Delta G = RT \ln K$$

$$\cong 1.4 \log P \, \text{kcal/mol at } 37°$$

or

$$-\Delta\Delta G \cong 1.4 \, \pi \, \text{kcal/mol}$$

σ is a measure of the change in the electronic nature of a compound caused by the introduction of a substituent X. It was first defined by Hammett (1940, 1970) as the difference in the pK_a of the benzoic acid bearing the substituent X and benzoic acid itself:

$$\sigma_X = pK_{a_{C_6H_5COOH}} - pK_{a_{X-C_6H_5COOH}}$$

Positive values of σ_X indicate that the substituent X overall is electron-withdrawing relative to hydrogen. As for the π value above, σ is proportional to a change in the Gibbs' free energy.

Taft (1956) and others established that there are two major contributions to the electronic effect. A *field* or *inductive* effect taking place through the σ-bond framework and a *resonance* or *mesomeric* effect exerted via the π-electron system. These separate effects have been estimated by two approaches. A chemical approach favoured by Taft's group measuring the electronic effects in systems in which there were no π-electrons gave the inductive effect as σ_I. The resonance component, σ_R, was defined such that for a *para* substituent on a benzene ring

$$\sigma = \sigma_I + \sigma_R$$

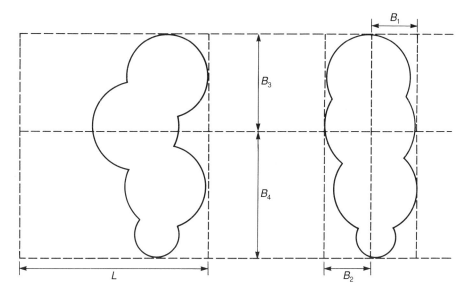

Fig. 5.7   Front and side views of the carboxyl substituent showing the Verloop shape parameters.

An alternative statistical approach, by Swain and Lupton (1986), using σ values for the *meta* and *para* substituents on aromatic systems gave rise to the $\mathscr{F}$(ield) and $\mathscr{R}$(esonance) scales.

L, B_1–B_4 are a series of parameters introduced by Verloop *et al.* (1976) to account for the shape of molecules. L represents the nominal length of a substituent based on CPK models of the corresponding substituted benzene measured along the axis of the bond between the substituent and the parent molecule. B_1 is the minimum van der Waals radius at right angles to this L-axis. The other radii are distances measured perpendicular to L at 90°, 180° and 270° rotation from B_1 and are labelled B_2–B_4 in order of increasing value. These are shown schematically in Fig. 5.7.

References

Austel, V. (1982). *Eur. J. Med. Chem.* **17**, 9.

Bradshaw, J. and Coates, I.H.C. (1983). In 'Quantitative Approaches to Drug Design' (J.C. Dearden, ed.), Elsevier, Oxford.

Bradshaw, J. and Taylor, P.J. (1989). *Quant. Struct. Act. Rel.* **8**, 279.

Cramer, R.D., Patterson, D.E. and Bunce, D.E. (1988). *J. Am. Chem. Soc.* **110**, 5959.

Free, S.M. and Wilson, J.W. (1964). *J. Med. Chem.* **7**, 395.

Hammett, L.P. (1940 and 1970). In 'Physical Organic Chemistry', McGraw-Hill, New York.

Heeres, J. (1985). In 'Medicinal Chemistry', 1st edn (S.M. Roberts and B.J. Price, eds), p. 249, Academic Press, London.

Hellberg, S. (1986). 'A Multivariate Approach to QSAR', University of Umea, Sweden.

Judd, D.B. (1980). In 'Histamine Antagonists', M. Sc. Thesis, CNAA, London.

Kenakin, T.P. (1987). 'Pharmacological Analysis of Drug–Receptor Interaction', Raven Press, New York.

Kier, L.B. and Hall, L.H. (1986). In 'Molecular Connectivity in Structure Activity Analysis', Research Studies Press, Letchworth.

Leo, A., Hansch, C. and Elkins, D. (1971). *Chem. Rev.* **71**, 525

Maliski, E., Bradshaw, J. and Latour, K. (1992) *Drug Design and Discovery* **9**, 1.

Martin, Y. (1978). 'Quantitative Drug Design', Marcel Dekker, New York.

Martin, Y.C. and Panas, H.N. (1979). *J. Med. Chem.* **22**, 784.

Plummer, E. (1990). In 'Reviews in Computational Chemistry' (Lipkowitz and Boyd, eds), VCH Publishers, New York.

Swain, C.G. and Lupton, E.C. (1968). *J. Am. Chem. Soc.* **90**, 4328.

Taft, R.W. (1956). 'Steric Effects in Organic Chemistry' (M.S. Newman, ed.), Wiley, New York.

Topliss, J.G. and Costello, R.G. (1972). *J. Med. Chem.* **15**, 1066.

Unger, S. (1984). 'QSAR in the Design of Bioactive Compounds', J.R. Prous, Barcelona.

Verloop, A., Hoogenstraaten, W. and Tipker, J. (1976). In 'Drug Design' (E.J. Ariens, ed.), Vol. 7, p. 165, Academic Press, New York.

Wold, S. (1991). *Quant. Struct. Act. Relat.* **10**, 191.

Wold, S., Dunn, W.J. III and Hellberg, S. (1984). In 'Drug Design, Fact or Fantasy' (Jolles and Wooldridge, eds), p. 95, Academic Press, London.

Yoshimura, Y., Hamaguchi, N., Kakeya, N. and Yashika, T. (1985). *Int. J. Pharm.* **26(3)**, 317.

–6–

The Current Status and Future Impact of Molecular Biology in Drug Discovery

T.J.R. HARRIS

Biotechnology,
Glaxo Group Research Ltd,
Greenford, Middx., U.K.

Dr Tim Harris, was educated at Uppingham School and at Birmingham University, where he took a Biochemistry degree, and a PhD in Molecular Virology. He worked at the Animal Virus Research Institute in Pirbright and at the State University of New York in Stony Brook (U.S.A.) before taking up a position as a molecular biologist at Celltech Limited. Since 1989 he has been Director of Biotechnology at Glaxo Group Research Limited, U.K.

I. Introduction

Molecular biology has now pervaded all areas of drug discovery from natural-product screening to receptor classification. The aim of this chapter is to review those areas of drug discovery where molecular biology is making a particularly important contribution and to try to put these into the proper perspective with respect to research and development in a pharmaceutical company. The areas that will be covered are listed in Table 6.1. There is something of a historical perspective to how the chapter develops. There is no doubt that molecular biology and genetic engineering are having their biggest impact as an enabling technology, that is providing the methods and techniques for understanding the molecular pathology of disease and for identifying points of intervention. The pharmaceutical industry was actually rather slow to adopt these powerful technologies.

MEDICINAL CHEMISTRY 2nd Edition
ISBN 0-12-274120-X

TABLE 6.1
Molecular Biology in Drug Discovery

Protein therapeutics
Protein engineering
Protein structure determination and modelling
Peptides as drugs
Novel assays and screens
Molecular basis of disease
Transgenic animals
Gene therapy

The lead was taken by the emergence of small 'biotech' companies specifically set up to exploit the developing science. Protein pharmaceuticals – the first products of these companies – provided the platform from which the other applications were developed.

II. Protein Therapeutics

The first protein produced for therapy by using genetic engineering was human insulin. Chemically synthesized genes coding for the A- and B-chains of the heterodimeric insulin molecule were expressed in the bacterium *Escherichia coli* (*E.coli*) as β-galactosidase fusion proteins* (Harris, 1983) and insulin reconstituted following cyanogen bromide (CNBr) cleavage and oxidation.

This method was necessary because direct expression of insulin A- or B-chains in *E.coli* led to instability of the peptides. Fortunately, neither the A- or B-chains contain methionine residues so a methionine residue could be incorporated into the fusion protein before the two peptide sequences allowing cleavage by CNBr (Fig. 6.1)

Human insulin is now prepared in *E.coli* and yeast (a fungus) from its precursor proinsulin. However, the production of insulin by the fusion route served as a paradigm for the expression of many other proteins including growth hormone of human and animal origin (Table 6.2). It is not always necessary to use the fusion protein approach as many larger proteins appear to be stable when made in *E.coli*. Note however, that agents may be needed to solubilize and refold the proteins (Marston, 1986). Despite the utility of *E.coli* expression systems, some proteins destined for therapeutic use (particularly large and post-translationally[†] modified ones) are better expressed in yeast (*Saccharomyces cerevisiae*) or in mammalian cells. A good example of the former is hepatitis B virus surface antigen, a protein capable of conferring protection against hepatitis B virus. The gene coding for the surface antigen protein was cloned into a yeast expression vector and transfected yeast cells were selected (Fig. 6.2). It was found that the yeast cells contained 22 nm particles of surface antigen protein resembling those present in the plasma of virus-infected individuals. These particles were purified, formu-

---

*Fusion proteins produced in *E.coli* consist of a portion of a bacterial polypeptide (in this case β-galactosidase) fused to a novel amino acid sequence (e.g. the A-chain of insulin). The fusion protein is translated from a chimeric gene cloned in a plasmid consisting of the bacterial protein coding sequence joined in frame to the gene coding for the new polypeptide.

[†]After polypeptide chains have been synthesized (translated from the mRNA) they are sometimes modified by cleavage or glycosylation or some other chemical modification of an amino acid side chain. This is termed post-translational modification.

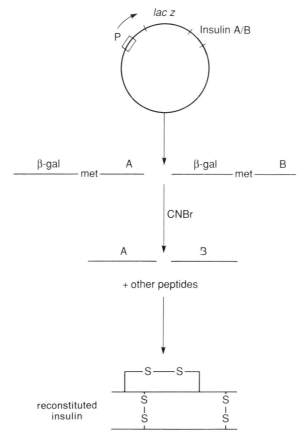

Fig. 6.1   Diagram showing the expression vector and fusion proteins for the production of insulin A- and B-chains in *E.coli*.

lated into a vaccine and shown to protect chimpanzees against challenge with the virus (McAleer *et al.*, 1984). Such yeast-derived vaccines are now available for use in humans.

Probably the best example of a human therapeutic protein made in mammalian cells is tissue plasminogen activator (t-PA). This large protein contains several glycosylation sites (points at which sugar residues are attached), and multiple disulphide bridges holding together the compact structure (Fig. 6.3)

The protein consists of a light chain containing the active site of the enzyme and a heavy chain which has other functions, e.g. mediating fibrin binding and degradation of the protein in the liver. The protein is involved in blood haemostasis, i.e. helping to control the balance between clotting and unclotting which goes on constantly in the bloodstream. It works by activating plasminogen, cleaving it to form plasmin, a potent fibrinolytic (Fig. 6.4) Its perceived advantage in a clinical setting over other bacterially derived plasminogen activators (e.g. streptokinase) was its inherent fibrin selectivity, so that plasminogen was activated only in the presence of fibrin (i.e. only after a blood clot was formed). Now that large quantities of the enzyme have been made, data are beginning

TABLE 6.2
Therapeutic Proteins Made Using Recombinant
DNA Techniques

Human insulin
Hepatitis B virus surface antigen
Human growth hormone
Interferon α
Interferon β
Interferon γ
Factor VIII
Tissue plasminogen activator (t-PA)
Erythropoietin
Granulocyte colony-stimulating factor (CSF)
Granulocyte–macrophage CSF

N.B. This list does not cover all potentially therapeutic proteins. It does, however, cover all those
that are made by recombinant DNA techniques
currently licensed for therapeutic use.

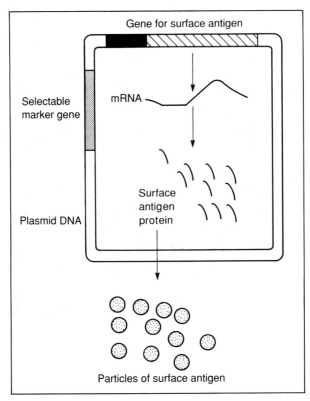

Fig. 6.2 Cartoon depicting the expression of hepatitis B virus surface antigen from a plasmid in a
recombinant yeast cell.

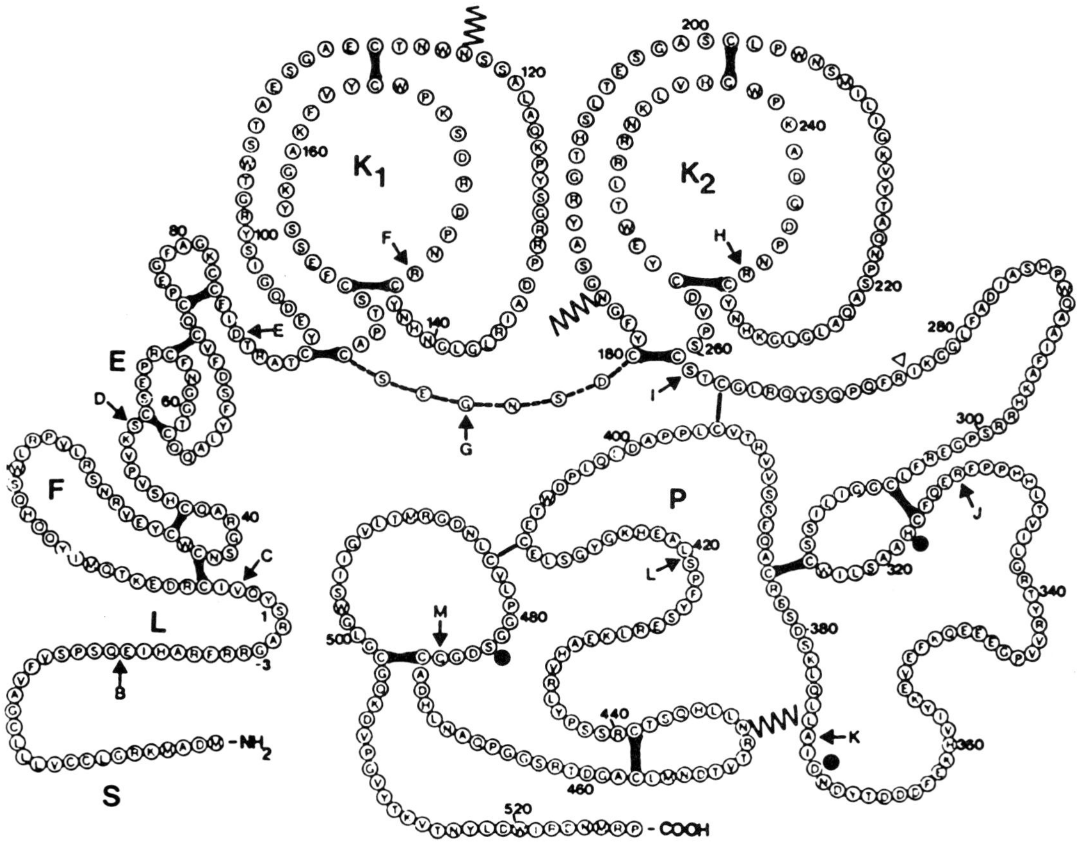

Fig. 6.3 The domain structure of t-PA. The disulphide bridges are marked [▮] as are the sites of glycosylation [≩]

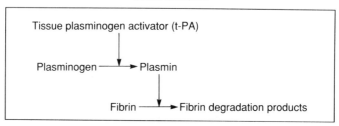

Fig. 6.4   Activation of plasminogen by tissue plasminogen activator.

to accumulate which compare t-PA with streptokinase in its ability to clear coronary thrombosis and to maintain arterial patency (see results of TIMI study group, 1989).

There are many other examples of therapeutic proteins (Table 6.2) which would not be available without the ability to clone their genes and express them in a variety of different systems. The colony-stimulating factors, granulocyte colony-stimulating factor, (G-CSF), granulocyte macrophage colony-stimulating factor (GM-CSF) and erythropoietin (EPO) are the most recent recombinant proteins to have been approved for human use (Cheson, 1990). G-CSF and GM-CSF stimulate white cell proliferation (particularly neutrophils) and have great potential in cancer chemotherapy where significant infection and morbidity result from chemically induced neutropenia (depletion of neutrophils). Erythropoietin stimulates red blood cells and is used to treat anaemia, particularly that occurring in patients requiring repeated kidney dialysis as a result of renal failure.

III.   Protein Engineering

Once a cloned gene or cDNA has been obtained, it is now quite straightforward to make site-specific modifications to the DNA, including point mutations, deletions or insertions. It is not the intention in this review to cover the basic technology of genetic engineering and protein engineering. This can be found in Old and Primrose (1991) or Brown (1991). Suffice to say, it is possible to alter the coding sequence of a cloned gene at will and generate new proteins, the function of which can be measured in various assays both *in vitro* and *in vivo*. This sort of protein engineering, coupled with structural analysis (see below) is leading to great insights into the mechanism of enzyme catalysis for several classes of enzyme and may well have implications for the design of transition-state analogues as potential drugs (Knowles, 1991). At present, however, the greatest impact that protein engineering is having in drug discovery is on the ability to generate new therapeutic entities. Antibodies provide a powerful example of this strategy.

Most monoclonal antibodies are derived from mice or rat cells and as a consequence will give rise to an immune response if injected into a human being. As there are several clinical indications – notably graft rejection, septic shock as well as tumour therapy and diagnosis – for which monoclonal antibodies are being developed, it is no surprise to find considerable interest in modifying the rodent antibodies so that they are less immunogenic; this process is referred to as 'humanization'. The antigen-combining site of an antibody is determined by the variable (V) domains of the heavy and light chains (Fig. 6.5). Thus, once genes coding for human immunoglobulin light and heavy chains were cloned, it was relatively straightforward to make genes where human V_L and V_H sequences are replaced by those coding for the murine V_L and V_H giving rise to chimeric genes (Winter and Milstein, 1991).

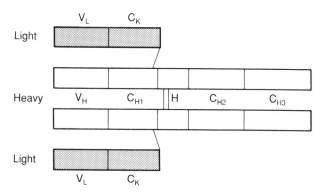

Fig. 6.5 Structure of an antibody molecule showing the association of heavy and light chains (held together by disulphide bridges) and the arrangement of variable (V) and constant (C) domains.

Expressing the chimeric genes gives rise to fusion proteins consisting of murine variable domains and human constant domains. Chimeric molecules have been made and expressed in several systems (Pluckthun, 1991). As the antigen-combining site is actually determined largely by hypervariable regions (complementarity determining regions, CDRs) within the variable domains, it has been possible to 'demousify' an antibody almost completely by 'CDR grafting'. Here the CDRs from murine heavy and light chain genes with the desired specificity are cloned into heavy- and light-chain variable domains of human origin (Reichmann *et al.*, 1988). The one CDR-grafted human antibody used clinically so far (CAMPATH-1) has shown considerable benefit to the patients and demonstrates the power of protein engineering to produce 'better' therapeutic proteins (Hale *et al.*, 1988). CAMPATH-1 is a rat monoclonal antibody specific for a human lymphocyte differentiation marker. The antigen is expressed on virtually all human lymphocytes and monocytes and the antibody has general anti-lymphocyte activity.

It is important to put therapeutic proteins and those modified by genetic engineering into the right perspective in the context of drug discovery. Table 6.3, which is compiled from several sources, indicates that the total market for drugs of this kind is unlikely to be much more than 8 billion pounds by year 2000. This is in an overall pharmaceutical market estimated to be over 200 billion pounds by this time.

Thus only about 4–5% of the market is likely to be made up by protein drugs produced by genetic engineering techniques. By far the biggest impact of molecular biology will be as an enabling technology which will allow biochemists, pharmacologists and chemists to do things that they could not do before and thus to improve the drug discovery process.

IV. Protein Structure Determination

The ability to produce proteins in quantity for determination of their structure is a good example of the above. There is no question that a detailed X-ray picture of a target enzyme (or receptor) with an inhibitor in place gives the medicinal chemist information which is useful for new compound design. The structure may not lead him directly to a new drug, because factors such as bioavailability and pharmacokinetics come into the equation, but it should help to refine an existing lead. There are now several examples

TABLE 6.3
Biotechnology Product Drug Market

Drug	Drug Market (Millions of Pounds)		
	1990/1991 (actual)	1996 (e)*	2001 (e)*
Cardiovascular	585	1145	2250
Erythropoietin	350	800	1300
Tissue plasminogen activator	210	225	450
Blood factors (e.g. Factor VIII)	25	50	100
Superoxide dismutase	0	20	100
Others	0	50	300
Cancer	350	975	2150
Colony-stimulating factors	50	350	950
Interferons (α, β, γ)	300	500	700
Interleukins	0	100	200
Others	0	25	300
Hormones/growth factors	750	1320	2050
Human growth hormone	200	470	650
Human insulin	550	750	1000
Others	0	100	400
Vaccines	200	320	550
Hepatitis B	200	300	500
Others	0	20	50
Monoclonal antibodies	20	200	500
Total	1905	3960	7500

*estimate

where structure determination by using either NMR or X-ray crystallography has aided drug design and one notable exception (so far) where enzyme structure has not really helped matters.

The enzyme renin is an aspartyl proteinase involved in the control of blood pressure where it catalyses the cleavage of angiotensinogen to angiotensin I which, after conversion into angiotensin II, mediates intracellular effects such as vasoconstriction by binding to a membrane receptor (see Chapter 9). Renin has a hexapeptide recognition sequence and, even with reasonable-resolution three-dimensional structural information, it has not been possible to design good renin inhibitors for use clinically. This is largely due to the fact that small peptide-based active-site inhibitors of three or four amino acids were not potent enough and those potent hexapeptide inhibitors that were synthesized were not orally bioavailable!

In contrast and ironically, the search for inhibitors of another proteinase of clinical importance, the proteinase from human immunodeficiency virus (HIV) [the human acquired immune deficiency syndrome (AIDS) virus], has been helped by structural analysis of enzyme inhibitor complexes (e.g. Swain et al., 1990). This has been done by either cloning the proteinase gene from the HIV genome or synthesizing the gene chemically and inserting it into an expression vector. After expression of the gene in

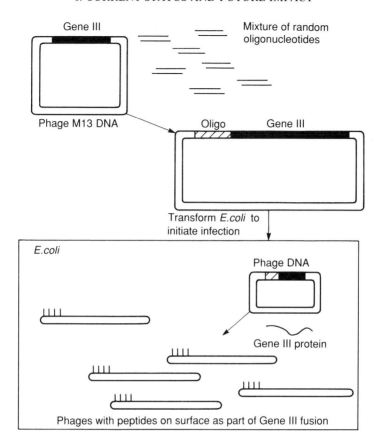

Fig. 6.6 Diagram to illustrate the construction of peptide fusion proteins on the surface of M13 bacteriophage.

E.coli, the protein was purified and crystallized with various inhibitors in place. The structure was then determined by X-ray diffraction (see Plate I). Without recourse to molecular biology (i.e. expressing the cloned gene) the structural work on HIV proteinase would not have been possible. For that matter, little of the knowledge that we have gained concerning the gene structure and function of the proteins of the virus itself would have been possible without molecular biology and recombinant DNA techniques.

V. Peptides as Drugs

The discussion above alludes to a larger issue concerning peptides as drugs. In terms of bioavailability, stability and pharmacokinetics, most peptides are as bad as proteins and, in general, do not make good drugs unless modified in some way.

Molecular biology is influencing the design of peptide mimetics in two ways. In the first place more pharmacologically active peptides are being discovered. Secondly, and probably of more importance, recent advances in molecular technology are allowing peptide ligands for any combining site (be it receptor or antibody) to be identified.

For this new method a degenerate oligonucleotide sequence coding for peptides from 6 to 15 residues is cloned into a bacteriophage M13 gene such that it is expressed as part of one of the coat proteins (protein III) of the phage on infecting *E.coli*. A population of phage is generated on infection with a random peptide sequence at the N-terminus of the coat protein III sequence (Fig. 6.6). It has been demonstrated that phage containing a specific peptide sequence can be selected from the population by affinity panning on columns or plates containing a specific antibody (Scott and Smith, 1990; Cwirla *et al.*, 1991). It should be possible in this way to identify a peptide-recognition sequence for any receptor binding a peptide or protein, which might then provide a starting point for a peptide mimetic programme to lead to a drug candidate [for complementary strategies, see Geysen *et al.* (1986) and Fodor *et al.* (1991)]. This has yet to be demonstrated in anything but a model system but the potential is clearly there. These methods have boosted attempts to make random libraries of peptides synthetically as a pool of shape-recognizing molecules (Houghton *et al.*, 1991; Lamb *et al.*, 1991).

VI. Receptor Structure and Function

Several important drugs act at receptor molecules as agonists or antagonists. For example, the histamine H_2 blockers (e.g. ranitidine and cimetidine) act to antagonize the activity of H_2 receptors in the stomach (Chapter 12), whereas the β_2-stimulants (e.g. the anti-asthmatic, salbutamol) stimulate β_2-adrenoreceptors in bronchiolar smooth muscle causing their relaxation (Chapter 11). The molecular nature of these and many other receptors has been clarified by cloning the genes coding for them and determining their amino acid sequences. Several classes of receptor have been identified (Fig. 6.7, Table 6.4). Those involved in catecholamine [(nor) adrenaline]-mediated functions have been shown to belong to the class of receptors called G-protein-linked or seven transmembrane (heptahelical) receptors, because the receptor protein chain appears to cross the cytoplasmic membrane seven times and because G-proteins, interacting with the portions of the molecule on the cytoplasmic side of the membrane, mediate intracellular

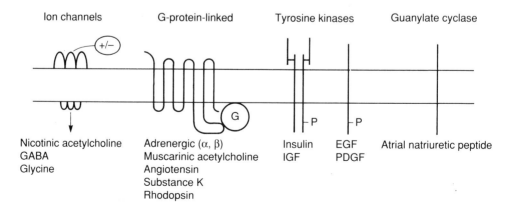

Ion channels	G-protein-linked	Tyrosine kinases	Guanylate cyclase
Nicotinic acetylcholine	Adrenergic (α, β)	Insulin EGF	Atrial natriuretic peptide
GABA	Muscarinic acetylcholine	IGF PDGF	
Glycine	Angiotensin		
	Substance K		
	Rhodopsin		

Fig. 6.7 Cartoon showing the organization of various receptor classes. Note the extracellular, intra-membrane and intracellular domains allowing binding on the outside of the cell to trigger intracellular events. Examples of the various different receptor classes are also shown. GABA, γ-aminobutyric acid; IGF, insulin-like growth factor; EGF, epidermal growth factor; PDGF, platelet-derived growth factor.

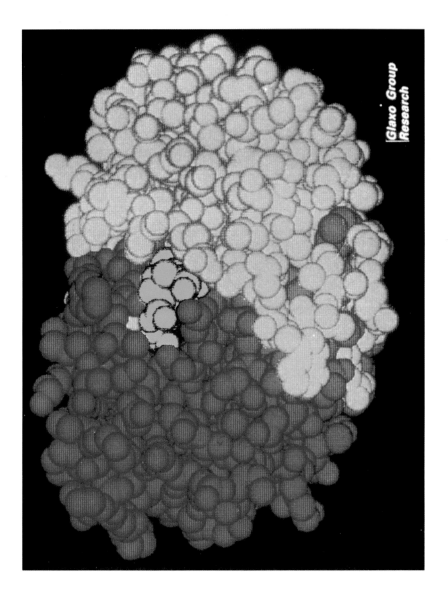

Plate I Raster graphics representation of the three-dimensional structure of HIV proteinase. The dimeric nature of the protein is clearly visible (one subunit in yellow, the other in red). The substrate is shown filling the active site of the enzyme (green).

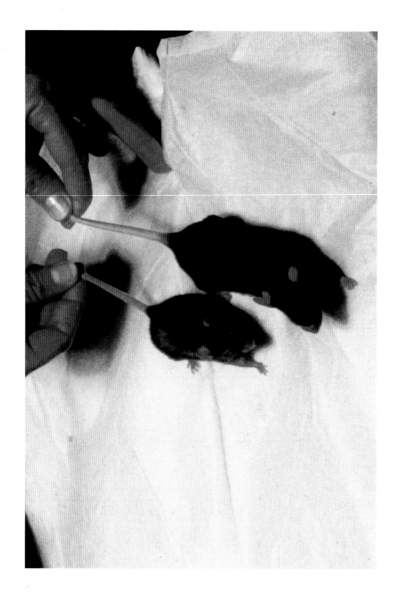

Plate II Photograph of a transgenic mouse containing the human growth hormone gene. The mouse on the left is a non-transgenic littermate.

TABLE 6.4
Receptor Classes

Class	Type	Ligand
Protein kinases	Cytoplasmic (intracellular) domain has tyrosine kinase activity	Insulin, PDGF, FGF, EGF
Major antigen receptor	Eight-chain complex consisting of four dimers. May couple to non-receptor-linked tyrosine kinases	T-cell antigen receptor
Heptahelical G-protein-linked	Protein crosses membrane seven times. Cytoplasmic loops connect to G-proteins	Neurokinins, dopamine, serotonin, thrombin, cannabinoid, rhodopsin, many others
Ion channels	Complex multisubunit transmembrane proteins make up the ion channel	Nicotinic acetylcholine glutamate, GABA, glycine (EAA), serotonin (5HT3)
Guanylate cyclases	Cell membrane form is a homodimer, the cytoplasmic domain of which has guanylate cyclase activity	Atrial natriuretic peptide Heat-stable enterotoxin
Protein tyrosine phosphatases*	Cytoplasmic domain has tyrosine phosphatase activity	Unknown at present
Lymphokine family†	Heterodimeric proteins in high-affinity states. Share subunits	IL2, IL3, IL4, IL6, GM-CSF, EPO, GH
Steroid receptors	Cytoplasmic proteins binding steroid hormones. Hormone–protein complex affects gene expression directly by binding to responsive elements in DNA	Oestrogen, androgen, glucocorticoids, TSH

*   Structure of some of these proteins is consistent with them having a receptor function.
†   Structurally related proteins regulating cell growth and differentiation. Signalling mechanism currently unknown.
Abbreviations: FGF, fibroblast growth factor; IL, interleukin; GH, growth hormone; PL, plasminogen; TSH, thyrotrophin; EAA, excitatory amino acid.

signalling (see also Chapter 7). There are many classes of molecule that have now been shown to interact with receptors of this type (Dohlman *et al.*, 1991) (Table 6.5).

Another receptor class with a substantial membership is the tyrosine kinase family. The receptors for insulin, platelet-derived growth factor (PDGF) and epidermal growth

TABLE 6.5
Diversity of Ligands for Cloned G-Protein-Coupled Receptors

Ligand	Example
Biogenic amines	Adrenaline Dopamine Serotonin (5-hydroxytryptamine)
Neurokinins	Substance P Substance K
Glycoprotein hormones	Thyrotrophin (TSH) Follicle-stimulating hormone
Polypeptides and peptide hormones	Angiotensin Bombesin Parathyroid hormone Endothelin
Sensory stimuli	Light Odorants
Other	Thrombin N-Formyl-Met-Leu-Phe α-Factor C_5A Thromboxane A_2

factor (EGF) belong to this class (Table 6.4). For these receptors it is thought that the extracellular domain binds the protein ligand which transduces a conformational change, leading to receptor dimerization and activation of the tyrosine kinase domain on the cytoplasmic side of the membrane. Tyrosine kinase acts to phosphorylate tyrosine residues in peptides and proteins. Proteins which are thus phosphorylated on tyrosine residues are involved in signal transmission leading to induction of gene expression and cellular proliferation. Other receptor proteins have intrinsic ion channel activity or have other enzymatic functions such as guanylate cyclase activity (cf. adenylate cyclase, Chapter 1) (Table 6.4, Fig. 6.7). The ability to study structure/function relationships in these receptor molecules by expressing them in different systems will lead to a much more fundamental understanding of the molecular basis of antagonism and agonism. By coupling this with physical structural data, it may be possible to understand in molecular detail how receptors work in much the same way as has been done for enzymes. Whether this will, in the short term, help the drug discoverers directly is not clear, although they will benefit from the ready availability of the proteins (see below). If nothing else, structure/function analysis of (for example) the β-adrenoreceptor will provide the medicinal chemist with an elegant rationalization of why and how particular β-receptor-active compounds act in the way they do.

VII. Novel Assays and Screens

Natural product screening is assuming a greater importance in drug discovery as recombinant DNA techniques reveal novel points of intervention in various diseases through an understanding of macromolecule–macromolecule interactions (e.g. protein–

TABLE 6.6
Impact of Molecular Biology in Natural-product Screening

---

* Provision of recombinant proteins for assays, e.g. enzymes and growth factors

* Analysis of receptor structure and function and production of cell lines expressing individual receptors

* Construction of screening organisms containing genes which alter phenotype

* Construction of cell lines for measuring alteration of gene transcription

---

protein, protein–peptide, protein–nucleic acid or protein–carbohydrate) which are not amenable directly to attack by medicinal chemists. The finding of a chemical lead from a screen of a compound collection or a source of natural products is sometimes a more profitable approach despite its somewhat serendipitous nature. Natural product screening often begins with isolation of bacteria and fungi from soil samples. These are cultured and small samples of the broths and solvent extracts of the mycelia are assayed for biological activity (e.g. the presence of an enzyme inhibitor or receptor-active substance). An alternative source of chemical diversity is provided by solvent extracts of plant material or marine organisms.

Molecular biology is influencing this activity in several ways (Table 6.6). The simplest example is in the provision of recombinant proteins for use in assays in amounts that would not otherwise be possible. An apposite example is again HIV proteinase. Insufficient enzyme would be available for screening for inhibitors without material expressed in *E.coli*. Simple assays such as cleavage of radioactive or fluorescently labelled peptide substrates have been developed. Recombinant growth factors and lymphokines are being made available in a similar way. Receptors (see above) are also becoming available in native or engineered forms for inclusion into high-throughput screens. A simple example is the use of a baculovirus–insect cell culture system for making the cytoplasmic tyrosine kinase domain of the EGF receptor (Wedegaertner and Gill, 1989). For more complex receptors (e.g. the G-protein-linked seven transmembrane domain family), it is possible to use the same system (Reilander *et al.*, 1991) although it may be preferable to use mammalian cells for expression. Very recently, it has been shown that the β-adrenergic receptor can be expressed in a functional form in yeast cells, possibly providing a new way to screen for β-adrenergic receptor-active compounds (King *et al.*, 1990).

Recombinant yeasts are quite robust in a screening format (i.e. they are not as susceptible to the non-specific effects of broths and extracts as tissue culture cells) and they may well have considerable potential as screening organisms (as they are eukaryotes), where particular molecular control circuits (points of intervention) can be cloned into the yeast with an appropriate read out. An example is a recombinant yeast developed at Glaxo which contains the HIV *TAT* gene expressed from a cytochrome *c* promoter, acting on its cognate control sequence (*tar*) placed in between a promoter and a reporter gene (β-galactosidase) (Fig. 6.8). In this system, the TAT protein is made and stimulates β-galactosidase gene expression by interacting with *tar*.

Agents that disrupt the interaction will lead to a reduction in enzyme production.

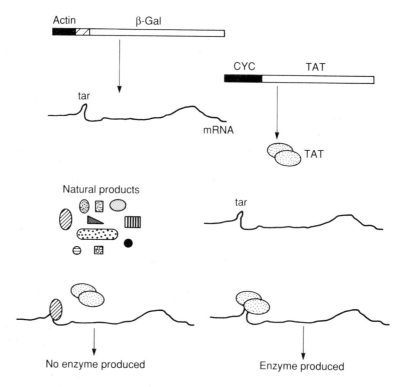

Fig. 6.8 The *TAT–tar* yeast screening organism is made by inserting two genes into the yeast cell. The first expresses β-galactosidase from an actin promoter with a *tar* sequence between the two elements. The *TAT* gene is separately expressed off a different promoter. When the *TAT* gene interacts with its cognate recognition sequence (*tar* in the mRNA) β-galactosidase is produced. In the presence of a natural product that disrupts *TAT–tar* binding no enzyme would be produced.

Control lines are also prepared where β-galactosidase is expressed directly (to check for non-specific inhibitors of the enzyme).

Recombinant yeast cells are also being used to screen for inhibitors of human phosphodiesterases (PDEases) – enzymes involved in signal transmission by modulating cyclic adenosine monophosphate (cAMP) levels. Yeast cells contain two genes encoding cAMP PDEases. PDEase-deficient mutants display properties associated with elevated levels of cAMP, e.g. inability to grow on acetate as the sole carbon source and sensitivity to heat shock. Yeast cells can be complemented by human PDEase genes and reverse the lesion. This assay forms the basis of a high-throughput screen to search for PDEase inhibitors (McHale *et al.*, 1990).

Natural-product screening is a *bona fide* part of any drug discovery operation in every therapeutic area. Molecular biology is clearly influencing the way this activity is carried out, just as it is changing the way biological research as a whole is undertaken.

VIII.   Molecular Basis of Disease

In such a short chapter as this, it is difficult to do justice to the way genetic engineering has altered fundamentally the way biological research is carried out and how the search for new medicines is practised. The description of disease in terms of specific alterations in gene structure and function has heralded the era of 'reverse genetics'. Knowledge of alterations in particular genes now provides a logical base for understanding disease in terms of underlying molecular pathology. This will result in therapeutic strategies directed against defined biochemical pathways and molecular targets.

There are some specific examples which can be listed briefly here which serve to illustrate what is meant by this. Oncogenes and several suppressor genes have been found and analysed at a molecular level over the past few years (Bishop, 1991). Oncogenes are genes that have oncogenic or cancer-causing potential. Oncogenes were initially found in retroviruses but were later discovered to be mutated forms of genes picked up from the host cell chromosome (proto-oncogenes). The possession of an oncogene gives the virus the potential to transform infected cells and give them the capacity to form tumours. Suppressor genes are normally involved in some undefined way in controlling the cell cycle (i.e. cell division and growth). When mutated or deleted from a cell, the cell undergoes division (mitosis) and DNA synthesis in an uncontrolled manner leading to transformation. These genes are intimately associated with the uncontrolled cellular proliferation which is the hallmark of most neoplasms. [A neoplasm is a new growth or tumour – it may be benign or malignant – displaying properties of uncontrolled cell growth, invasiveness and metastatic spread (if malignant)]. Understanding the genetic basis of the control of cell division will influence drug discovery because new points of intervention in this process will become apparent. For example there is, currently, a great deal of interest in trying to find inhibitors of oncogene function in the hope that this might switch off the molecular drive leading to cellular proliferation. Alternatively, reversing the 'switching off' of suppressor gene function – which also occurs in tumour cells – may have the same effect.

The disease of cystic fibrosis (CF) affects several million people. It is one of the most common genetic disorders (one in twenty of the population are carriers of the disease). Until recently the genetic defect responsible for the disease had not been determined. Isolation of the gene for CF has provided considerable insight into the molecular basis of the disease (Kerem et al., 1989). The gene for CF was located on the human genome by family-linkage studies. It codes for a protein of 1480 amino acids referred to as the cystic fibrosis transmembrane conductance regulator (CFTR). Evidence supporting a role for this gene in the disease now comes from several sources. Sequence analysis has revealed a variety of mutations in the genes of affected individuals. Extensive studies have indicated that the most common mutation (70% of affected individuals) in CF is a 3 base pair deletion encoding phenylalanine at codon 508, a mutation never found in normal individuals. Expression of a cDNA clone of the mRNA coding for the CFTR in mammalian cells has shown that it influences chloride ion conductance and specifically complements the defective gene (Drumm et al., 1990).

These studies are a paradigm for dissecting the genetic basis of disease, particularly genetic defects due to a single gene. In future, similar 'from gene outwards' investigations may lead to a better understanding of 'genetic diseases', such as Alzheimer's disease, and more complex inherited diseases, such as heart disease and schizophrenia.

The final examples were alluded to at the beginning of this chapter. Substantial progress has been made in understanding the immune response and the role of protein factors (lymphokines) affecting the activity of B- and T-cells of the immune system. Similarly a number of those proteins involved in haematopoiesis (that is the differentia-

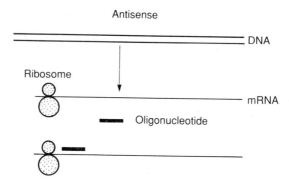

Oligonucleotides prevent translation of the mRNA in the cytoplasm

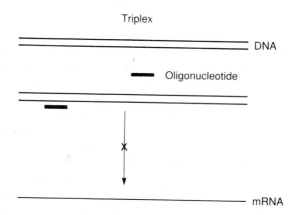

Triplex oligonucleotides bind directly to DNA in the nucleus forming
a triple helix. No mRNA is made

Fig. 6.9 Cartoon indicating the mode of action of antisense and triplex oligonucleotides.

tion of pluripotent stem cells into the cells found in blood) have been cloned and
expressed and their functions analysed in detail.

IX. Affecting the Genes Themselves

In addition to its role in improving current approaches to the treatment of disease,
molecular biology is creating completely novel opportunities for disease modulation.
The ultimate form of disease therapy will come in the ability to control gene expression
selectively. This may give rise to the practice of therapeutic intervention by switching
genes off and on in a predictable fashion. The first tentative steps down this route have
been taken with the construction of modified antisense and triplex oligonucleotides,
which recognize specific nucleotide sequences in either nucleus or cytoplasm (Fig. 6.9).
Rapid progress is being made in the development of the chemistry required to make

bioavailable and selective molecules with desirable pharmacokinetics (Crooke, 1991). There is also considerable excitement in the area of gene therapy, that is the correction of a genetic defect by the introduction of a 'normal' gene into the correct chromosomal location in a somatic cell (Culver *et al.*, 1991). Significant progress has been made and a number of clinical trials of gene therapy in specific human disease situations are underway, following on from experiments in mice. There is a long way to go, however, before this becomes anything other than a purely experimental technique.

X. Transgenic Animals

The prospect of human gene therapy would not be a reality if it had not been for the development of transgenic animals, particularly transgenic mice. Large amounts of a gene coding for a particular protein are microinjected into fertilized mouse eggs which are then implanted into foster mothers. In some of the offspring the gene is incorporated into the chromosome and in some cases the gene is expressed but usually in an un-controlled way. Mice that incorporate the gene such that it is inherited by future offspring are 'transgenic' (Palmiter and Brinster, 1985). (This is in contrast with human gene therapy where the intention is only to change the genetic make up of the somatic cells *not* the germ cells). The function of many genes is being investigated by expressing them in whole animals as transgenes.

The most notable are mice expressing the *myc* oncogene (an example of an oncogene derived from a retrovirus that causes cancer in birds) which are patented in the U.S.A. These are mice (so-called Harvard mouse or 'mick' mouse) with a predetermined susceptibility to cancer owing to expression of the oncogene. Transgenic mice containing a growth hormone gene are much larger than those littermates which are not transgenic, giving a visual example of the power of the technique (Plate II). With the ability now to delete as well as to add functional genes using embryonic stem cells, a whole new set of animal models will be available. These will be used to test compounds in relevant models as well as to disentangle some of the interactions that lead to certain diseases.

XI. Conclusions

As can be seen from Table 6.1, there are many overlapping parts of the drug discovery process that are being touched by molecular biology and biotechnology. Hardly a day goes by without a new gene being discovered or the function of another one better defined. The power of this technology is pervasive, as evidenced by the number of small 'biotech' companies set up specifically to utilize it. Some of these companies will become major players in the pharmaceutical industry, others will not. The fact that they were set up, however, has helped to increase the awareness of the importance of molecular biology and genetic engineering in drug discovery. The art of the exercise, however, is the utilization of the resources once you have them and the integration of the disciplines of medicinal chemistry, cell biology, pharmacology and biochemistry, all of which are needed for successful drug research. It is the companies that get this balance right that will be finding and developing the new medicines of the future.

References

Bishop, J.M. (1991). *Cell* **64**, 235.

Brown, T.A. (1991). 'Gene cloning – An introduction,' 2nd edn, Van Nostrand Rheinhold, London.

Cheson, B.D. (1990). *Drug News Perspect.* **3**, 154.

Crooke, S.T. (1991). *Curr. Opin. Biotechnol.* **2**, 282.

Culver, K., Cornetta, K., Morgan, R., Morecki, S., Abersold, P., Kasid, A., Lotze, M., Rosenberg, S., Anderson, W.F. and Blaese, R.M. (1991). *Proc. Natl. Acad. Sci. U.S.A.*, **88**, 3155.

Cwirla, S.E., Peters, E.A., Barrett, R.W. and Dower, W.J. (1991). *Proc. Natl. Acad. Sci. U.S.A.* **87**, 6378.

Dohlman, H.G., Thorner, J., Caron, M.G. and Lefkowitz, R.J. (1991). *Annu. Rev. Biochem.* **60**, 653.

Drumm, M.L., Pope, H.A., Cliff, W.H., Rommens, J.M., Marvin, S.A., Tsui, L.C., Collins, F.S., Frizzell, R.A. and Wilson, J.M. (1990). *Cell* **62** 1227.

Fodor, S.P.A., Read, J.L., Pirrung, M.C., Stryer, L., Tsai Lu, A. and Solas, D. (1991). *Science* **251**, 767.

Geysen, M., Rodda, S.J. and Mason, T.J. (1986). *Mol. Immunol.* **23**, 709.

Hale, G., Clark, M., Marcus, R., Winter, G., Dyer, M.J.S., Philips, J.M., Reichmann, L. and Waldmann, H. (1988). *Lancet* **ii** 1394.

Harris, T.J.R. (1983). *Genet. Eng.* **4**, 127.

Houghten, R.A., Pinella, C., Blondelle, S.E., Appel, J.R., Dooley, C.T. and Cuervo, J.H. (1991). *Nature* **354**, 84.

Kerem, B.A., Rommens, J.M., Buchanan, J.A., Markiewicz, D., Cox, T.K., Chakravati, A., Buckwald, M. and Tsui, L.C. (1989). *Science* **245**, 1073.

King, K., Dolhman, H.G., Thorner, J., Caron, M.G. and Lefkowitz, R.J. (1990). *Science* **250**, 121.

Knowles, J.R. (1991). *Nature* **350**, 121.

Lam, K.S., Salmon, S.E., Hersh, E.M., Hruby, V.J., Kazmierski, W.M. and Knapp, R.J. (1991). *Nature* **354**, 82.

Marston, F.A.O. (1986). *Biochem J.* **240**, 1.

McAleer, W.J., Buynak, E.B., Maigetter, R.Z., Wampler, D.E., Miller, W.J. and Hilleman, M.R. (1984). *Nature* **307**, 178.

McHale, M.M., Cieslinski, L.R., Eng, W.K., Johnson, R.K., Torphy, T.J. and Livi, G.P. (1991). *Mol. Pharmacol.* **39** 109.

Old, R.W. and Primrose, S.B. (1991). 'Principles of Gene Manipulation. An Introduction to Genetic Engineering' 4th end, Blackwell, Oxford.

Palmiter, R.D. and Brinster, R.L. (1985). *Cell* **41**, 343.

Pluckthun, A. (1991). *Biotechnology* **9**, 545.

Reichmann, L., Clark, M., Waldmann, H. and Winter, G. (1988). *Nature* **332**, 323.

Reilander, H., Boege, F., Vasudevan, S., Maul, G., Hakman, M., Dees, C., Hampe, W., Helmreich, E.J.M. and Michel, H. (1991). *FEBS Lett.* **282**, 441.

Saiki, R.K., Gelfand, D.H., Stoffel, S., Scharf, S.J., Higuchi, R., Horn, G.T., Mullis, K.B. and Erlich, H.A. (1988) *Science* **239**, 487.

Scott, J.K. and Smith, G.P. (1990). *Science* **249**, 386.

Shokat, K.M. and Schultz, P.G. (1990). *Annu. Rev. Immunol.*, **8** 335.

Swain, A.L., Miller, M.M., Green, J., Rich, D.H., Schneider, J., Kent, S.B.H. and Wlodawer, A. (1990). *Proc. Natl. Acad. Sci. U.S.A.* **87**, 8805.

TIMI Study Group (1989). *New Engl. J. Med.* **320**, 618.

Wedegaertner, P.B. and Gill, G.N. (1989). *J. Biol. Chem.* **264**, 11346.

Winter, G. and Milstein, C. (1991). *Nature* **349**, 293.

–7–

General Approaches to Discovering New Drugs: An Historical Perspective

C.R. GANELLIN
Department of Chemistry,
University College London,
London, U.K.

Professor C. Robin Ganellin, FRS read chemistry at Queen Mary College London and studied under M.J.S. Dewar for a Ph.D. in Organic Chemistry. He conducted postdoctoral research with A.C. Cope at MIT (1959) and subsequently joined Smith Kline and French Laboratories as a medicinal chemist, eventually becoming Vice-President for Research at Welwyn. In 1986 he was elected as a Fellow of the Royal Society and was appointed to the SK&F Chair of Medicinal Chemistry in the Chemistry Department of University College London. Co-discoverer of the H_2-receptor histamine antagonists and of the drug cimetidine, he has special interest in applying the principles of physical–organic chemistry to structure–activity analysis. He has been accorded wide recognition for his work and, in particular, has received awards from the Royal Society of Chemistry, American Chemical Society, La Societé de Chimie Thérapeutique, the Sociedad Espanola de Quimica Therapeutica, Society of Chemical Industry and Society for Drug Research, and in 1990 was elected into the USA National Inventors Hall of Fame.

I. Introduction

The development of new medicines takes place mainly in the modern pharmaceutical industry and most of the new drugs are also discovered in the research laboratories of pharmaceutical companies.

To gain an historical perspective it is of interest to consider the main approaches to drug discovery which have been taken in the past, particularly during the twentieth century with the rise of the modern multinational pharmaceutical industry. For excellent accounts of many examples of drug discovery, see Sneader, (1985) and Bindra and Lednicer, (1982).

MEDICINAL CHEMISTRY 2nd Edition
ISBN 0-12-274120-X

Broadly speaking, one can discern four main sources for new drugs. These are:

(1) Natural products
(2) Existing drugs
(3) Screens
(4) Physiological mechanisms

II. Natural Products as a Source for New Drugs

Historically, natural products provide the oldest source for new medicines. Natural selection during evolution, and competition between the species, has produced powerful biologically active natural products which can serve as chemical leads, to be refined by the chemist to give more specifically acting drugs.

For example, moulds and bacteria produce substances that prevent other organisms from growing in their vicinity. The famous penicillium mould led to the considerable range of semisynthetic penicillins, and gave rise to the concept of seeking naturally occurring antibiotics. This was taken up by the microbiologists who argued that bacteria which cause infections in humans do not survive for long in soil because they are destroyed by other soil-inhabiting microbes. Extensive soil-screening research programmes have led to a considerable array of antibiotics which have provided some very potent life-saving drugs, e.g. streptomycin, chloramphenicol, chlortetracycline, erythromycin, neomycin, bacitracin, polymixin.

Microbial fermentation products may also provide a source for leads to other types of drug, e.g. the discovery of a novel cholecystokinin (CCK) antagonist, from *Aspergillus alliaceus* (Chang *et al.*, 1985), served as the starting point for scientists at Merck Sharpe and Dohme to develop very specific and potent non-peptide antagonists at CCK-A and CCK-B receptors respectively. The fermentation broth contained a substance that inhibited the binding of ^{125}I-labelled CCK-33 to a preparation from rat pancreas. The substance, named Asperlicin, had $IC_{50} = 1.4\,\mu M$; its structure was determined and the chemists recognized two parts to the molecule: a benzodiazepine and a tryptophan-derived group (Fig. 7.1). Structure–activity exploration led to a very potent synthetic inhibitor (MK-329, Devazepide), having $IC_{50} = 0.08\,nM$, i.e. over 10 000-fold increase in potency; a non-peptide antagonist of a peptide. Such fermentation broths contain hundreds, if not thousands, of chemicals and are a potentially rich source of novel inhibitors.

A variety of venoms and toxins are used by animals as protection or to paralyse their prey; some are extremely potent, requiring only minute doses, e.g. α-bungarotoxin (from snake venom) which combines with acetylcholine receptors, tetrodotoxin (from puffer fish) which blocks sodium channels, apamin (from bee venom) which blocks Ca^{2+}-activated potassium channels, and batroxobin (from the venom of a pit viper) which is a thrombin-like enzyme; they have served as starting points for investigation of hormone receptors, ion channels and enzymes. Such natural products have provided important leads to drugs such as muscle relaxants and haemostatics. Recent subjects for study have included spiders, frogs and sponges. Indeed, marine life offers a vast untapped resource for future investigation (Hall and Strichartz, 1990).

Another fruitful means of identifying pharmacologically active natural products has been the folk law remedies, which are mainly plant products (Fig. 7.2): alkaloids such

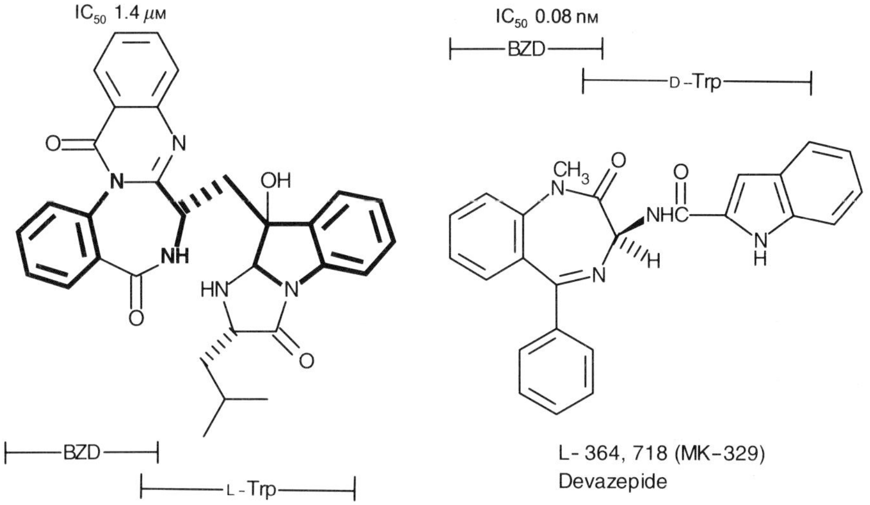

Fig. 7.1   Asperlicin, a natural product lead from *Aspergillus alliaceus*, was the starting point for designing potent non-peptide inhibitors at CCK-A receptors (Chang *et al.*, 1985). Two substructures were noticed, a benzodiazepine (BZD) and a tryptophan-derived group (Trp).

as morphine (from the opium poppy known in ancient Egypt), atropine and hyoscine (from plants of the Solanaceae family known to the ancient Greeks) and reserpine (from *Rauwolfia serpentina*, the snakeroot, popular in India as a herbal remedy), and non-nitrogenous natural products such as salicylates, e.g. salicin from the willow tree (genus *Salix*, botanical sources known to Hippocrates) and the glycosides, e.g. digitoxin and digoxin in digitalis from the foxglove (in folk use in England for centuries).

Natural products continue to provide a very fruitful source of drug leads especially when coupled with screening methods based on modern pharmacological or biochemical procedures, e.g. for enzyme inhibitors or in receptor-binding assays.

III.   Existing Drugs as a Basis for New Drug Discovery

The most fruitful basis for the discovery of a new drug is to start with an old drug. This has been the most common and reliable route to new products. Thus, existing drugs may need to be improved, e.g. to obtain a better dosage form, to improve drug absorption or duration, to increase potency to reduce the daily dose, or to avoid certain side effects.

For example, antihistamines, useful for treating hay fever, often make people feel sleepy which is a nuisance when they need to stay alert such as when driving a car. This has stimulated the development of new antihistamines which were introduced in the 1980s (e.g. terfenadine, astemizole, mequitazine, acrivastine; Fig. 7.3) and have a much lower tendency to cause sedation.

On the other hand there may be side effects that can be exploited, arising from astute observation during pharmacological studies in animals or from clinical investigation in patients.

The discovery of sulphonamide diuretics such as chlorothiazide in the 1950s followed from an observation involving sulphanilamide (Fig. 7.4), i.e. *p*-aminobenzene-sulphonamide, the active ingredient of Prontosil® Rubrum (the first antibacterial

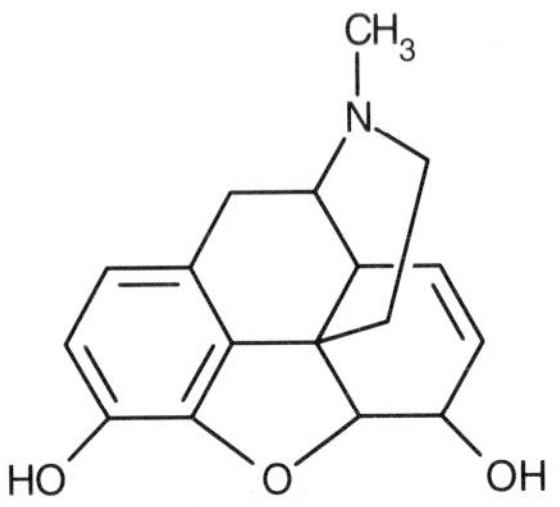

Morphine

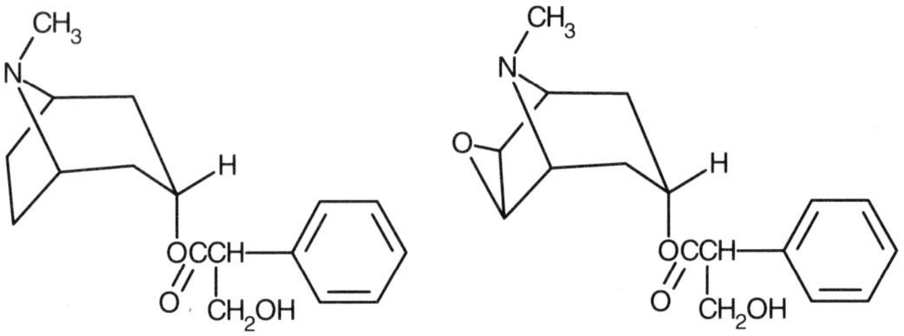

Atropine and hyoscine

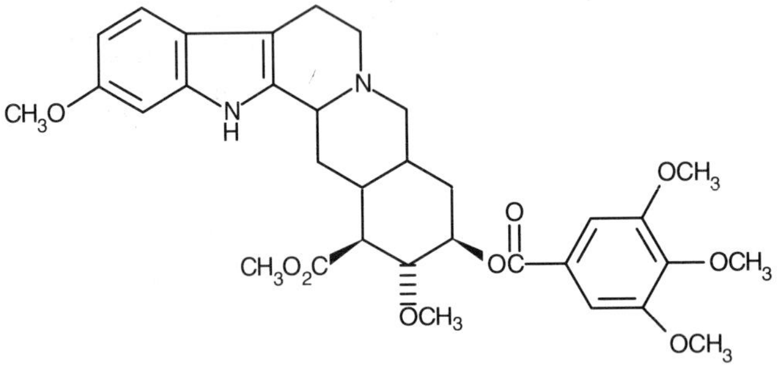

Reserpine

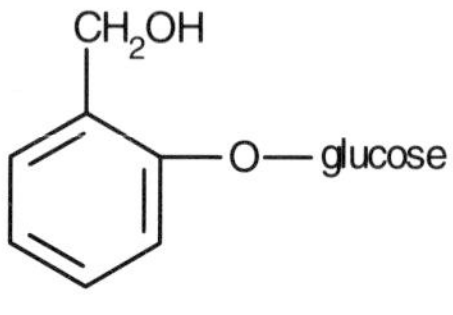

Salicin

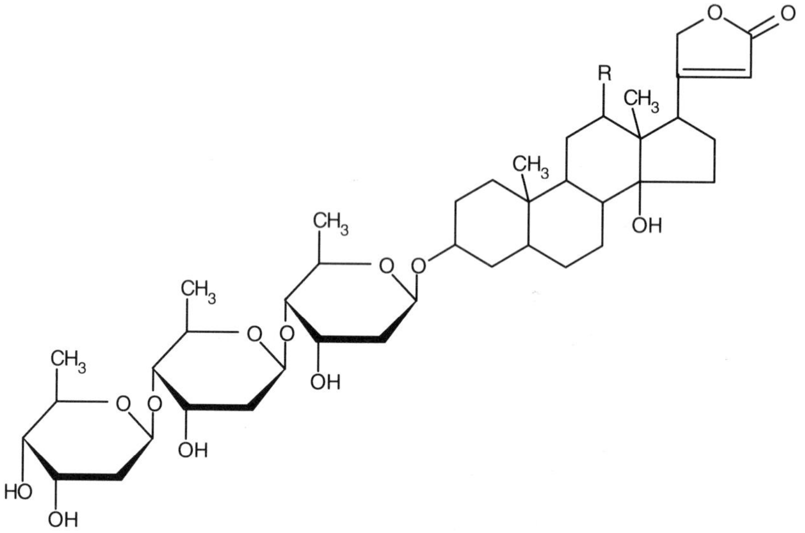

Digitoxin (R = H), digoxin (R = OH)

Fig. 7.2   Examples of active constituents of natural product medicines in use since ancient times.

sulphonamide; see later). Sulphanilamide was found to cause alkaline diuresis in patients who had been given massive doses and this was later shown to be due to inhibition of the enzyme, carbonic anhydrase (Schwartz, 1949). Eventually the lead was used by chemists at Sharp and Dohme to make other benzene sulphonamides and this led to

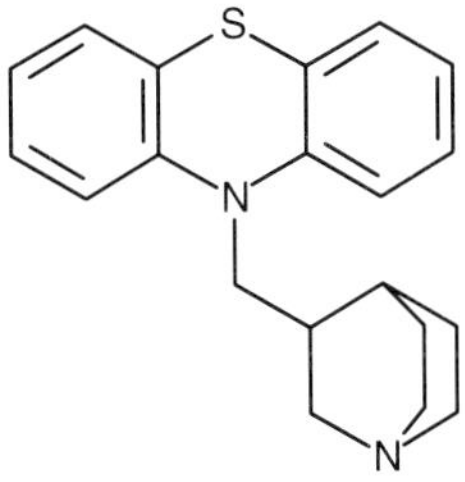

Mequitazine

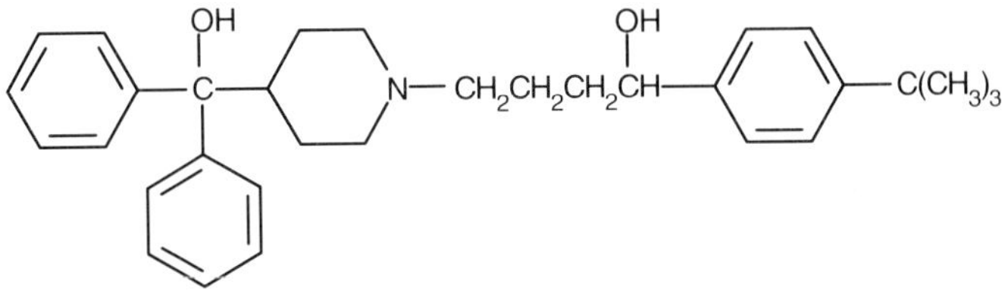

Terfenadine

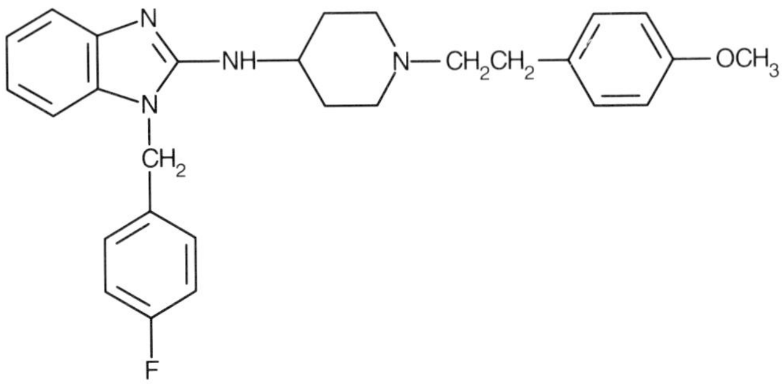

Astemizole

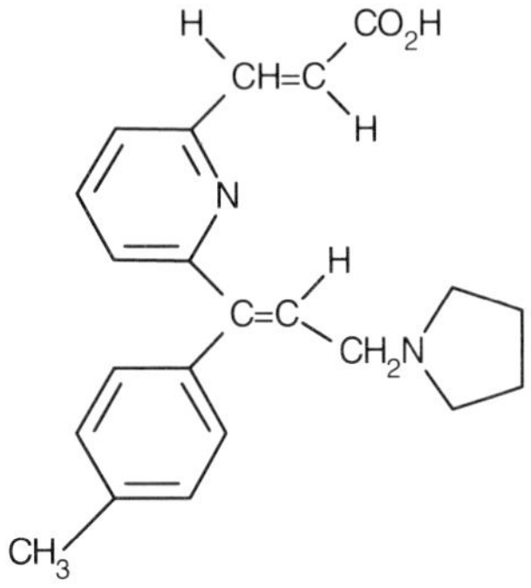

Acrivastine

Fig. 7.3   Newer antihistamines claimed to have a low liability to cause sedation.

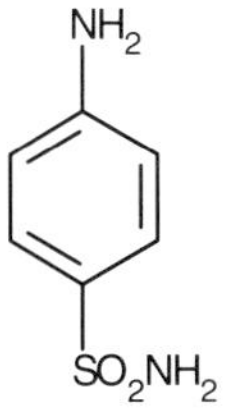

Sulphanilamide

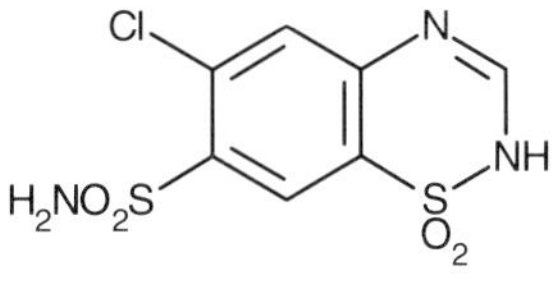

Chlorothiazide

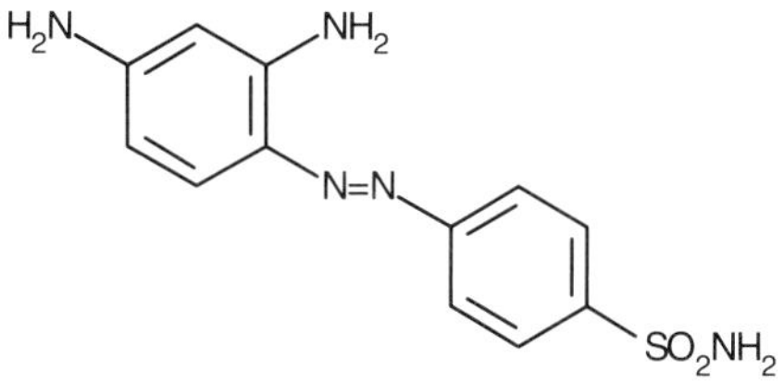

Prontosil Rubrum (Sulphamidochrysoidine)

Fig. 7.4   Sulphanilamide, an antibacterial drug, was the lead to chlorothiazide and the start of the new diuretics. Sulphanilamide was originally shown to be the active constituent of the first life-saving anti-bacterial sulphonamide drug, Prontosil Rubrum.

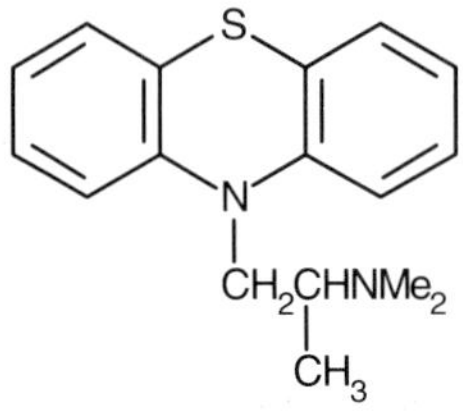

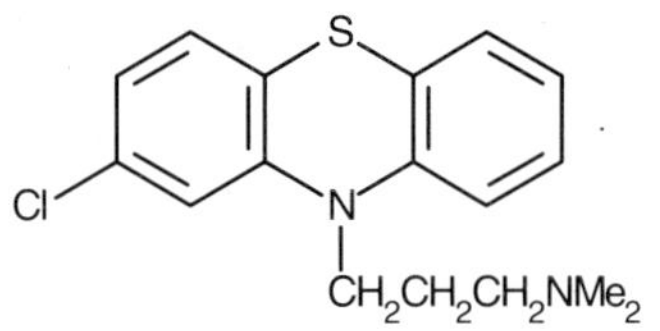

Promethazine   Chlorpromazine

Fig. 7.5 The antihistamine, promethazine, which provided a lead to chlorpromazine and so ushered in the revolution in psychiatric medicine.

chlorothiazide (1957; Fig. 7.4) the first of many thiazide diuretics. Overnight these rendered the mercurial diuretics obsolete.

The phenothiazine tranquilizers, which revolutionized the treatment of psychiatric patients, resulted from an astute observation of the effects of the antihistamine, promethazine (Laborit, 1954). Antihistamines were being studied by Laborit, a French Navy surgeon, for possibly preventing surgical shock. Promethazine seemed to be better than the others, but it became apparent that it had unusual effects on the central nervous system (CNS). Rhone-Poulenc (the manufacturers) became interested and, since promethazine is a phenothiazine, other phenothiazines were tested to enhance the CNS-depressant effects in the belief that this would improve the utility in surgical shock; this led to chlorpromazine (Fig. 7.5). When tested on patients undergoing surgery they seemed relaxed and unconcerned. The significance was not lost on the investigators and the drug was tried on a manic patient (January 1952). So started the revolution in the treatment of schizophrenia and the transformation of psychiatric medicine. The use of chlorpromazine was approved in the Autumn of 1952 and resulted in nearly 70% of the patients suffering from schizophrenia in France being able to return to the mainstream of society.

A much more recent example is metoclopramide (Fig. 7.6), an antagonist of dopamine (see Fig. 7.9) at its D_2-receptors, which was found to be useful in patients as an antiemetic (Gralla et al., 1981). More potent D_2-antagonists were not so effective, so that it was realized that some other property was likely to be important. It was known that

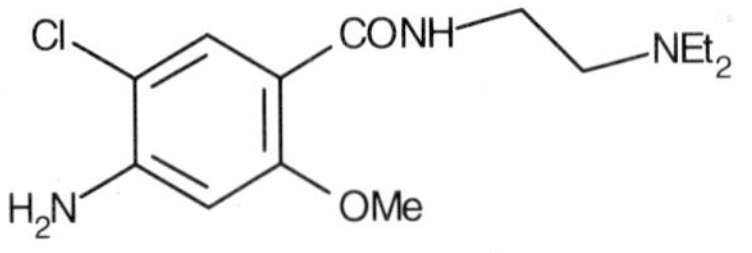

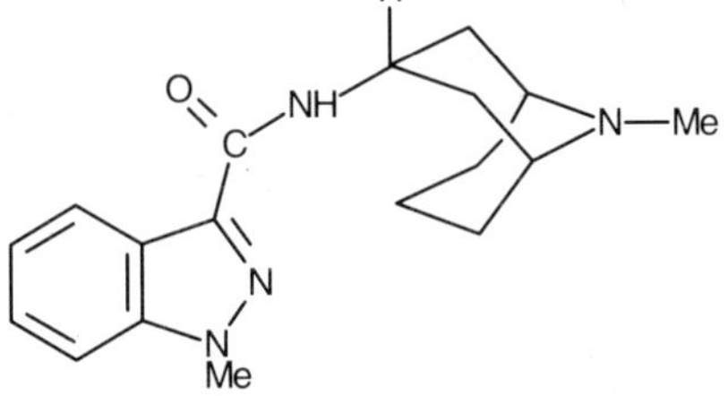

Metoclopramide

Granisetron

Fig. 7.6 Metoclopramide, a dopamine antagonist, was later shown to act at 5-hydroxytryptamine 5-HT$_3$ receptors and provided the lead for the development of potent antiemetic drugs, ondansetron and granisetron.

metoclopramide was also an antagonist of 5-hydroxytryptamine (5-HT) (Fozard and Mobarok Ali, 1981) (see Fig. 7.9) and later this was characterized as acting at 5-HT$_3$ receptors and led to the development of the new compounds granisetron (Bermudez *et al.*, 1990) and ondansetron.

Many new products arise, however, which may only represent minor improvements over the old one. These are the so-called 'Me-Too' products. 'Me-Too' is a term often used perjoratively by those who wish to attack the pharmaceutical industry. Actually, historically, the process has provided the main route whereby a particular type of drug action has been optimized in terms of both selectivity (to avoid side effects) and application (for a particular patient population). Eventually, to realize the full potential of a new 'Me-Too' drug and to reveal its advantages it is necessary to market it in order to gain access to a sufficiently wide patient population. But, of course this can lead to the proliferation of products and it can take many years for clinicians to determine the most suitable drug treatment.

IV. Disease Models as Screens for New Drugs

The screening approach with natural products for new antibiotics, antimetabolites and enzyme inhibitors has had its counterpart with unnatural products, i.e. synthetic chemicals. The idea has been to test large numbers of compounds on a relatively simple system to reveal the required activity. This has been a third main source of new drugs.

The background for this approach lies in the dyestuff's industry: Paul Ehrlich, founder and prophet of chemotherapy, discovered that synthetic dyes were absorbed differentially into tissues and that they could kill parasites and bacteria without affecting mammalian cells. From this work came Salvarsan in 1910, an arsenic compound for treating syphilis (Burger, 1954). Several large chemical companies followed up this discovery by establishing their own research programmes seeking drugs against venereal diseases. From then on, a key to a new drug discovery was seen to be the systematic examination of hundreds of synthetic chemicals.

In 1931, Gerhard Domagk, working for I.G. Farbenindustrie, turned to screening sulphonamide derivatives of azo dyes. This work gave rise to Prontosil Rubrum (red), published in 1935, the first truly effective chemotherapeutic agent for any generalized bacterial infection. This discovery led, via sulphanilamide (Fig. 7.4), to the development of the sulphonamide class of antibacterials (Rose, 1964).

The success of these discoveries in chemotherapy dominated research thinking in the pharmaceutical industry for many years and screens were also established for non-infectious diseases, e.g. for anticonvulsants (useful in epilepsy), analgesics (in the hope of finding non-addictive ones), antihypertensives, antiinflammatory, antiulcer, etc.

There is, however, a fundamental distinction between the antiinfective screens, and the screens that seek a treatment for a disease which we can call 'metabolically based'. In the antiinfective screens (antibacterial, antifungal, antihelmintic, antiprotozoal, antiviral) a drug is sought which is lethal to the pathogen, but leaves the host unharmed. It is a search for selective toxicity between species.

By contrast, in the metabolically based diseases, e.g. allergy, asthma, cancer, duodenal ulcers, epilepsy, gout, hypertension, inflammation and so on, the cause is usually unknown, and we seek selectivity within the same being. In these latter situations, an animal model is used as a screening test, in which a clinical condition is induced in a laboratory animal such as a rabbit or rat, and compounds are tested to see whether they alleviate it.

The model often simulates the disease by presenting similar symptoms, but it may be misleading if the underlying causes are quite different; then the procedure throws up false leads, e.g. compounds that protect the laboratory animal, but when tested clinically are found not to be active in man. Nevertheless, successes have been achieved. The non-steroidal antiinflammatory drugs are a very good example; they were discovered by screening in animals in which various forms of inflammation had been artificially induced. Many other types of drug have been discovered by such relatively simplistic screening procedures. Other searches have been less successful, e.g. pharmaceutical companies spent some 30 years screening for antiulcer agents without much progress being made.

A major drawback is how to generate a lead compound, because the model offers no chemical guidance, and in the past the main source of potential leads was the vast number of compounds held on company files available for screening. The approach is now mainly historical since the outlook for the future is based on more rational grounds as indicated below.

V. Physiological Mechanisms; the Modern 'Rational Approach' to Drug Design

With the advent of greater understanding of physiological mechanisms it has become possible to take a more mechanistic approach to research and start from a rationally argued hypothesis to design drugs. Progress depends largely upon the current state of understanding of physiology in relation to diseases. However, the target diseases selected for study in the pharmaceutical industry are generally those prevalent in Western society. This is the modern 'rational approach' to drug design which is becoming increasingly important with the development of information in cell biochemistry and cell biology, especially where this is understood at the molecular level.

The modern approach to drug discovery requires very close collaboration between chemists and biologists and, reflecting this, most large pharmaceutical companies now organize their research in project teams instead of by scientific discipline.

The 'rational approach' to new drug discovery requires several essential ingredients for success:

(1) evidence of a physiological basis for understanding a disease, so that one may hypothesize that a drug with a particular action should be therapeutically beneficial;
(2) an explicit chemical starting point;
(3) bioassay systems which measure the desired drug activity in the laboratory;
(4) a test which measures the activity of the drug in humans that can be related to a potential therapeutic treatment.

One may discern, broadly, five main sites for drug action:

(1) *Enzymes* – where new molecules are made in tissues (the basis of metabolic activity).
(2) *Receptors* – where circulating messengers, e.g. biogenic amines and peptides, act to alter cellular activity.

(3) *Transport systems* – that selectively permit access through membranes into and out of cells, e.g. ion channels, transporter molecules.

(4) *Cell replication and protein synthesis* – controlled by DNA and RNA.

(5) *Storage sites* – where molecules are kept in an inactive form for later use, e.g. mast cells, blood platelets, neurones.

The body is controlled by chemical messengers (physiological mediators). There is a very complex communication system and each messenger has specific functions and is recognized at specialized sites where it acts. In disease, something has got out of balance, and in drug therapy the aim is to redress that balance.

Thus enzymes have active sites which specifically recognize the appropriate substrates which they can process. If we wish to interrupt this process we may design enzyme inhibitors, using the chemistry of the substrate as a starting point. Hitchings (1980) describes how in the 1940s at Burroughs Wellcome their proposed research effort was based on the exploration of the enzymes and metabolic pathways of nucleic acid anabolism . . . 'It was our thought that metabolites which were analogs of purine and pyrimidine bases could be used as probes of metabolic events, and that the findings could be used to design more specific antimetabolites* which eventually might show sufficient selectivity to be useful in chemotherapy'. The research was extraordinarily productive and led to such antimetabolite drugs as 6-mercaptopurine, azathioprine, pyrimethamine, trimethoprim and allopurinol (Table 7.1).

If the enzymic activity can be adequately assayed, it may be possible to characterize the active site. For example, angiotensin-converting enzyme (ACE) was characterized as a metalloenzyme containing zinc at the active site and, following a natural-point lead, this led to the design of the inhibitor captopril for treatment of hypertension (Ondetti *et al.*, 1977) (see Chapter 9).

The substrate specificity may be studied using small peptides as probes (Llorens *et al.*, 1980) which may then be converted into potent inhibitors, as was effected in the design of the enkephalinase inhibitor, thiorphan. Enkephalinase is a dipeptide hydrolase which specifically cleaves the Gly-Phe bond in the opioid, enkephalin (Fig. 7.7), hence an inhibitor might be an analgesic. Llorens *et al.* (1980) screened commercially available dipeptides looking for an inhibitor of enkephalinase activity in mouse brain (striatum) (Table 7.2). They then chose the best inhibitors as leads. The structure–activity studies suggested the presence of a hydrophobic binding pocket in the enzyme and they selected phenylalanylglycine (Phe-Gly-OH) and incorporated a group to bind to Zn; by analogy with captopril, they included a thiol group and obtained thiorphan (Fig. 7.7) and an increase in potency of 1000-fold ($K_1 = 2$ nM).

The work of Llorens *et al.* suggests an important approach which may have a wider application, namely the use of dipeptides and small oligopeptides to characterize binding to the enzymically active site of peptidases. In future, when a new peptidase is isolated, it should be possible to characterize its binding site using dipeptides, and then incorporate suitable groups to increase affinity and stability to give inhibitors. Looking

---

*Antimetabolites are structural analogues of intermediates (substrates or coenzymes) in the physiologically occurring metabolic pathways; they act by inhibiting a particular enzyme and so may block the biosynthesis of a physiologically important substance.

TABLE 7.1

Marketed Drugs Resulting from the Search for Useful Antimetabolites as
Enzyme Inhibitors of Nucleic Acid Metabolic Pathways (Hitchings, 1980)

Product	Year first synthesized	Application
6-Thioguanine	1950	Antileukaemic
6-Mercaptopurine	1951	Immunosuppressive
Azathioprine	1957	Immunosuppressive
Allopurinol	1956	Antihyperuricaemic
Diaveridine	1949	Coccidiostat
Pyrimethamine	1950	Antimalarial
Trimethoprim	1956	Antibacterial
		Antiprotozoal

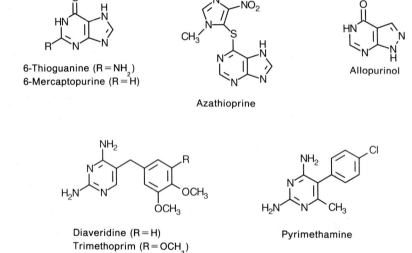

6-Thioguanine (R = NH$_2$)
6-Mercaptopurine (R = H)

Azathioprine

Allopurinol

Diaveridine (R = H)
Trimethoprim (R = OCH$_3$)

Pyrimethamine

further into the future, as the structures of enzymes become known one would hope to design inhibitors using knowledge of the molecular structure of the active site.

Many different substances circulate as chemical messengers and combine with their own receptors, e.g. amino acids, biogenic amines, peptides, prostaglandins, purines, steroids; they are remarkably specific since the messengers are not normally recognized by other receptors, i.e. the receptors discriminate between different messengers. In the case of biogenic amines (see Chapter 3), it has been possible to differentiate between sites for the same messenger, suggesting different subpopulations of receptors, to provide further scope for introducing selectivity of drug action (Jack, 1976, 1991). If too many

[Met]enkephalin   Tyr-Gly-Gly-Phe-Met-OH
[Leu]enkephalin   Tyr-Gly-Gly-Phe-Leu-OH
Enkephalinase cleaves the Gly-Phe bond
Inhibitors: H$_2$NCHCONHCH$_2$CO$_2$H   HSCH$_2$CHCONHCH$_2$CO$_2$H
                                                              | |
                                                           CH$_2$Ph   CH$_2$Ph

                                              Phe-Gly-OH   Thiorphan K_i = 2nM

Fig. 7.7 Enkephalin and the enkephalinase inhibitors, Phe-Gly-OH (the dipeptide lead) and thiorphan.

TABLE 7.2
IC$_{50}$ Results from Screening of Various Dipeptides as Inhibitors of Enkephalinase Activity from Mouse Striatum (Llorens *et al.*, 1980) which led to the Design of Thiorphan

X-L-Ala	IC$_{50}$ (μM)	L-Phe-Y or L-Tyr-Y	IC$_{50}$ (μM)
L-Phe-L-Ala	1.0 $\pm$ 0.2	L-Tyr-L-Ala	1.2 $\pm$ 0.1
L-Tyr-L-Ala	1.3 $\pm$ 0.1	L-Tyr-Gly	1.6 $\pm$ 0.9
L-Trp-L-Ala	2.4 $\pm$ 0.3	L-Phe-L-Ala	1.0 $\pm$ 0.2
L-Ile-L-Ala	10.0 $\pm$ 2.0	L-Phe-Gly	5.0 $\pm$ 1.0
L-Pro-L-Ala	10.0 $\pm$ 2.0	L-Phe-L-Phe	5.0 $\pm$ 1.0
L-Val-L-Ala	36.0 $\pm$ 1.0	L-Phe-L-Trp	5.0 $\pm$ 1.0
L-Met-L-Ala	$\sim$ 100.0	L-Phe-L-Val	$\sim$ 10.0
L-Thr-L-Ala	$\sim$ 100.0	L-Phe-L-Ile	$\sim$ 10.0
L-Ala-L-Ala	$\sim$ 100.0	L-Phe-L-Leu	$\sim$ 20.0
L-Ser-L-Ala	$\sim$ 100.0	L-Phe-L-Ser	$\sim$ 30.0
L-Glu-L-Ala	$>$ 100.0	L-Phe-L-Asp	$>$ 100.0
L-Asp-L-Ala	$>$ 100.0	L-Phe-L-Pro	$>$ 100.0

messages are getting through, then we can seek to design an antagonist to block the receptor, using the chemistry of the natural messenger as a lead (Black, 1976, 1989).

There are many potential sites for drug intervention to affect biogenic amine action especially in their role as neurotransmitters (Fig. 7.8). In addition to blocking postsynaptic receptors (R^1, R^2 etc.), there are presynaptic autoreceptors which may modulate transmitter release or synthesis, storage sites and reuptake mechanisms which modulate free transmitter concentration, and enzymes which can be blocked to alter the rate of transmitter turnover, e.g. inhibit biosynthesis to deplete the transmitter or inhibit metabolism to prolong its existence. Furthermore, the amine receptor is usually coupled to a second messenger enzyme (protein kinase) or an ion channel, so that its influence may be modified by drug interference with the amplifying transducing mechanism.

Undoubtedly, some of the best examples of rational drug design have been based on using the chemical structure of the natural transmitter (i.e. the biogenic amine) as a

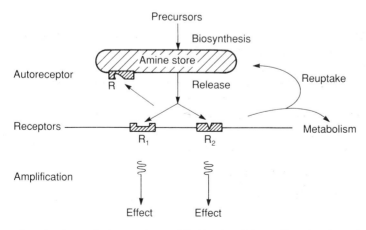

Fig. 7.8 Biogenic amine transmission can be modified by drug intervention at various stages and sites using enzyme inhibitors, or selective receptor blockers or stimulants. Drugs may also affect amine release or uptake.

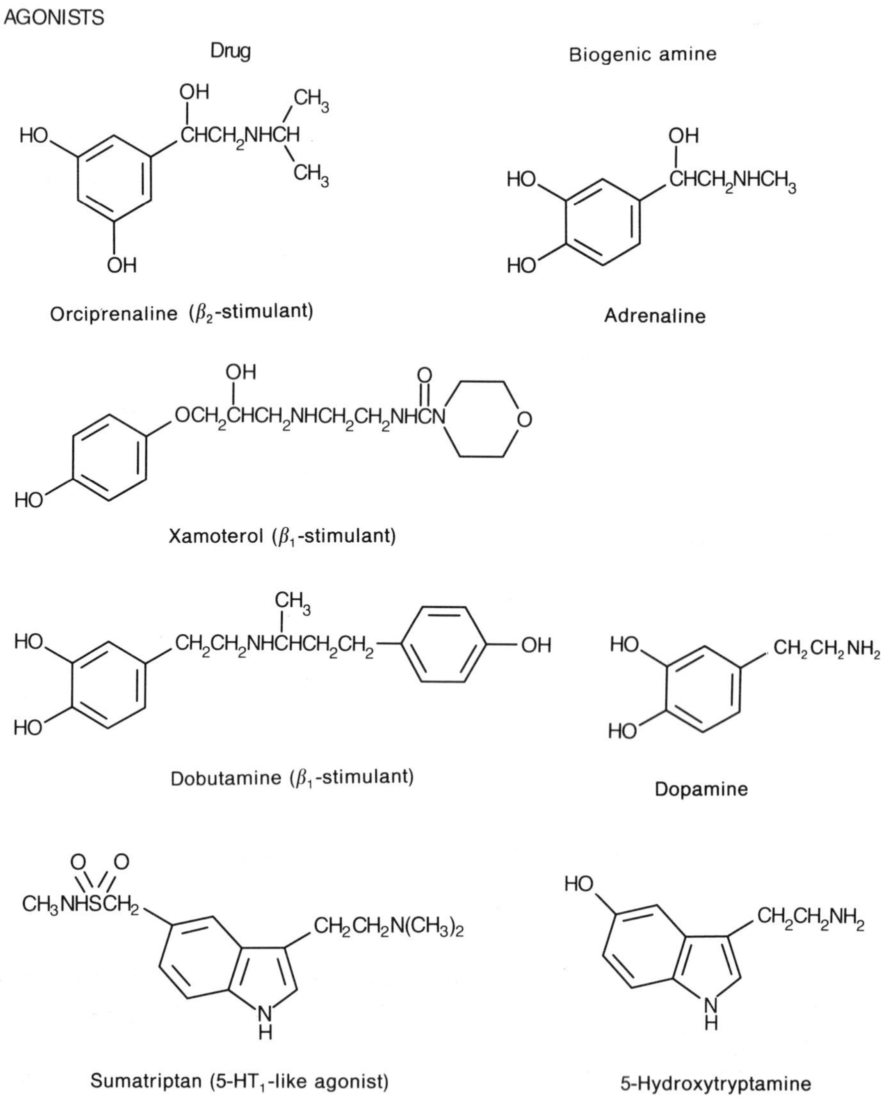

AGONISTS

Fig. 7.9 Chemical structures of some agonists and antagonists at biogenic amine receptors where the design was based on the amine structure as a template.

template. Thus selective agonists for postsynaptic receptors have been made for adrenergic β-receptors, e.g. salbutamol (Hartley *et al.*, 1968) (see Chapter 11), dobutamine (Tuttle and Mills, 1975) and xamoterol (Barlow *et al.*, 1981), and 5-hydroxy-tryptamine receptors, e.g. sumatriptan (Brittain *et al.*, 1987), by the modification and addition of appropriate substituents to the biogenic amine structure (Fig. 7.9).

Antagonists for biogenic amine receptors have been designed by retaining a partial structure of the amine (for receptor recognition) and then incorporating additional groups to increase the binding affinity between drug and receptor (Black, 1989) e.g. pronethalol (Black and Stephenson, 1962) for adrenergic β-receptors (see Chapter 10),

ANTAGONISTS

Drug
Biogenic amine

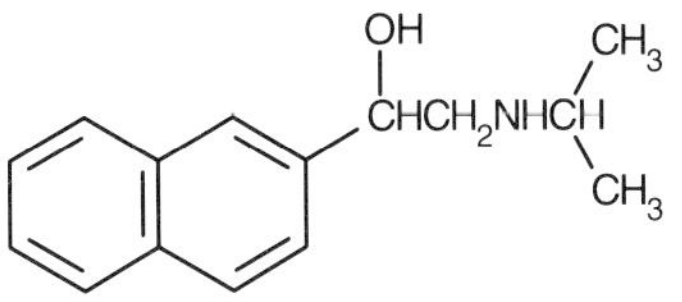

Pronethalol (β-blocker)

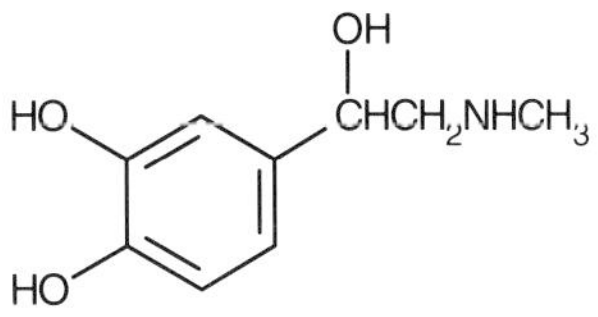

Adrenaline

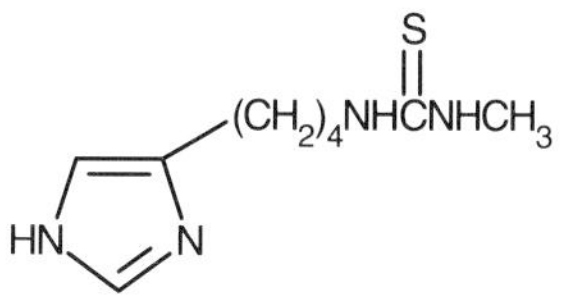

Burimamide (H_2-antagonist)

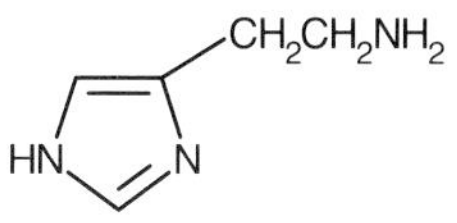

Histamine

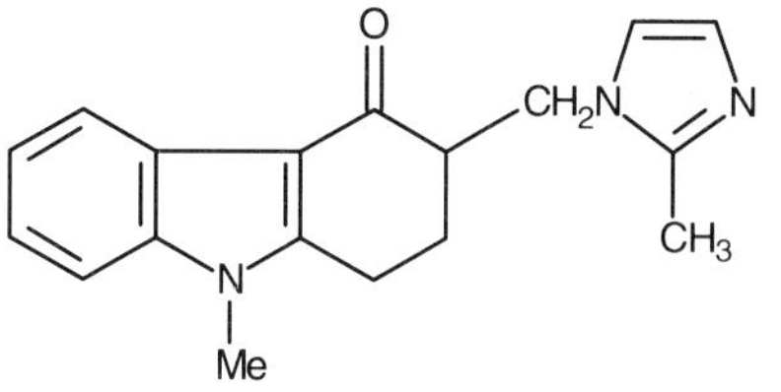

Ondansetron (5-HT_3-antagonist)

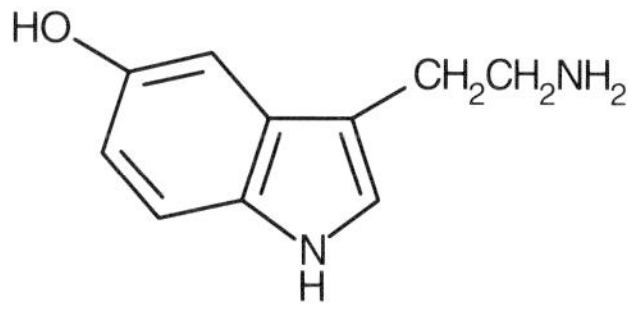

5-Hydroxytryptamine

Fig. 7.9 (Continued).

burimamide (Black *et al.*, 1972) for histamine H_2-receptors (see Chapter 12) and ondansetron and granisetron for 5-hydroxytryptamine 5-HT_3-receptors (Bermudez *et al.*, 1990) (Fig. 7.9).

Thus both enzymes and receptors are highly specialized sites involved in molecular recognition, whereby only the appropriate substances interact with them in a productive manner. This leads to a very high degree of selectivity of action and hence they provide very good chemical starting points for drug design. Since the key to a successful drug lies in its selectivity of action, the rational approach has been especially effective in modern drug discovery when based on enzymes or hormone receptors. This is reflected in the award of the Nobel Prize for Physiology and Medicine in 1988 to Black, Elion and Hitchings for their contributions to the discoveries of adrenergic β-blockers, histamine H_2-antagonists and antimetabolite enzyme inhibitors.

The past three decades have seen some notable drug design successes (Table 7.3) based on the approach of using as leads enzymes and substrates, or receptors for biogenic amines.

With the development of biochemistry and cell biology has come a greater understanding of cellular mechanisms and control, and many of the drugs discovered serendipi-

TABLE 7.3
Some Examples of Drugs Designed to Act on Enzymes or Receptors

Drug	Use	Mechanism of action
Enzyme inhibitors		
Pargyline	Antihypertensive	Inhibits monoamine oxidase-B
Captopril	Antihypertensive	Inhibits angiotensin-converting enzyme (ACE)
Carbidopa	Potentiates use of L-dopa in Parkinson's disease	Inhibits dopa decarboxylase peripherally
Clavulanic acid	Potentiates antibiotic action of penicillins	Inhibits β-lactamase
Methotrexate	Antitumour	Inhibits dihydrofolate reductase
Cytarabine	Antileukaemia	Inhibits DNA polymerase
Lovastatin	Lowers cholesterol level	Inhibits HMG-CoA* reductase
Omeprazole	Antiulcer	Inhibits H^+/K^+ ATPase ('proton pump')
Receptor stimulants or blockers		
Atracurium	Neuromuscular blockade	Blocks acetylcholine nicotinic receptors
Pirenzepine	Antiulcer	Selectively blocks acetylcholine muscarinic M_1 receptors
Fenoldopam	Congestive heart failure	Stimulate dopamine D_1-receptors
Butaclamol	Antipsychotic	Blocks dopamine D_2-receptors
Prazosin	Antihypertensive	Blocks adrenergic α_1-receptors
Labetolol	Antihypertensive	Blocks adrenergic α- and β- receptors
Propranolol	Antihypertensive	Blocks adrenergic β-receptors
Xamoterol	Heart failure	Stimulates adrenergic β_1-receptors
Salbutamol	Antiasthma	Stimulates adrenergic β_2-receptors
Ondansetron	Antiemetic	Blocks serotonin $5HT_3$-receptors
Sumatriptan	Antimigraine	Stimulates serotonin '$5HT_1$-like' receptors
Mepyramine	Antiallergy	Blocks histamine H_1-receptors
Cimetidine	Antiulcer	Blocks histamine H_2-receptors

*HMG-CoA, hydroxymethylglutaryl-coenzyme A.

Chemical structures

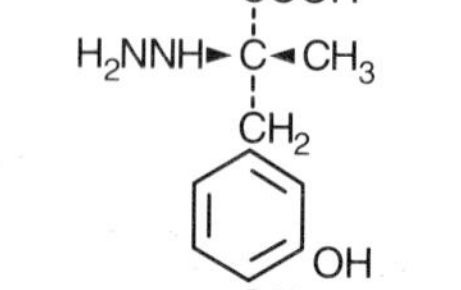

Pargyline

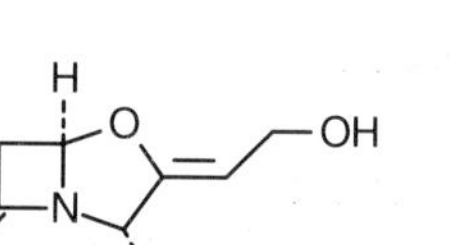

Captopril

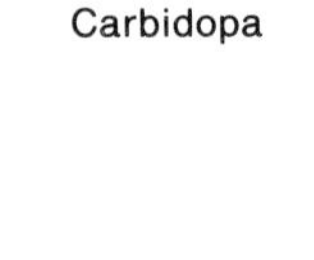

Carbidopa

Clavulanic acid

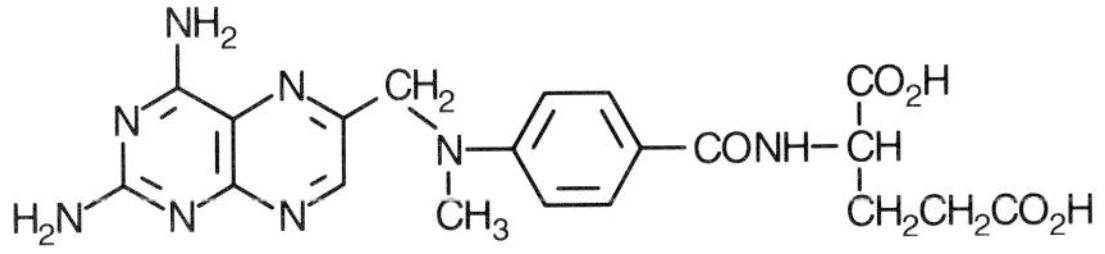

Methotrexate

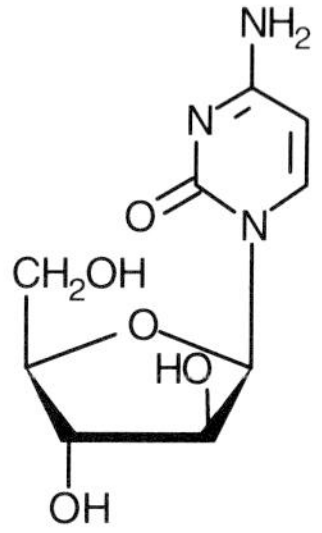

Cytarabine

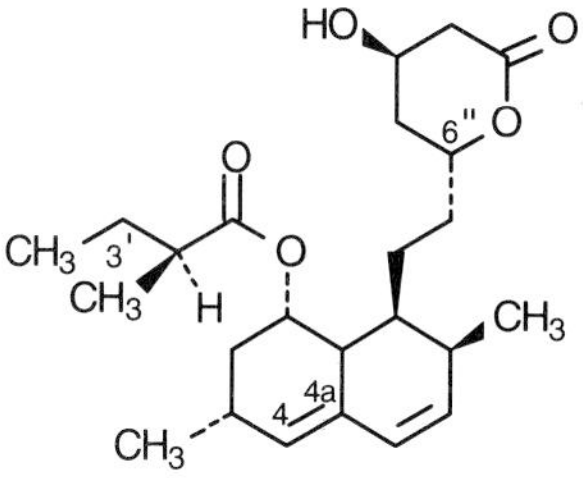

Lovastatin

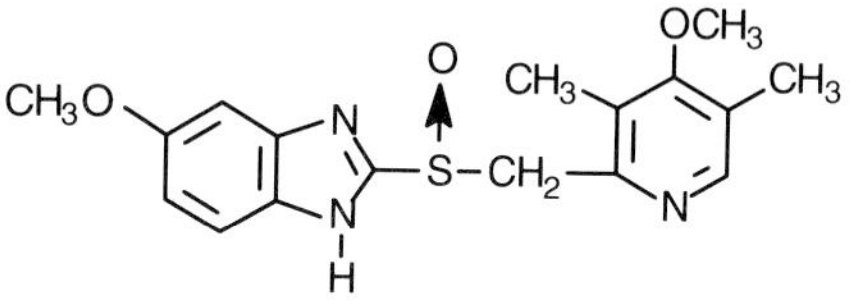

Omeprazole

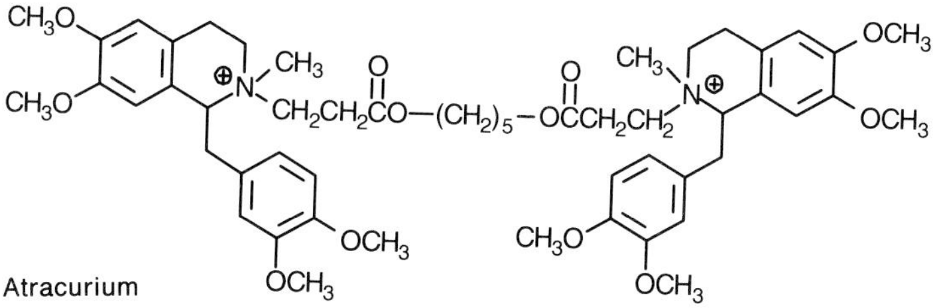

Atracurium

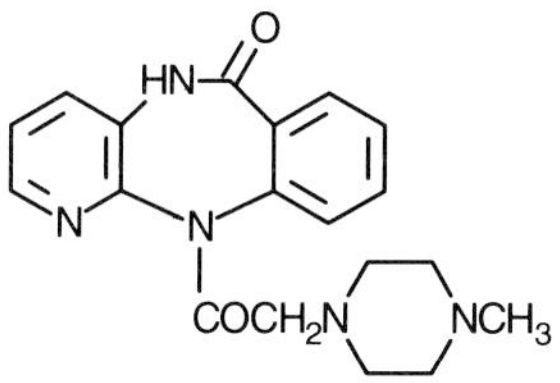

Pirenzepine

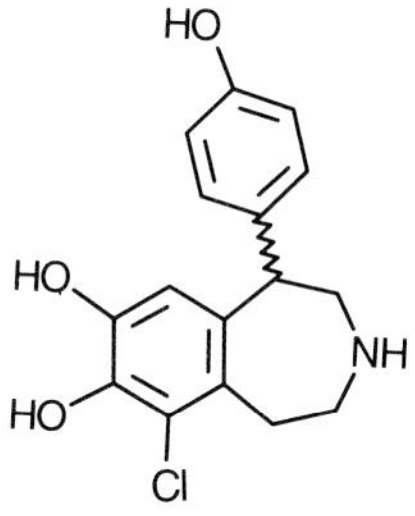

Fenoldopam

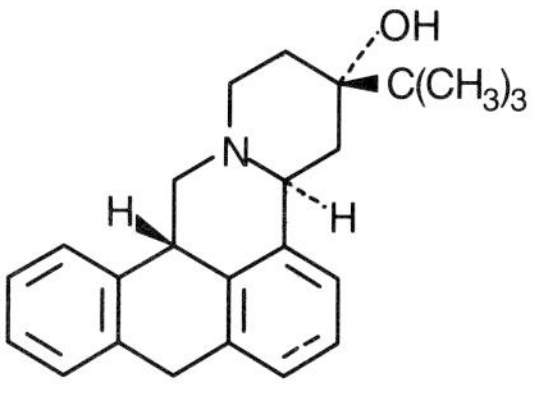

Butaclamol

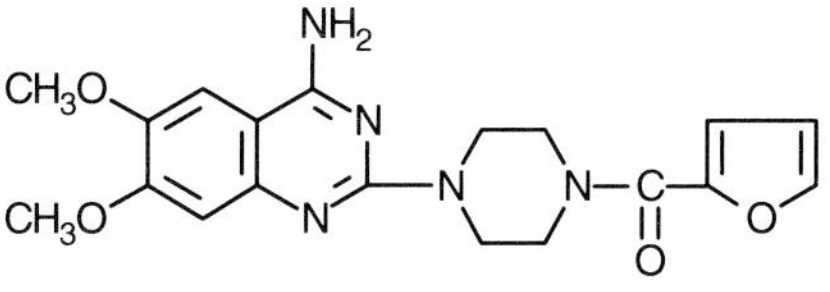

Prazosin

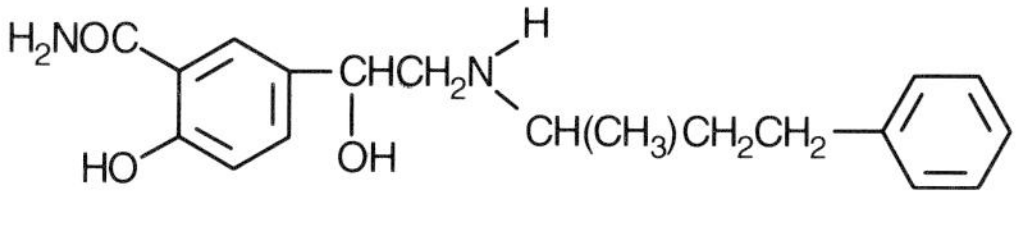

Labetolol

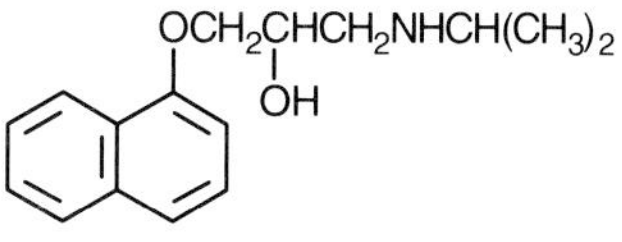

Propranolol

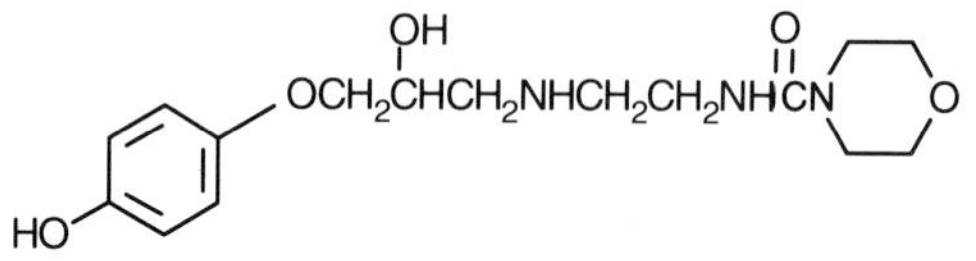

Xamoterol

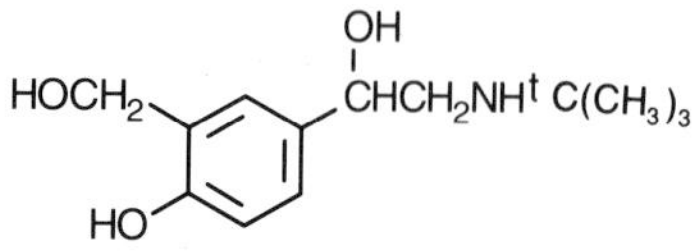

Salbutamol

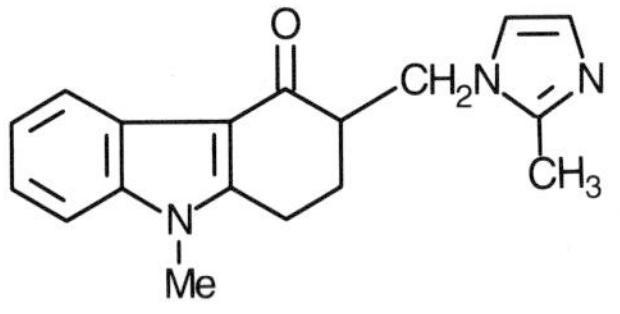

Ondansetron

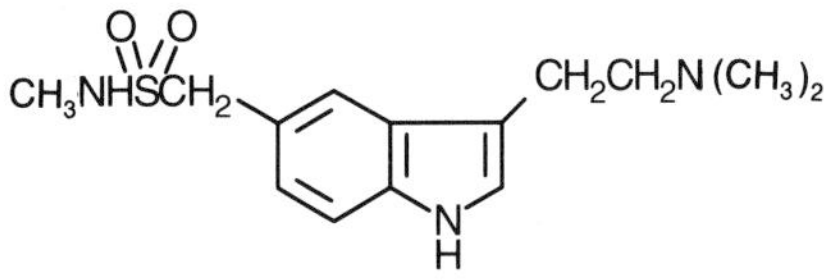

Sumatriptan

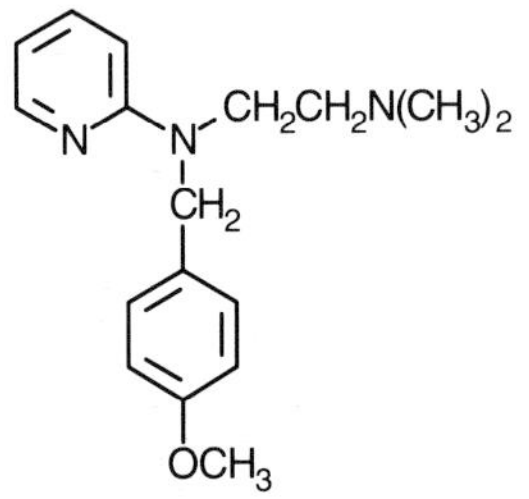

Mepyramine

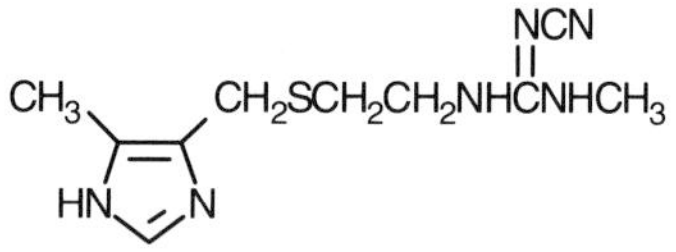

Cimetidine

tously or by screening procedures over the past three decades have subsequently been shown to act by interfering with particular mechanisms; some examples are shown in Table 7.4. Such examples have provided a further stimulus to the development of agents rationally, from mechanistic considerations. Thus there is a strong interplay between drug discovery and retrospective rationalization. A semi-empirical discovery of a useful drug provides the stimulus for deeper probing into how and why it works and thus to a deeper understanding of what underlies the disease. This in turn may give rise to new concepts and new discoveries.

VI.   Uncertainties in drug design and development

This chapter has been concerned with the design and discovery of new compounds with interesting biological properties as potential drugs. There has been a phenomenal increase in recent years in the amount of information about cellular mechanisms and, therefore, as we approach the twenty-first century, the outlook for rational drug design ought to be very optimistic. Unfortunately, however, there has been a key factor missing from our understanding.

   The body is under multifactorial control. There are many natural checks and balances. For any given function, there are usually several messengers and several types of receptor; there are also amplification systems, modulating systems, feedback inhibitory mechanisms, various ion fluxes and so on; if we block one pathway by drug action, another pathway is likely to take over. The consequence is that we cannot be sure at the outset that designing a drug to act on a particular receptor or enzyme will necessarily

TABLE 7.4
Some Examples of Old Drugs and New Mechanisms

Drug	Use	Mechanism of action
Aspirin	Antiinflammatory	Inhibits the enzyme prostaglandin synthetase
Theophylline	Antiasthma (bronchodilator)	Inhibits the enzyme phosphodiesterase
Atropine	Mydriatic (dilates pupil of the eye)	Blocks the muscarinic receptors for acetylcholine
Haloperidol	Neuroleptic	Blocks the receptors for dopamine in the brain
Chlordiazepoxide	Antianxiety	Binds allosterically to the receptors for γ-aminobutyric acid (GABA) and opens the channels for the transport of Cl^- in brain neurones
Tolbutamide	Hypoglycaemic (antidiabetic)	Blocks the transport of K^+ through its channels in the insulin-secreting β-cells of the pancreas
Nifedipine	Coronary vasodilator used to treat angina	Blocks the channels for transport of Ca^{2+} in vascular smooth muscle
Aminacrine	Antibacterial	Intercalates into the DNA of bacterial nucleus to inhibit growth
Cromoglycate	Antiasthma	Prevents histamine release from its storage sites in mast cells in the lung
Nicorandil	Vasodilator	Stimulates K^+ transport through its channels in vascular smooth muscle

Chemical structures

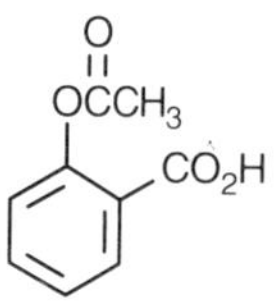

Aspirin

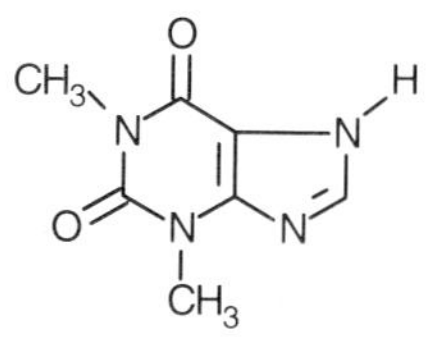

Theophylline

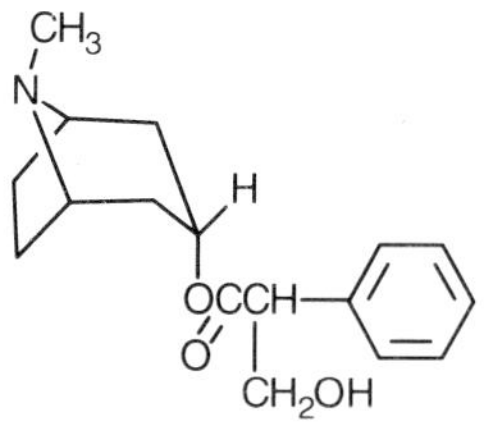

Atropine

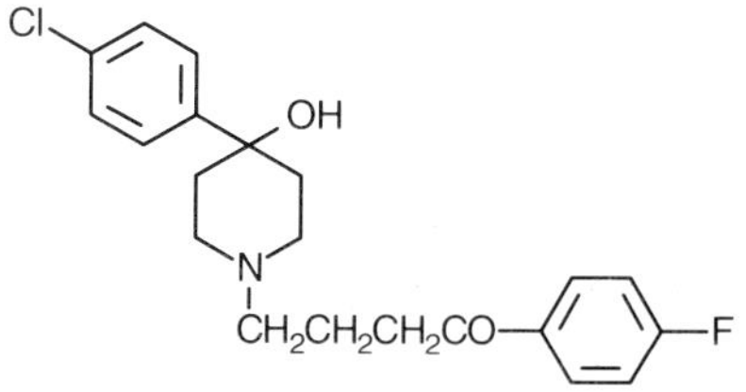

Haloperidol

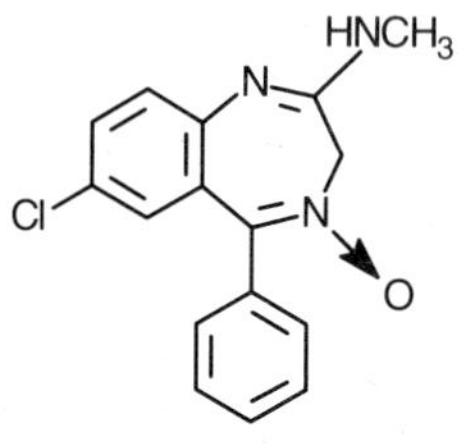

Chlordiazepoxide

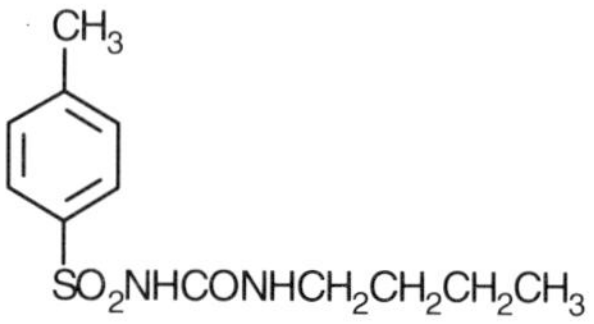

Tolbutamide

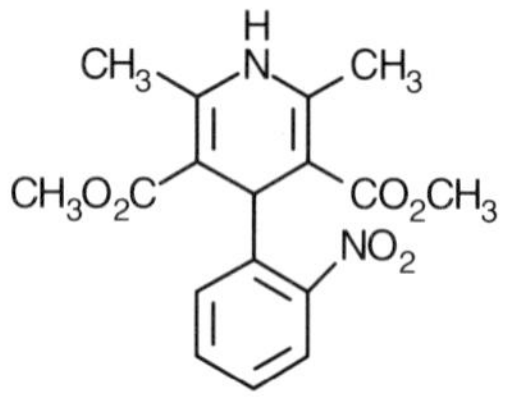

Nifedipine

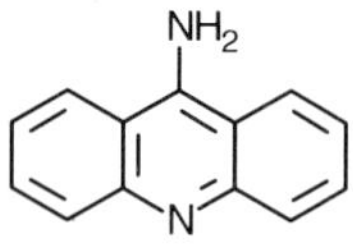

Aminacrine

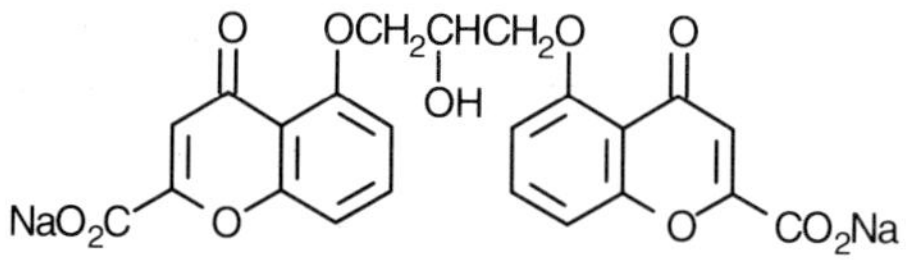

Sodium cromoglycate

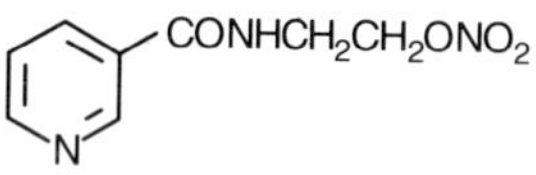

Nicorandil

provide treatment for a given medical condition, even though we may know that it is involved in the physiological controlling mechanisms. Indeed, there are now many examples of nicely designed 'drugs' which are still looking for a suitable disease.

Enzymes and receptors are ubiquitous and occur in many different tissue systems. Blocking them at a tissue site involved in a disease may be therapeutically effective but blockade concurrently in other tissue sites (not involved in the disease) may be thoroughly undesirable.

For the above reasons, there is still a strong element of speculation in drug design, and a considerable uncertainty in achieving success. Thus only a small proportion of drug discoveries are destined to become useful therapeutic agents.

Complications also arise during drug design because a drug has to be administered and find its way to the desired site of action, whereas the natural messenger may be generated locally or stored nearby to its required site of action. Also, after it has acted, the natural transmitter is removed by specific enzymes or reuptake mechanisms, whereas a drug has to be disposed of by being excreted. We have to balance the desired pharmacology with the biochemical needs to achieve drug access and elimination. In altering the chemistry of a drug to achieve adequate disposition one may inadvertently introduce other pharmacological properties, thereby reducing the selectivity of action.

The discovery of a potential new drug as a possible therapeutic agent is followed by its development: the studies that have to be carried out in the laboratory and on animals to characterize the properties of any new drug and to assess its safety, before it can be tried in the human body, often referred to as 'preclinical R&D'. Then comes the third stage, 'clinical R&D' namely human studies progressing from healthy volunteers through to clinical trials on patients. Although one can differentiate between these three stages, there is usually considerable overlap in timing.

Safety assessment often starts with testing *in vitro* on cell cultures to examine for potential effects on cell reproduction. This is usually followed by repeated daily doses (at different dose levels) to groups of laboratory animals for 7–14 days to see what signs of toxicity might be revealed. Eventually this will be repeated for longer periods, e.g. daily dosing for 1 month, 3 months or 6 months and in at least two species of animal. During this period various indicators of biochemical function will be assessed and at the end of the period the animals will be killed and tissue samples from all the main organs will be examined microscopically for possible damage or unusual effects.

These safety studies are enormously time-consuming and involve many skilled people. A study *in vivo* to show that a potential drug is not likely to cause cancer can take more than 3 years to complete; it will involve daily dosing to groups of laboratory rats or mice for almost their lifetime (18 months to 2 years) followed by extensive microscopic examination of the tissues by qualified pathologists.

The scale of the safety assessment is considerable and can require 500–750 kg of compound to be synthesized before it reaches the manufacturing stage. In addition, the synthetic route used to generate the material for safety assessment must be close to that used in the intended production process. This requires that considerable process investigation will have taken place, and there is great pressure to do this rapidly because the generation of sufficient chemical at this juncture can be rate-limiting. This investment in time, effort and money has to be made before it is known whether the new drug will be therapeutically effective or whether it may show an unacceptable 'side effect' problem. None of the product is sold during this time. This phase is very costly and there is a high chance that there will be no financial return on the investment, i.e. it is a risky business.

It takes many years to invent and develop a new drug, typically 10–15 years. Today's new research programmes will not come to fruition until we have already entered the twenty-first century!

References

Barlow, J.J., Main, B.G. and Snow H.M. (1981). *J. Med. Chem.* **24**, 315.

Bermudez, J., Fake, C.S., Joiner, C.F., Joiner, K.A., King, F.D., Miner, W.D. and Sanger, G.J. (1990). *J. Med. Chem* **33**, 1924.

Bindra, J.S. and Lednicer, D. (eds) (1982). 'Chronicles of Drug Discovery', Vol. 1 *et seq.* J. Wiley, New York.

Black, J.W., (1976). *Pharm. J.* **217** 303.

Black, J. (1989). *Science* **245**, 486.

Black, J.W. and Stephenson, J.S. (1962). *Lancet* **2**, 311.

Black, J.W., Duncan, W.A.M., Durant, G.J., Ganellin, C.R. and Parsons, M.E. (1972). *Nature (London)* **236**, 385.

Brittain, R.T., Butina, D., Coates, I.H., Feniuk, W., P.P.A. Humphrey, Jack, D., Oxford, A.W., and Perren, M.J. (1987). *Br. J. Pharmacol.* **92** (Suppl), 618P.

Burger, A., (1954). *Chem. Eng. News* **32**, 4172.

Chang, R.S.L., Lotti, V.J., Monaghan, R.L., Birnbaum, J., Stapley, E.O., Goetz, M.A., Albers-Schönberg, G., Patchett, A.A., Liesch, J.M., Hensens, O.D. and Springer, J.P. (1985). *Science* **230**, 177.

Fozard, J.R. and Mobarok Ali, A.T.M. (1981) *Eur. J. Pharmacol.* **49**, 109.

Gralla, R.J., Itri, L.M., Pisko, S.E., Squillante, A.E., Kelsen, D.P., Braun, D.W., Bordin, L.A., Braun, J.J. and Young, C.W. (1981) *New Eng. J. Med.* **305**, 905.

Hall, S. and Strichartz, G. (eds) (1990). *ACS Symposium Series*, **418**. 'Marine Toxins: Origin, Structure and Molecular Pharmacology,' American Chemical Society, Washington D.C.

Hartley, D., Jack, D., Lunts, L. and Ritchie, A.C.H. (1968). *Nature (London)* **219**, 861.

Hitchings, G.H. (1980). *Trends Pharm. Sci.* **1**, 167.

Jack, D. (1976). *Pharm. J.* **217**, 229.

Jack, D. (1991). *Br. J. Clin. Pharmacol.* **31**, 501.

Laborit, H. (1954). *Presse Med.* **62**, 359.

Llorens, C., Gacel, G., Swerts, J.P., Perdrisot, R., M.C. Fournie-Zaluski, Schwartz, J.C. and Roques, B.P. (1980). *Biochem. Biophys. Res. Commun.* **96**, 1710.

Ondetti, M.A., Rubin, B. and Cushman, D.W. (1977). *Science* **196**, 441.

Rose, F.L. (1964). *Chem. Ind.*, 858.

Schwartz, W.B. (1949). *New Engl. J. Med.* **240**, 173.

Sneader, W. (1985). 'Drug Discovery: The Evolution of Modern Medicines', J. Wiley, Chichester.

Tuttle, R.R. and Mills, J. (1975). *Circ. Res.* **36**, 185.

– 8 –

Discovery and Development of Cromakalim and Related Potassium Channel Activators

G. STEMP and J.M. EVANS

SmithKline Beecham Pharmaceuticals,
Medicinal Research Centre,
Coldharbour Road,
The Pinnacles,
Harlow, Essex CM19 5AD, U.K.

Dr Geoffrey Stemp graduated in chemistry from Nottingham University and completed his Ph.D. there in 1979. He then joined Beecham Pharmaceuticals and is currently a Senior Medicinal Chemist at SmithKline Beecham. His research interests include potassium channel activators and dopamine antagonists.

Dr John Evans obtained his degrees in chemistry (B.Sc. and Ph.D.) at University College of Wales, Aberystwyth, 1960–1966. After taking up Postdoctoral Research Fellowships with S.M. Kupchan at the University of Wisconsin and Sir E.R.H. Jones at Oxford, he became a Lecturer at Escuela de Quimica, Universidad Central de Venezuela, 1969–1973. He joined Beecham Pharmaceuticals in 1974 and is currently a Senior Medicinal Chemist at SmithKline Beecham. His research interests include potassium channels and their modulators.

MEDICINAL CHEMISTRY 2nd Edition
ISBN 0-12-274120-X

I. Introduction

Cardiovascular disease is the major cause of morbidity and mortality in the Western world, and high blood pressure (hypertension) is a significant risk factor for heart attacks and strokes. The causes of hypertension – which occurs in 15–20% of the population – are not fully understood, but a number of factors such as age, genetic history, stress, diet and smoking are believed to be involved. Early treatment for high blood pressure often involves a reappraisal of the patient's lifestyle as this can lead to a normalization of blood pressure. However, moderate to severe hypertension requires drug treatment, for, although no symptoms are associated with high blood pressure, the consequences of non-treatment such as an increased risk of stroke, kidney failure or impaired vision are serious.

An individual's blood pressure depends upon many factors but, in general, diastolic blood pressure in a normal individual should be below 90 mmHg. Mild hypertension is defined as a diastolic pressure of 90–100 mmHg, and between 110 and 130 mmHg is classified as severe hypertension.

During the last two decades the treatment of high blood pressure has, like that of many other disease areas, undergone a remarkable revolution with the introduction of several classes of drugs such as the β-adrenoceptor antagonists (β-blockers) (see Chapter 10), calcium channel blockers and angiotensin-converting enzyme (ACE) inhibitors (see Chapter 9). More recently a new class of compounds has been discovered which lowers blood pressure by a novel mechanism of action, involving an increase in the outward movement of potassium ions through channels in the membranes of vascular smooth muscle cells, leading to relaxation of the smooth muscle. We have therefore called these compounds potassium channel activators (KCAs). Cromakalim (1) was the first antihypertensive agent to be shown to act exclusively by this mechanism (Hamilton et al., 1986) although interest is now focused on the biologically active ($-$)-enantiomer, BRL 38227 (2).

(1) Cromakalim
(2) BRL 38227 (3S, 4R)

Although the physiology of potassium channels has been studied for several decades, a major contribution to the spectacular increase in the investigation of these channels in recent years can be ascribed to the discovery and elucidation of the mode of action of cromakalim. This chapter describes the discovery of cromakalim and highlights some further structure–activity developments in potassium channel activators.

II.   Discovery of Cromakalim

In the early 1970s, ICI introduced the β-blocker class of drug for the treatment of angina

(3) $ArOCH_2CHOHCH_2NHCH(Me)_2$

(3a)   Ar =

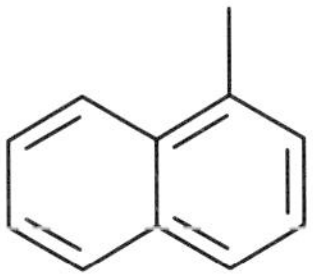

(3b)   Ar = H_2NCOCH_2

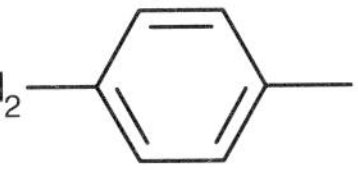

pectoris and hypertension, examples of which are propranolol (3a) and, more recently, atenolol (3b). However, at that time there was some doubt that β-blockade was responsible for their antihypertensive activity. These drugs are characterized by an aryloxy substituent attached to one terminal of a flexible propanolamine chain, which can adopt many conformations.

It was suggested by one of our colleagues at Harlow – Eric Watts – that by restricting their conformation by cyclizing the carbon atom bearing the terminal amino group on to the aromatic ring (compounds 4, Scheme 8.1), β-blocking activity would be lost but antihypertensive activity might be retained. Such antihypertensive agents would lack the side effects associated with β-blockers.

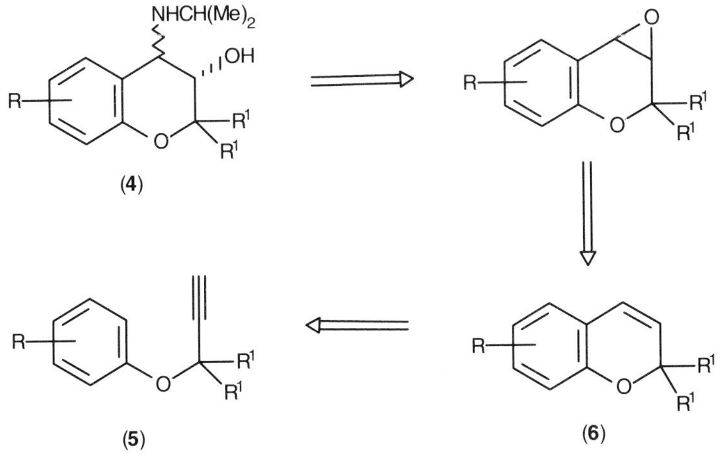

Scheme 8.1   Retrosynthetic analysis of aminochromanols (4).

A retrosynthetic scheme for the preparation of compounds (4) was devised (Scheme 8.1), the initial step being the Claisen rearrangement of arylpropargyl ethers (5) to chromenes (6).

The yield of this reaction is greatly enhanced (Harfenist and Thom, 1972] by the presence of a *gem*-dimethyl group (R^1 = Me) at C-2, so we decided to carry out a model synthesis (Scheme 8.2) employing *p*-nitrophenylpropargyl ether (5), the aromatic substituent being available for further manipulation as required.

The nitrochromene (6) was converted to bromohydrin (7) and then to epoxide (8) which was opened with isopropylamine to give the *trans*-isopropylamino alcohol (9)

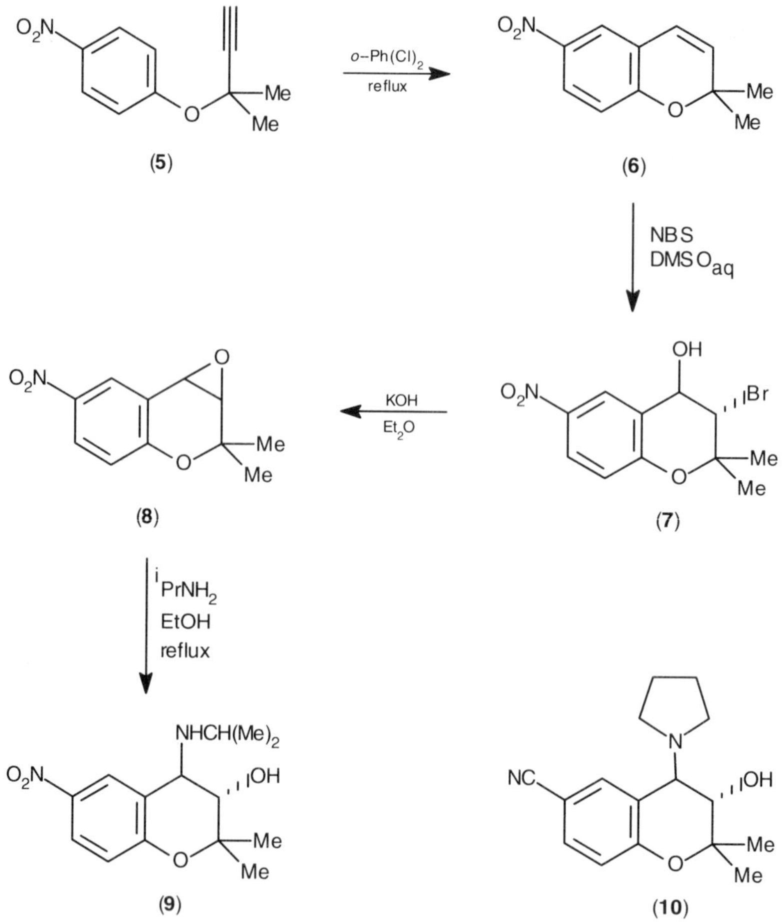

Scheme 8.2   Model synthesis of 2,2-dimethyl-6-nitroaminochromanol (9).

*In this and subsequent schemes, all compounds are racemic unless otherwise stated.

(Evans *et al.*, 1983). This compound was indeed found to lower blood pressure in hypertensive rats (Table 8.1), and to our great surprise by a direct peripheral vasodilator mechanism. No β-blocking activity was observed. Systematic variation of the substituents around the benzopyran ring of compound (9) indicated that the *gem*-dimethyl group, and an electron-withdrawing aromatic substituent were essential for high potency, while cyclic amino groups were preferred to the original isopropylamino groups. This optimization of activity led to the 6-cyano-4-pyrrolidinylbenzopyran (10) which was more than a hundred-fold more potent than compound (9) (Evans *et al.*, 1983). However, when tested for vasorelaxant activity *in vitro*, compound (10) showed only moderate activity, indicating that metabolism of this compound might occur *in vivo*. Thus, as part of a programme investigating the potential metabolites of compound (10), it was decided to prepare the compound (1) in which a carbonyl group had been incorporated α to the amine nitrogen, since α-oxidation of amines to amides is a well-known metabolic process. The replacement of pyrrolidine by pyrrolidinone resulted in a further 3-fold increase in potency in the hypertensive rat.

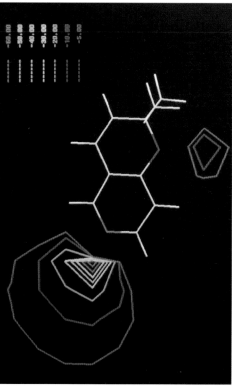

Plates I and II   Two-dimensional electrostatic potential energy maps of 6-cyano-2,2-dimethyl-2H-1-benzopyran (Plate I) and 2,2-dimethyl-2H-pyrano[3,2-c]pyridine (Plate II).

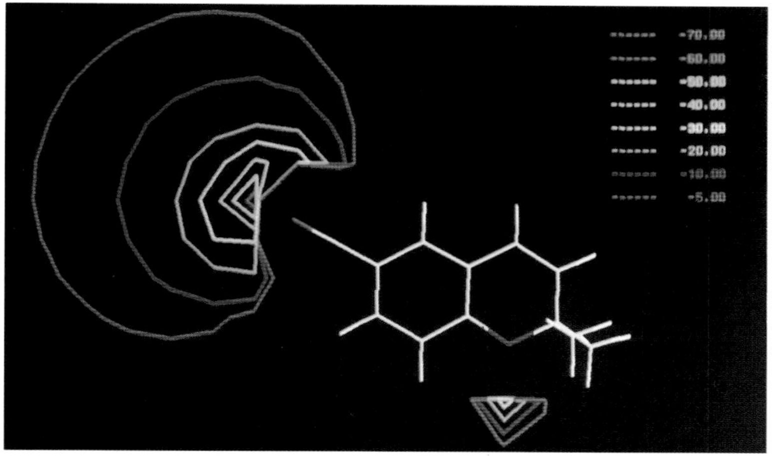

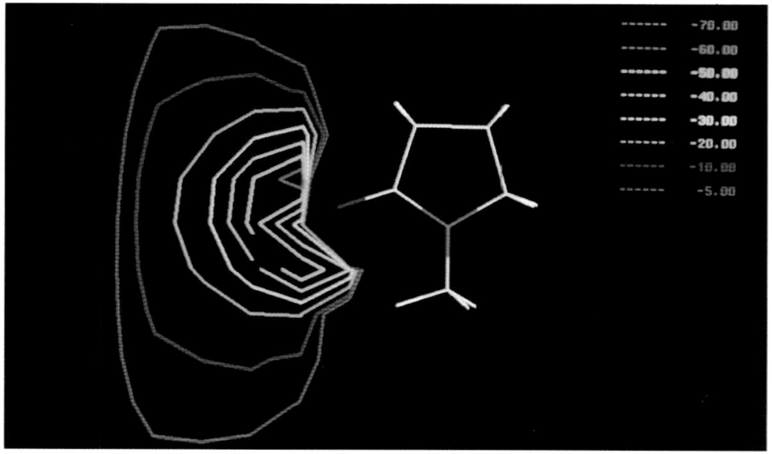

Plates III–VI   Two-dimensional electrostatic potential energy maps of N-methyl-2-pyrrolidinone (Plate III), NN'-dimethylurea (Plate IV) (shown overleaf), NN'-dimethylcyanoguanidine (Plate V) and 1-methyl-3-amino-5-methylamino-1,2,4-triazole (Plate VI).

These maps were generated using the ab initio program GAMESS and displayed on a Silicon Graphics Iris workstation. The contour levels are in kcal/mol and show the areas around each molecule that are attractive towards a point positive charge; they are colour coded as indicated at the side of each photographic plate. The areas of negative potential thus indicate possible hydrogen-bonding sites around each molecule. The structural formulae (in white) are coloured red and blue to represent O and N atoms respectively.

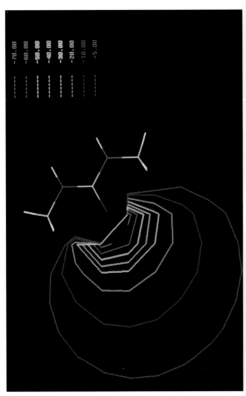

IV

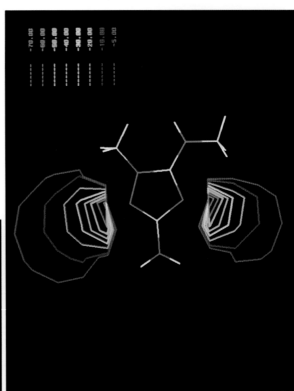

VI

V

TABLE 8.1
Antihypertensive Activity of 4-Substituted Benzopyran-3-ols

Compund	R¹	R²	Dose (p.o.) (mg/kg)	Max. decrease in blood pressure (%)
			Hypertensive rats	
9	NO₂	NHCH(Me)₂	100	36 ± 4*
10	CN	pyrrolidine	1.0	29 ± 2†
1	CN	pyrrolidinone	0.3	39 ± 4†

Results are means ± S.E.M.

*Deoxycorticosterone acetate (DOCA)/saline rats (Evans *et al.*, 1983, 1984). Hypertension was induced by subcutaneous implantation of DOCA, together with unilateral nephrectomy and replacement of the drinking water with 1% NaCl solution for 5 weeks after surgery. Two months later animals were considered hypertensive if their systolic blood pressure was greater than 160 mmHg. Compounds were dosed orally to groups of at least three rats.

†Spontaneously hypertensive rat (SHR) (Evans *et al.*, 1984; Ashwood *et al.*, 1986). Compounds were dosed orally to groups of at least five of these genetically hypertensive rats, derived from the Japanese (Okamoto) strain. Animals with a systolic blood pressure greater than 180 mmHg were considered hypertensive.

The synthesis of compound (**1**) was achieved (Scheme 8.3) by either direct reaction of epoxide (**11**), obtained by an analogous route to that shown in Scheme 8.2, with the anion of 2-pyrrolidinone, or acylation of amino alcohol (**12**) with 4-chlorobutyryl chloride to give compound (**13**), followed by ring closure. Resolution of (**1**) was achieved via the *S*-α-methylbenzylcarbamate and the activity found to reside almost exclusively in the (−)-3*S*, 4*R* enantiomer (**2**) (Ashwood *et al.*, 1986). The absolute stereochemistry of (**2**) was determined from X-ray crystallography, using the corresponding *S*-α-methylbenzylcarbamate. Amino alcohol (**12**) could also be resolved via its bromo-camphor sulphonate salt to give an alternative synthesis of (**2**).

III. Structure–Activity Studies

Following the discovery of the high antihypertensive potency of compound (**1**), we embarked upon a systematic investigation designed to optimize activity in this structure, using the fall in blood pressure after oral dosing in the spontaneously hypertensive rat (SHR) as our primary screen (see footnote in Table 8.1).

The optimal position for substitution in the aromatic ring was investigated (see Table 8.2) in the nitro series (**14, 15, 16**), and the 6-nitro analogue (**14**) was found to be the most potent. At position 6 the nature of the substituent required was investigated and we

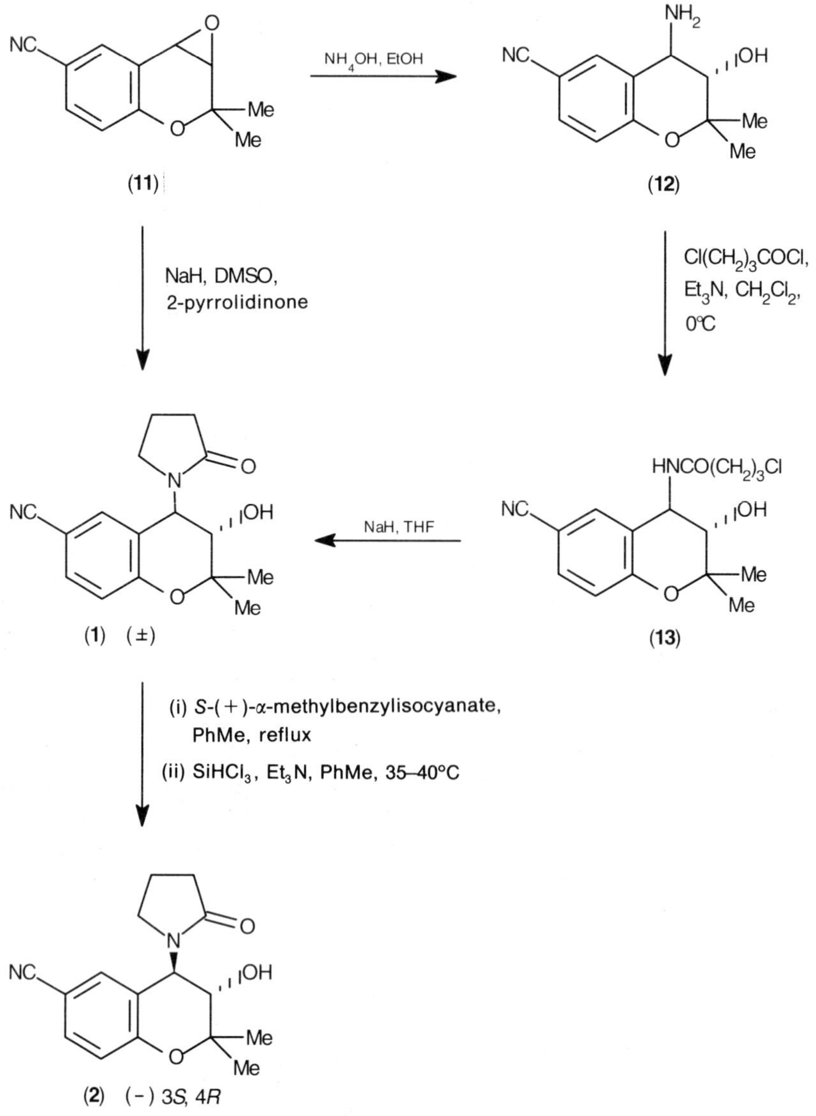

Scheme 8.3   Synthesis and resolution of cromakalin (1).

found that strong electron-withdrawing groups such as cyano (1), nitro (14) or acetyl (17) gave the most potent compounds, whereas a chloro substituent (18) resulted in only modest potency, and the analogue (19) lacking a substituent at position 6 showed little activity. It was also found that combining a strong electron-withdrawing group (NO$_2$) at position 6 with an electron-donating group (NH$_2$) at position 7, as in compound (20), enhanced potency. As noted previously, the *gem*-dimethyl group greatly enhanced the yields of the Claisen rearrangement, and it was fortunate from a synthetic point of view to find that the presence of this grouping at position 2 of the benzopyran ring was critical for activity; compare compounds (14), (21) and (22) (Ashwood *et al.*, 1986).

During the epoxide-opening reaction [(11)→(1)] shown in Scheme 8.3, small amounts

TABLE 8.2

Effect on Antihypertensive Activity of Substitution around the Benzopyran Ring

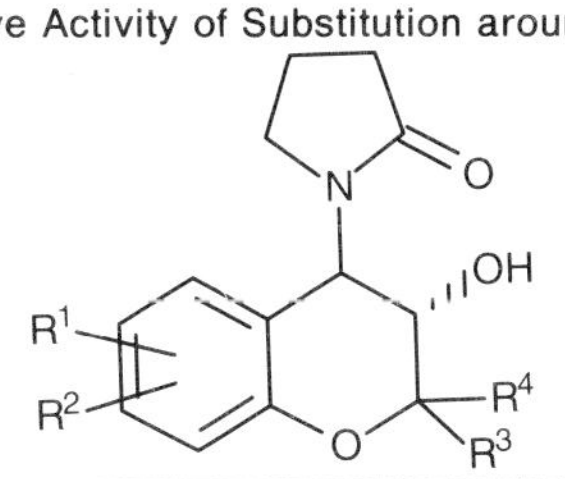

Compound	R^1	R^2	R^3	R^4	SHR Dose (p.o.) (mg/kg)	Max. decrease in blood pressure (%)
1	6-CN	H	Me	Me	0.3	39 ± 4
14	6-NO$_2$	H	Me	Me	0.3	27 ± 5
15	7-NO$_2$	H	Me	Me	0.3	12 ± 5
16	8-NO$_2$	H	Me	Me	1.0	7 ± 4
17	6-COMe	H	Me	Me	0.3	33 ± 5
18	6-Cl	H	Me	Me	10.0	37 ± 7
19	H	H	Me	Me	10.0	9 ± 4
20	6-NO$_2$	7-NH$_2$	Me	Me	0.1	51 ± 4
21	6-NO$_2$	H	Me	H	10.0	34 ± 3
22	6-NO$_2$	H	H	H	10.0	5 ± 1

Results are means ± S.E.M.

of the achiral chromene (23) were obtained, and it was found that treatment of (1) with sodium hydride in tetrahydrofuran under reflux gave (23) in good yield. Interestingly, (23) was found to be equipotent with (1) in the spontaneously hypertensive rat. Hydrogenation of (23) gave chroman (24) but this compound was much less potent than (1). The corresponding *cis*-analogue of compound (1) was also found to be less potent (Ashwood *et al.*, 1986).

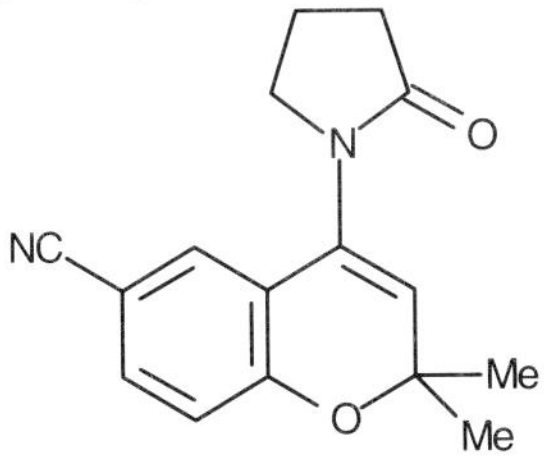

(23) 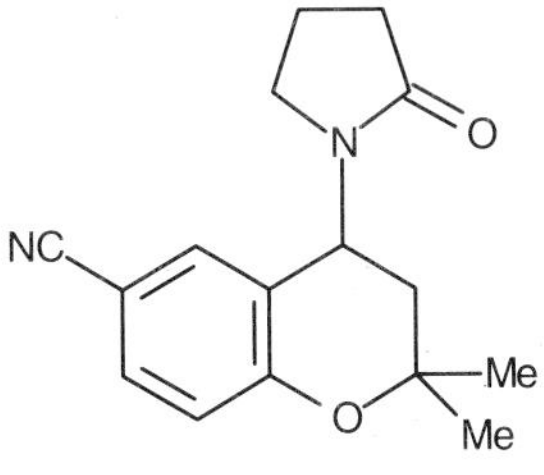   (24)

The size of the lactam ring at position 4 was found to have an important effect on potency (Table 8.3), with the six-membered ring (25) conferring greater potency than the five-membered ring. Increasing the lactam ring size beyond six led to a marked decrease in potency (Ashwood *et al.*, 1986). Incorporation of heteroatoms into five- or six-membered lactams also provided compounds of high potency. The oxazolidinone (26) and thiazolidinone (27) were slightly less potent, but the piperazinone (30) was slightly more potent than the pyrrolidinone (1). Interestingly, thiomorpholinone (29) was much less potent than morpholinone (28), presumably because the effective ring size of this substituent is greater than the optimum of five or six [Ashwood *et al.*, 1986). The

TABLE 8.3

Effect on Antihypertensive Activity of Lactam Ring Size and Heteroatom Substitution

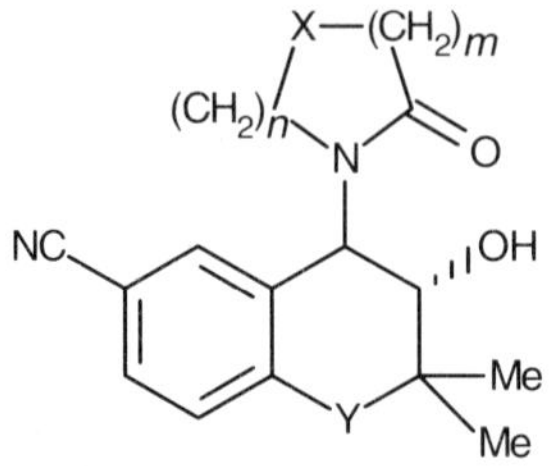

					SHR	
Compound	X	Y	m	n	Dose (p.o.) (mg/kg)	Max. decrease in blood pressure (%)
1	CH_2	O	1	1	0.3	39 ± 3
25	CH_2	O	1	2	0.1	46 ± 7
26	O	O	0	2	0.3	20 ± 4
27	S	O	0	2	0.3	27 ± 3
28	O	O	1	2	0.3	26 ± 3
29	S	O	1	2	10.0	12 ± 4
30	NH	O	1	2	0.1	28 ± 3
31	CH_2	NH	1	1	3.0	26 ± 5
32	CH_2	CH_2	1	1	3.0	29 ± 2

Results are means ± S.E.M.

tetrahydroquinoline (31) and tetrahydronaphthalene (32) analogues of compound (1) showed that the pyran oxygen was an important feature as these compounds were approximately 10-fold less potent in the spontaneously hypertensive rat (Ashwood *et al.*, 1991). The thiolactam corresponding to compound (1), obtained by treating (1) with Lawesson's reagent, proved to be equipotent to compound (1).

It was surprising to find (Ashwood *et al.*, 1990) that the acyclic analogue of compound (1), the N-acetylethylamino compound (33), prepared as a more flexible pyrrolidinone replacement, retained about a third of its potency in the spontaneously hypertensive rat since, in the previous aminochromanol series, acyclic compounds [e.g. (9)] were much less potent than the cyclic amines [e.g. (10)]. However, an investigation of acyclic amides related to (33) revealed that removal of the N-ethyl group of compound (33) caused a dramatic increase in potency, with the acetylamino compound (34) being slightly more potent than compound (1).

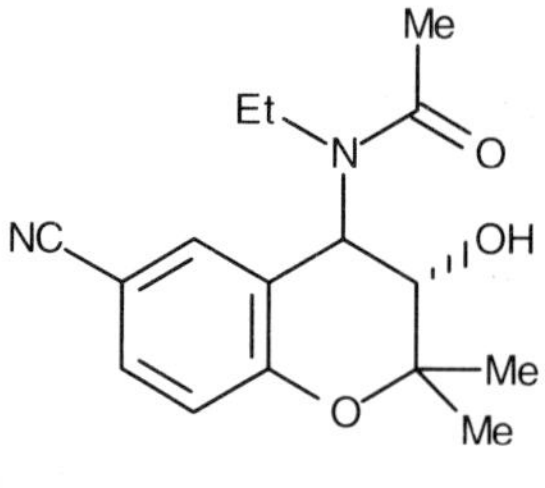

(33)

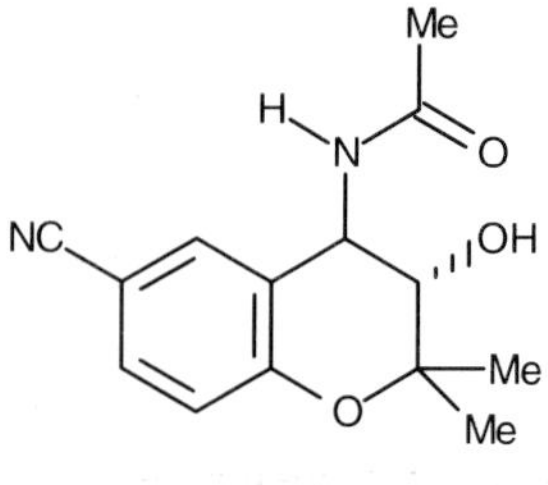

(34)

TABLE 8.4
Antihypertensive Activity of 4-(Acyclic amido)-Benzopyran-3-ols

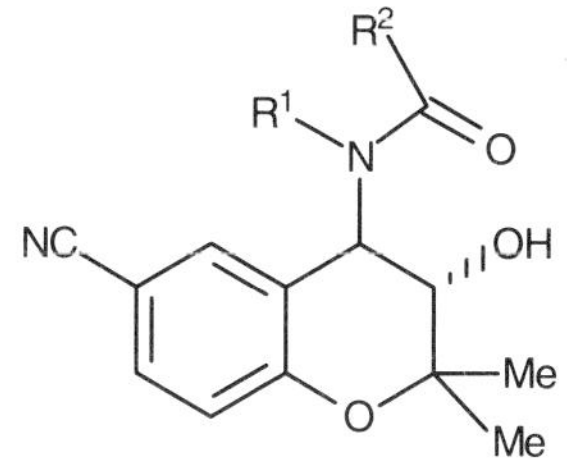

Compound	R^1	R^2	Dose (p.o.) (mg/kg)	Max. decrease in blood pressure (%)
1	-(CH$_2$)$_3$-		0.3	39 ± 4
33	Et	Me	1.0	22 ± 5
34	H	Me	1.0	67 ± 4
35	H	Ph	1.0	41 ± 8
36	H	NHMe	1.0	61 ± 3

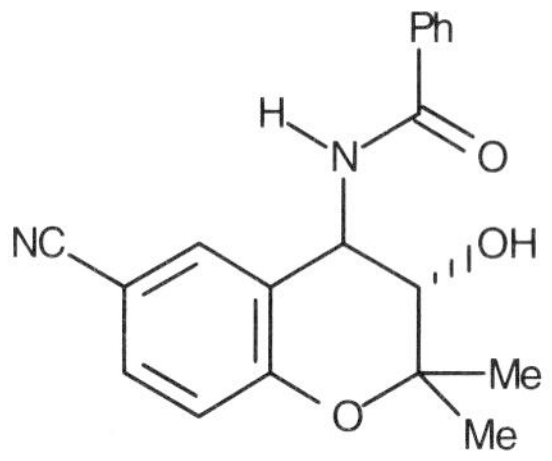

(35)

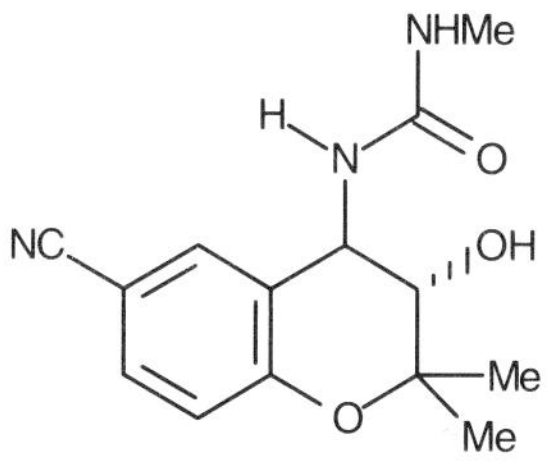

(36)

Replacing the methyl group in compound (34) by phenyl to give benzamide (35) or by methylamino to give methylurea (36) also retained activity (Table 8.4).

From the compounds discussed in this section, a number were selected to be taken through to secondary screening in both conscious and anaesthetized spontaneously hypertensive rats and renal hypertensive cat models. From the data obtained from these models, compound (1), cromakalim, was selected for progression to the clinic based on an improved haemodynamic profile (less tachycardia and increased renal blood flow) and greater potency compared with the calcium antagonist nifedipine (37) (Buckingham et al., 1986a).

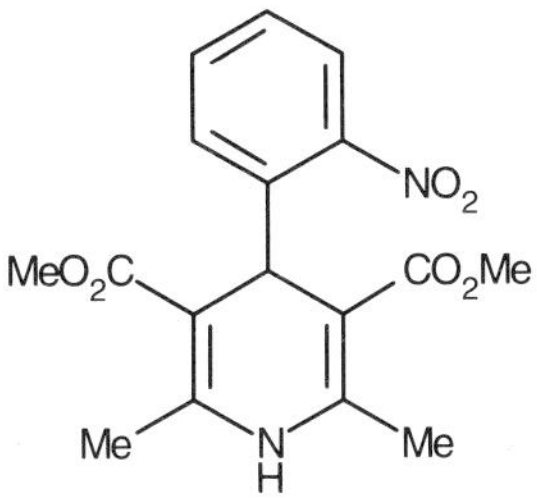

Nifedipine (37)

Cromakalim has since been superseded in clinical trials by its 3S, 4R-enantiomer BRL 38227.

IV.   Mode of Action of Cromakalim

The ability of cromakalim to relax isolated blood vessels (Hamilton *et al.*, 1986; Bucking-ham *et al.*, 1986b) demonstrated that the antihypertensive activity of the compound was probably due to a mechanism involving direct vasodilation, and this was confirmed in whole animals (Buckingham *et al.*, 1986a). Electrophysiological studies in cardiac tissue (Cain and Metzler, 1985) suggested that cromakalim enhanced the outward conductance of potassium ions, and this work was extended to microelectrode studies in rat portal vein (Hamilton *et al.*, 1986) to record cell membrane potential differences, together with measurement of mechanical tension, performed using various drug concentrations. After administration of cromakalim, the spontaneous, or natural, electrical and mechanical activities were attenuated. Also, at the highest concentration (5 μM) a significant hyper-polarization of the cell membrane was observed, until a value of approximately -90 mV was attained (Fig. 8.1). This is close to the theoretical equilibrium potential for potas-sium ions, suggesting that cromakalim is able to open a group of potassium ion channels which are essentially closed at the resting membrane potential, with the consequent outward movement of potassium ions and relaxation of the tissue.

Further evidence for this potassium ion efflux has been obtained using ^{86}Rb as a marker for potassium ions in isolated blood vessels such as rat portal vein (Hamilton *et al.*, 1986) and rabbit mesenteric artery (Coldwell and Howlett, 1987). In these experi-ments the tissue was incubated in a buffered physiological solution containing ^{86}Rb, and the basal efflux of the ions measured by the number of counts for ^{86}Rb released in fresh replacement buffer solution bathing the tissue. When the efflux of ^{86}Rb had reached a steady state (see Fig. 8.2), cromakalim (5 μM) was introduced and a doubling of the ^{86}Rb released from the tissue was observed. Similar results have also been obtained using ^{42}K and ^{43}K, although the short half-lives of these isotopes compared with ^{86}Rb preclude their use on a routine basis.

More recently it has been shown that the vasorelaxant action of cromakalim, both *in vitro* and *in vivo*, can be inhibited by the antidiabetic sulphonylurea glibenclamide (**38**), a potassium channel blocker (Buckingham *et al.*, 1989).

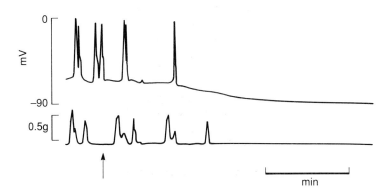

Fig. 8.1   Cell membrane potential difference (upper panel) and mechanical tension (lower panel) in rat isolated portal vein following administration of cromakalim (5 μM) ($\uparrow$). [Reproduced, with permission, from Hamilton, *et al.*, (1986).]

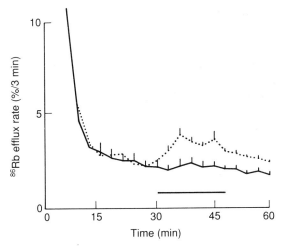

Fig. 8.2 A typical mean efflux curve from rabbit mesenteric artery (solid line) and the effect of cromakalim (5 μM) on such a curve (dotted line). Cromakalim, when present, is indicated by the horizontal bar. [Reproduced with permission from Coldwell and Howlett (1987).]

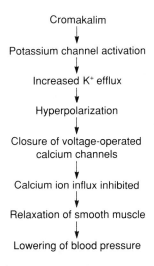

Glibenclamide (38)

The probable sequence of events (Fig. 8.3) culminating in drug-induced relaxation of the blood vessel is potassium ion channel activation, followed by increased potassium ion efflux and hyperpolarization of the cell membrane.

Cromakalim
↓
Potassium channel activation
↓
Increased K⁺ efflux
↓
Hyperpolarization
↓
Closure of voltage-operated
calcium channels
↓
Calcium ion influx inhibited
↓
Relaxation of smooth muscle
↓
Lowering of blood pressure

Fig. 8.3 Probable cascade of events leading to smooth muscle relaxation.

This hyperpolarization prevents the opening of the voltage-operated calcium channels through which calcium ions move into the cells. Since calcium ions are known to be involved in the process of cell contraction, a lowering of the intracellular calcium ion concentration via this mechanism leads to smooth muscle relaxation, with a consequent fall in blood pressure.

V.   Recent Developments

Following the selection of cromakalim for progression to the clinic, our investigations into the structural modification of this molecule continued. We were particularly interested in preparing compounds with different physicochemical properties to cromakalim, and also in preparing as wide a variety of structures as possible for evaluation by our pharmacologists.

A.   [3,2-c]-Pyranopyridines

As part of the strategy outlined above, we investigated the activity of compounds where an additional heteroatom had been incorporated into the aromatic ring. A number of candidate structures were generated and their two-dimensional electrostatic potential maps compared with that of the 6-cyanobenzopyran (Plate I). From these candidates, the [3,2-c]pyranopyridine ring system (39) was selected for synthesis based on the similarity of its potential map (Plate II) to that of the 6-cyanobenzopyran. As a test of this electrostatic potential model the isomeric [3,2-b]-, [2,3-c]- and [2,3-b]-pyranopyridines (40–42) were also targeted for synthesis.

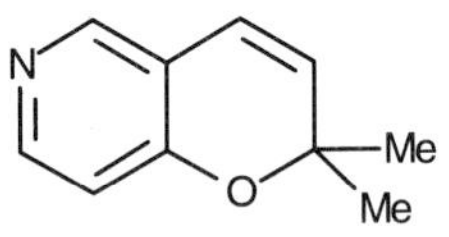

3, 2-c (39) 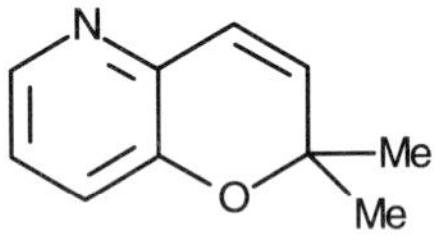   3, 2-b (40)

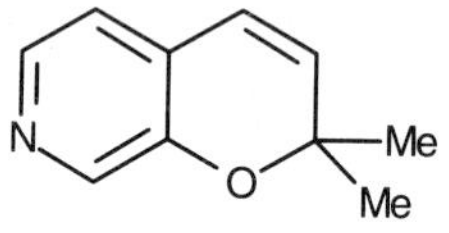

2, 3-c (41) 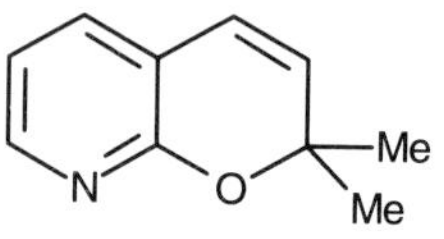   2, 3-b (42)

The syntheses of these 2,2-dimethylpyranopyridines had not been reported previously, so it was decided to attempt this via the Claisen rearrangement. Alkylation of 4-hydroxypyridine (43) with 3-chloro-3-methylbut-1-yne under phase transfer conditions (PTC), followed by heating the propargyl ether (44) in o-dichlorobenzene gave the desired [3,2-c]-pyranopyridine (39) in good yield.

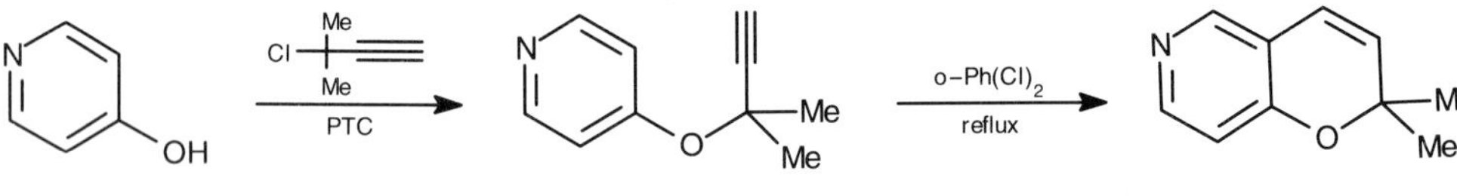

(43)   (44)   3, 2-c (39)

Scheme 8.4   Synthesis of 2,2-dimethyl-[3,2-c]-pyranopyridine ring.

In contrast with the cyclization of *meta*-substituted arylpropargyl ethers where substantial amounts of both possible isomers were obtained (Ashwood *et al.*, 1986), a similar sequence starting from 3-hydroxypyridine (45) did not lead to the expected mixture of [3,2-*b*]- and [2,3-*c*]- isomers. Inspection of the crude reaction mixture by ^{1}H NMR indicated that almost exclusive formation of the [3,2-*b*]- isomer (40) had occurred, with less than 5% of the [2,3-*c*]- isomer (41) present.

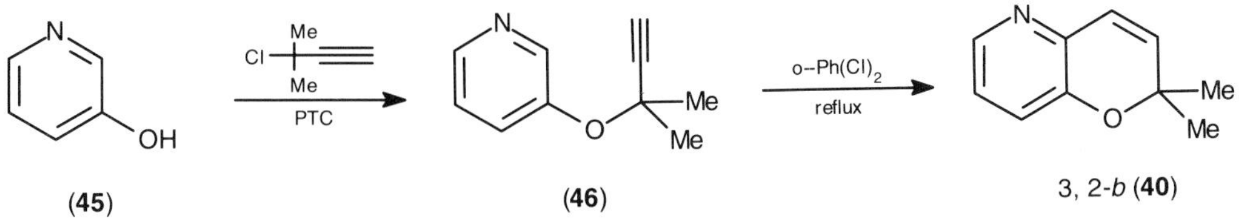

Scheme 8.5 Synthesis of 2,2-dimethyl[3,2-*b*]-pyranopyridine ring.

In order to redirect cyclization of the propargyl ether to give the [2,3-*c*]- isomer (41) we decided to block the 2-position by using 2-bromo-3-hydroxypyridine (47) as starting material. This time, cyclization of the intermediate ether (48) gave the required [2,3-*c*]pyranopyridine (49), which was debrominated by treatment with *n*-BuLi followed by quenching with water.

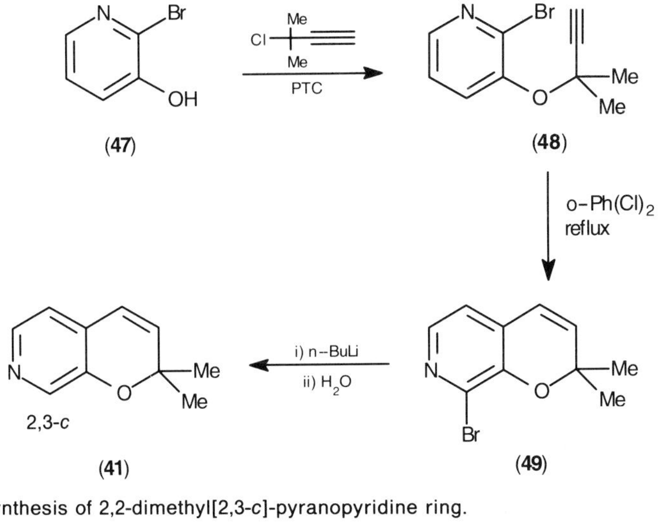

Scheme 8.6 Synthesis of 2,2-dimethyl[2,3-*c*]-pyranopyridine ring.

The synthesis of the remaining [2,3-*b*]- isomer (42) required a change in strategy, as *O*-alkylation of 2-pyridone with 3-chloro-3-methylbutyne could not be achieved. Thus, lithiation of 3,5-dibromo-2-methoxypyridine (50), followed by reaction with 3-methylcrotonaldehyde gave the allylic alcohol (51) which on brief treatment with HBr in acetic acid gave the [2,3-*b*]-pyranopyridine (52). Debromination with *t*-BuLi, followed by quenching with methanol gave the parent [2,3-*b*]- isomer (42) in good yield (Evans and Stemp, 1988).

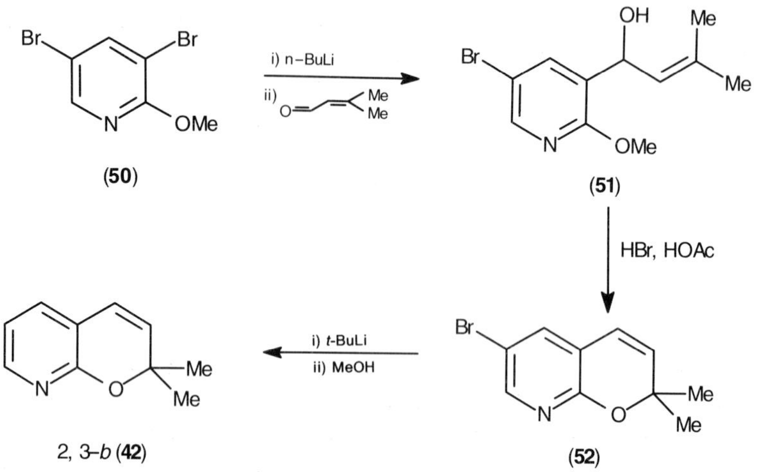

Scheme 8.7   Synthesis of 2,2-dimethyl[2,3-*b*]-pyranopyridine ring.

All of these pyranopyridines were converted to the corresponding pyrrolidinones (**53–56**) using the route described in Schemes 8.2 and 8.3, and in accordance with the electrostatic model, the most potent compound was the [3,2-*c*]-pyranopyridine (**54**), where the 6-cyano substituent of cromakalim had been replaced by the pyridine nitrogen (Table 8.5) (Burrell *et al.*, 1990).

TABLE 8.5
Antihypertensive Activity of Pyranopyridine-3-ols

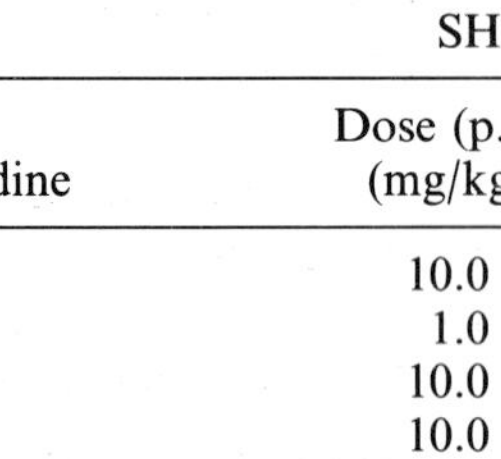

| | | SHR | |
| | | | |
Compound	Pyranopyridine	Dose (p.o.) (mg/kg)	Max. decrease in blood pressure (%)
53	[3,2-*b*]	10.0	15 ± 2
54	[3,2-*c*]	1.0	39 ± 7
55	[2,3-*c*]	10.0	28 ± 2
56	[2,3-*b*]	10.0	5 ± 2

Results are means ± S.E.M.

B.   6-Alkyl Benzopyrans

In accordance with the requirement for a strong electron-withdrawing group, the 6-CF$_3$ analogue of cromakalim showed high potency, while, as pointed out earlier, the unsubstituted analogue (**19**) did not lower blood pressure at the top dose investigated of 10 mg/kg (Burrell *et al.*, 1990). In order to confirm that the activity of the 6-substituted analogues reflected the electron-withdrawing character of the substituent, the 6-methyl analogue (**57**) was synthesized. The 6-methyl substituent would fulfil any steric requirement at this position, being similar in size to CF_3, but since methyl is weakly electron-donating, antihypertensive activity was not anticipated.

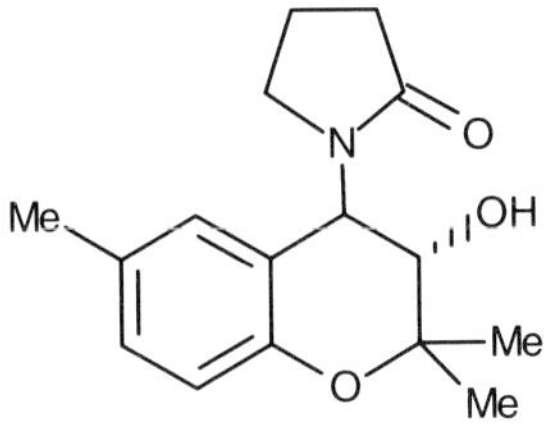

(57)

Surprisingly, compound (57) showed good activity at 10 mg/kg in the spontaneously hypertensive rat, lowering blood pressure by approximately 50%. This result prompted an investigation of a series of 6-alkyl analogues of compound (57) (Burrell *et al.*, 1990). Optimal activity was found for the 6-ethyl benzopyran (58) which retains approximately one-third of the activity of cromakalim. It therefore appears that a strong electron-withdrawing group at the 6-position is not obligatory for good antihypertensive activity!

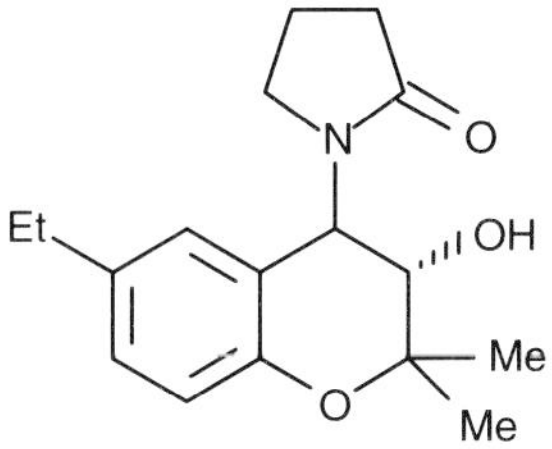

(58)

C. Amide Replacements at C-4

The use of two-dimensional electrostatic potential mapping in the design of novel potassium channel activators has also been extended to the development of novel replacements for the pyrrolidinone substituent at the 4-position of cromakalim. Following the observation that activity was retained when the pyrrolidinone ring of cromakalim (1) was replaced by an N-methylurea (compound 36), comparison of the electrostatic potential energy maps of N-methyl-2-pyrrolidinone (Plate III) and NN'-dimethylurea (Plate IV) with that of NN'-dimethylcyanoguanidine (Plate V) indicated that this urea replacement was worthy of synthesis in the cromakalim series, since the negative potential areas for each of these substituents would occupy similar regions in space when attached to the benzopyran nucleus. The cyanoguanidine (60), synthesized via the route shown in Scheme 8.8, was found to be approximately equipotent with cromakalim in the spontaneously hypertensive rat.

This result with compound (60) prompted the investigation of a series of amide and urea isosteres. From this study, the N'-methyltriazole (61), which can be considered as a cyclized form of the cyanoguanidine (60), was found to be approximately 3-fold more potent than cromakalim (Burrell and Stemp, 1990). Compound (61) was synthesized from the methylthiocyanoimidate (59) by treatment with methylhydrazine, and was isolated as the major product together with the N"-methyl isomer (62).

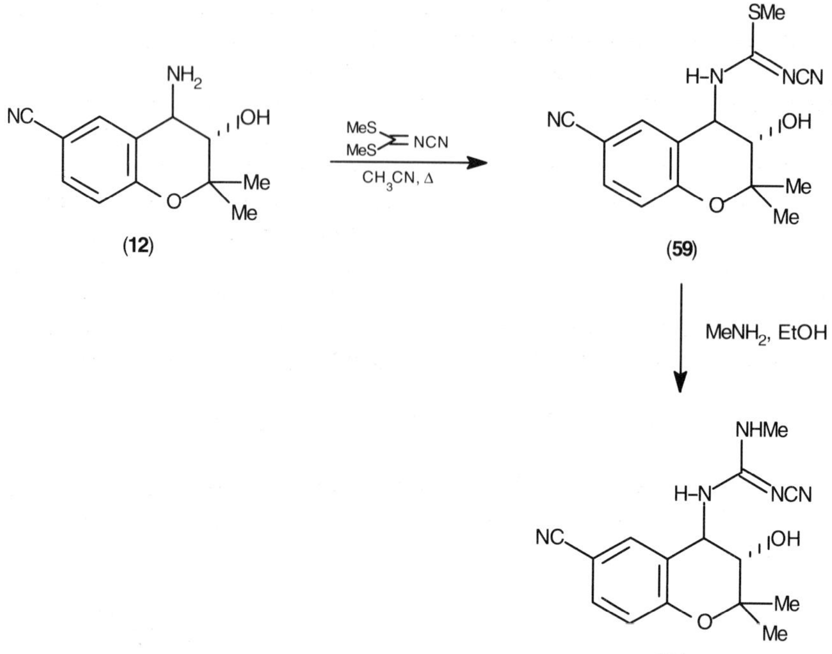

(12) (59) (60)

Scheme 8.8 Synthesis of cyanoguanidine analogue (60).

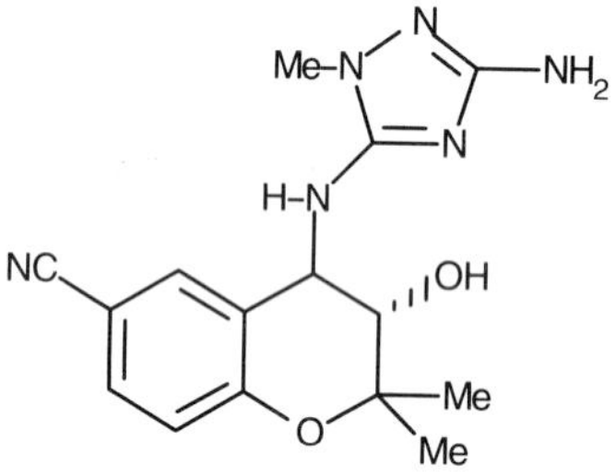

(61)

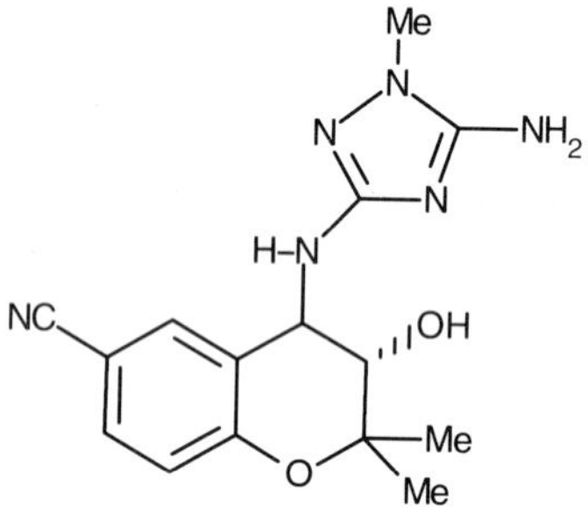

(62)

Interestingly, compound (**62**) was at least 10-fold less potent than compound (**61**), as was the desmethyl analogue (**63**).

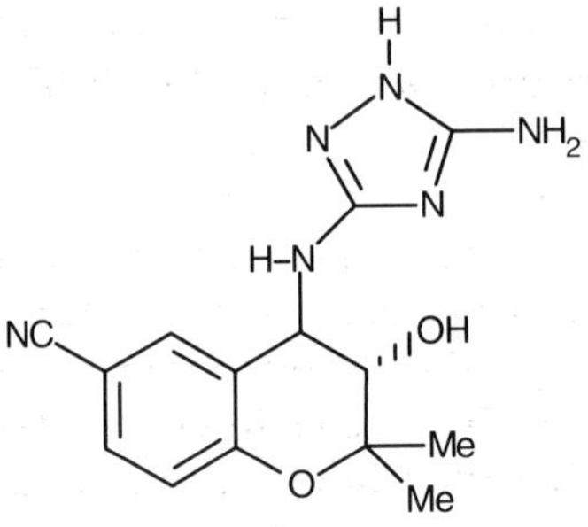

(63)

The triazole ring of compound (63) can exist in three possible tautomeric forms and *ab initio* calculations showed that the tautomer shown above predominates to the extent of 99% over the other two. Once identified, using long-range carbon–proton couplings in the NMR spectra, compounds (61) and (62) could be readily distinguished using the chemical shift of the proton at C-4 of the benzopyran ring. In the ^{1}H NMR spectrum of compound (63), the chemical shift of the C-4 proton is very close to that observed for compound (62), confirming the major tautomeric form as that shown above, and possibly explaining its lower potency. The two-dimensional electrostatic potential map of the triazole ring in compound (61) shows (Plate VI) an area of negative potential, associated with the N″ nitrogen atom, which can overlap with that of the cyanoguanidine (60), which would not be present in compounds (62) and (63), and this provides a possible explanation for the differences in potency between these triazoles.

D. 3-Amino- and 3-Amido-benzopyrans

Two other compounds which fulfilled the strategy outlined in the introduction to this section were the 3-amino analogue (64) of cromakalim and compound (65) where the amide and hydroxyl substituents had been transposed.

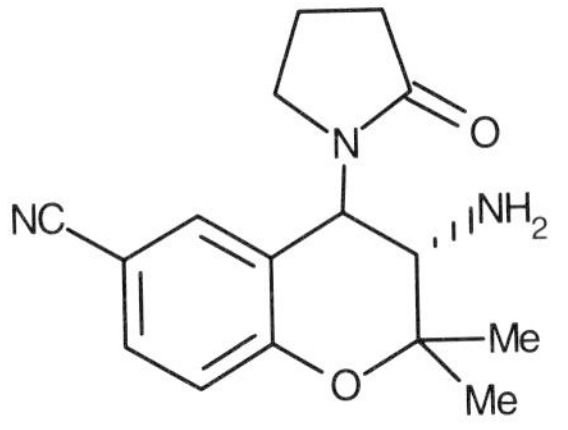

(64)

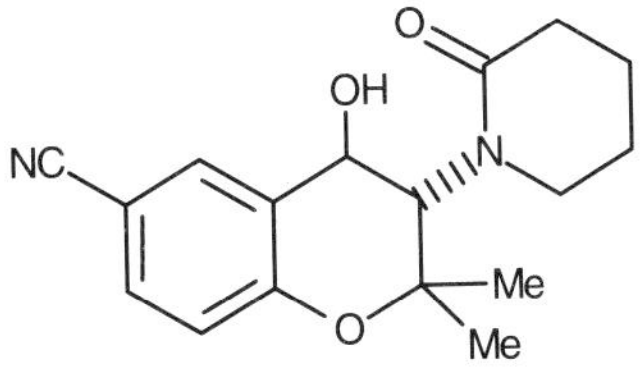

(65)

It was envisaged that both (64) and (65) would be available from a common intermediate, such as the protected aziridine (66).

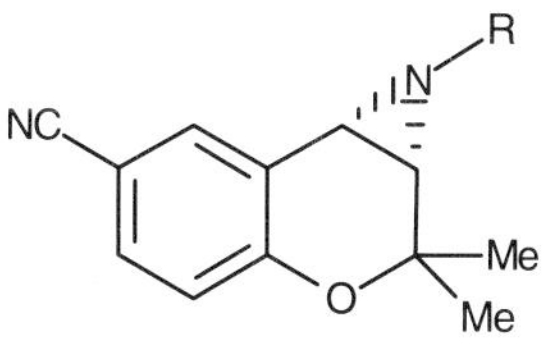

(66)

Traditional routes based on the preparation of aziridines from alkenes, epoxides or amino alcohols failed in the 6-cyanobenzopyran series, so the route shown in Scheme 8.9, based on the regio- and stereo-selective addition of *NN*-dichloro-*t*-butylcarbamate to chromenes was developed (Orlek and Stemp, 1991).

Treatment of 6-cyanochromene (67) with *NN*-dichloro-*t*-butylcarbamate, followed by *in situ* reduction with sodium metabisulphite gave the *trans* adduct (68) in good yield, and reaction of (68) with potassium carbonate in aqueous ethanol smoothly led to the Boc-protected aziridine (69). Opening of the aziridine ring with pyrrolidinone anion, and deprotection with trifluoroacetic acid (TFA) gave the desired 3-amino derivative (64) of cromakalim, and treatment of (69) with dilute acid followed by elaboration of the piperidinone ring gave the transposed analogue (65). Interestingly, both (64) and (65)

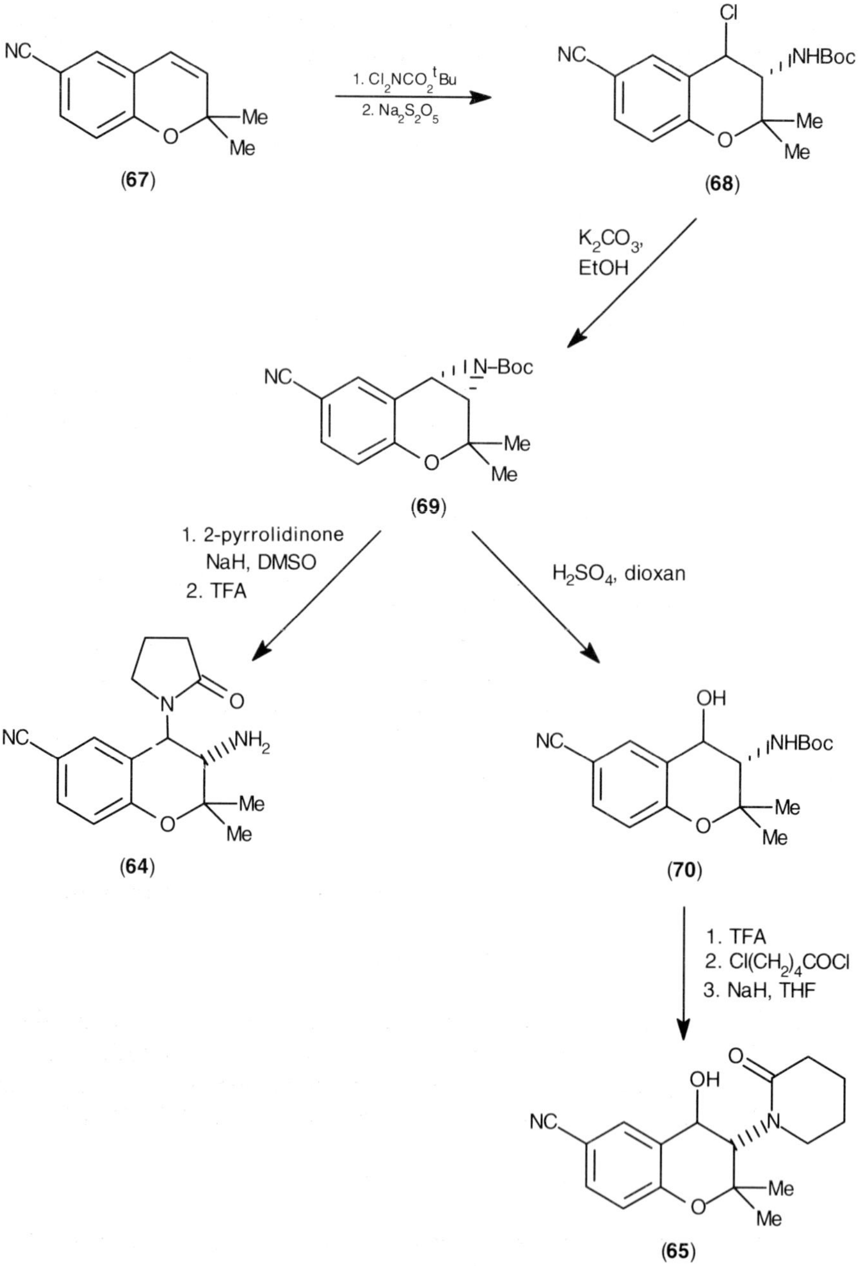

Scheme 8.9 Synthesis of 3-amino and 3-amido analogues of cromakalim via aziridine intermediate.

showed very little blood pressure lowering activity, but the hydroxycarbamate (70) was active at 3 mg/kg. This surprising activity provided the encouragement for further investigation and resulted in a series of compounds with a completely different structure–activity relationship about the amide carbonyl from that seen in the 4-amido (cromakalim) series. The *t*-butyl urea (71) was the most potent compound, being approximately equipotent with cromakalim (Stemp, 1990).

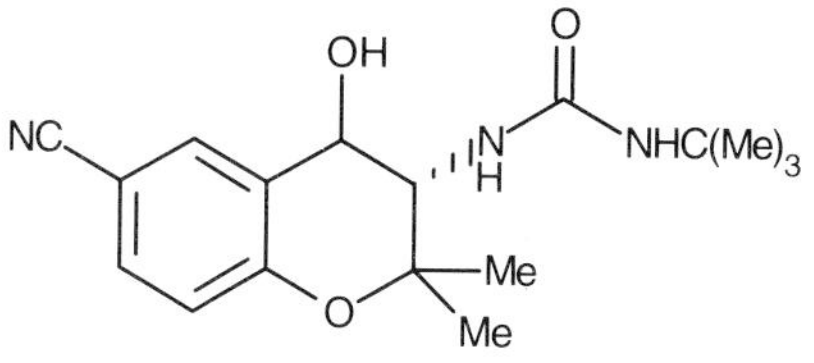

(71)

The antihypertensive activity of (71) was, like that of cromakalim, blocked by pretreatment with glibenclamide (38), confirming that this compound can be classified as a potassium channel activator. It remains to be seen whether this transposition of the amide and hydroxyl groups as in compound (71) will result in any changes in profile compared with cromakalim.

E.   Other Potassium Channel Activators

Since the discovery of the mechanism of action of cromakalim, the recently launched vasodilator pinacidil (72), diazoxide (73) and the sulphate metabolite of minoxidil (74) have also been classed as potassium channel activators, although they are less potent (see Fig. 8.4). The thioamide RP 49356 (75), developed from a series of compounds with antiulcer activity, has also been shown to belong to this class of compound. A number of other benzopyrans have also been described as potassium channel activators (see Evans *et al.*, 1992; and references therein) and a selection of these (76–79) are shown in Fig. 8.4.

VI. Therapeutic Potential of Potassium Channel Activators

A.   Hypertension

The potential of cromakalim as an antihypertensive agent was investigated in a double-blind study (Vanden Burg *et al.*, 1987) involving 73 mildly hypertensive patients. A single oral dose of 1.5 mg, given daily for 8 days, lowered diastolic blood pressure significantly and was well tolerated. Similar studies have been conducted with BRL 38227.

B.   Bronchial Asthma

In addition to the potential of potassium channel activators as a novel class of antihypertensive agents, there is potential in the treatment of other disease states where amelioration can be achieved by smooth muscle relaxation. Of particular importance is the treatment of asthma by relaxation of bronchial smooth muscle, and in this respect cromakalim was found to relax isolated guinea pig tracheal strips (Arch *et al.*, 1988), and given orally or by inhalation (Arch *et al.*, 1988; Bowring *et al.*, 1989) protected guinea pigs from histamine-induced bronchospasm. In comparison, pinacidil produced a similar protective effect at a much higher dose (Taylor *et al.*, 1988). The ability to relax human bronchial tissue has also been observed for cromakalim (Taylor *et al.*, 1988).

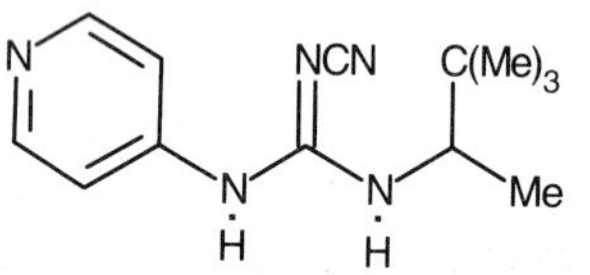

Pinacidil (**72**) 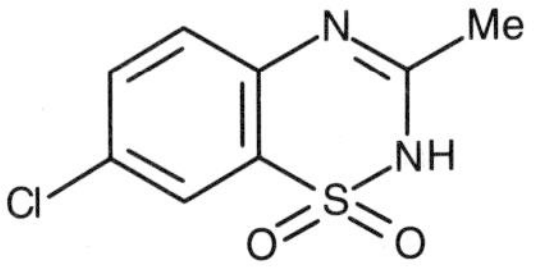   Diazoxide (**73**) 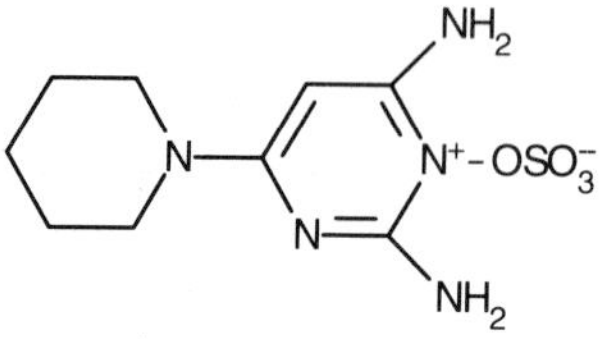   Minoxidil sulphate (**74**)

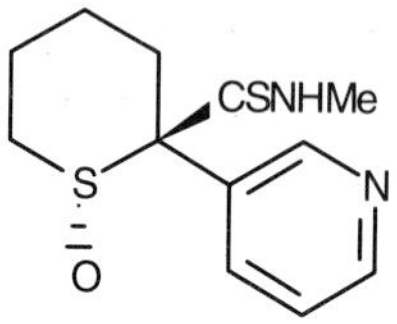

RP 49356 (**75**) 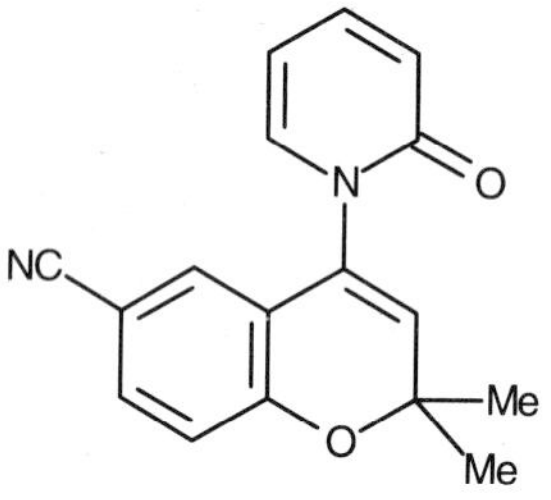   EMD 52692 (**76**) 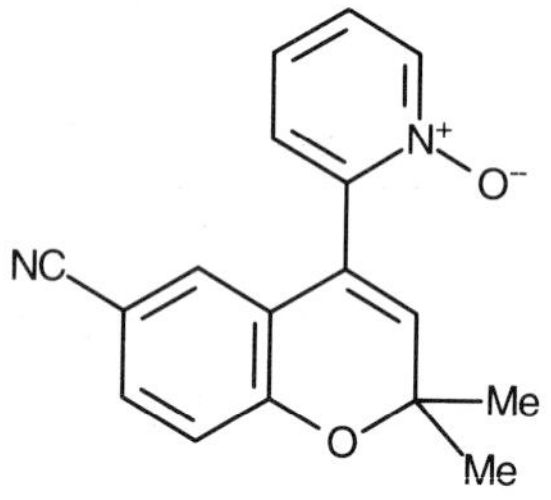   Ro 31-6930 (**77**)

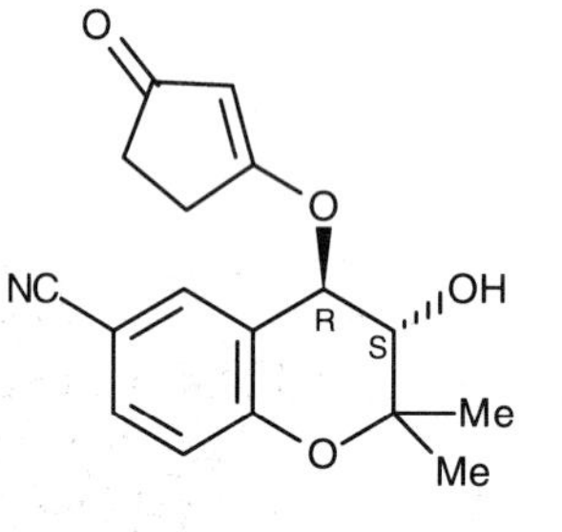

SDZ PCO-400 (**78**) 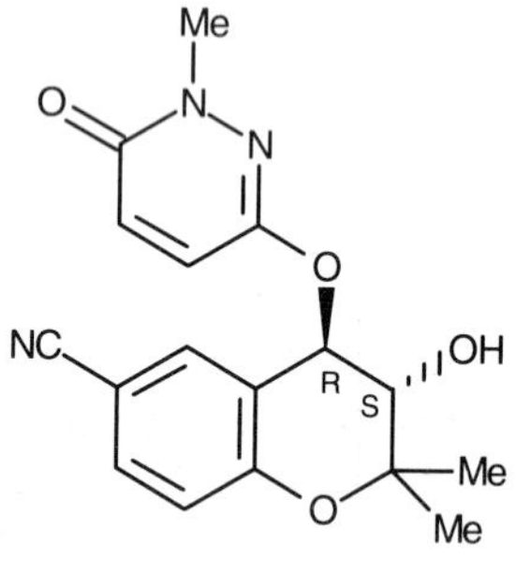   EMD 57283 (**79**)

Fig. 8.4 Structural diversity of potassium channel activators.

Extension of these studies to healthy volunteers demonstrated that cromakalim is an effective inhibitor of histamine-induced bronchoconstriction. The long plasma half-life of cromakalim in man of around 18–23 h [Gill *et al.*, 1988] suggested that it would be of particular value in the treatment of patients with nocturnal asthma. Indeed, in a clinical trial in such patients (Williams, *et al.*, 1990) cromakalim, administered as a single oral dose of 0.5 mg at night, prevented early morning bronchoconstriction. In a repeat dose study, administration of either 0.25 mg or 0.5 mg of cromakalim nightly, for 5 nights, also significantly prevented early morning bronchoconstriction (Williams *et al.*, 1990). Importantly, at the low oral doses used, no reduction in blood pressure was observed. The structure–activity relationships of benzopyran potassium channel activators and some close analogues for relaxing isolated guinea pig trachealis have subsequently been reported (Buckle *et al.*, 1990, 1991).

C. Urinary Incontinence

Another potential application of potassium channel activators is in the treatment of urinary incontinence. Certain abnormalities of the bladder which lead to incontinence are associated with the instability of the detrusor smooth muscle. In humans one cause is a partial obstruction of bladder outflow due to prostatic hypertrophy. Thus a requirement exists for a drug that can abolish the involuntary spontaneous contractility of the detrusor muscle, while maintaining the normal micturition reflex. Cromakalim (at concentrations $> 0.5\,\mu\text{M}$) was found (Foster *et al.*, 1989a) to abolish spontaneous mechanical activity in guinea pig, and, significantly, human detrusor muscle *in vitro* (Foster *et al.*, 1989b). Furthermore, in a pig model of urinary incontinence, cromakalim prevented unstable contractions during bladder filling without compromising micturition (Foster *et al.*, 1989b), a profile which would be highly desirable in the treatment of detrusor instability in humans.

D. Other Potential Applications

In addition to the therapeutic areas mentioned above there are a number of other possible indications for potassium channel activators based on their pharmacological properties, including peripheral vascular disease, congestive heart failure, angina and epilepsy.

VII. Conclusion

The discovery of the mechanism of action of cromakalim has stimulated worldwide interest in potassium channels. In this chapter we have described the ideas leading to the synthesis of cromakalim and some of the approaches we have adopted towards the discovery of novel potassium channel activators. Studies to date have demonstrated that these drugs may have potential in the treatment of a variety of diseases.

Acknowledgements

We thank our many colleagues who have contributed to the chemistry and pharmacology described in this chapter.

References

Arch, J.R.S., Buckle, D.R., Bumstead, J., Clarke, G.D., Taylor, J.F. and Taylor, S.G. (1988). *Br. J. Pharmacol.* **95**, 763.

Ashwood, V.A., Buckingham, R.E., Cassidy, F., Evans, J.M., Faruk, E.A., Hamilton, T.C., Nash, D.J., Stemp, G. and Willcocks, K. (1986). *J. Med. Chem.* **29**, 2194.

Ashwood, V.A., Cassidy, F., Coldwell, M.C., Evans, J.M., Hamilton, T.C., Howlett, D.R., Smith, D.M. and Stemp, G. (1990). *J. Med. Chem.* **33**, 2667.

Ashwood, V.A., Cassidy, F., Evans, J.M., Gagliardi, S. and Stemp, G. (1991). *J. Med. Chem.* **34**, 3261.

Bowring, N.E., Taylor, J.F., Francis, G.F. and Arch, J.R.S. (1989). *Br. J. Pharmacol.* **98**, 805P.

Buckingham, R.E., Clapham, J.C., Hamilton, T.C., Longman, S.D., Norton, J. and Poyser, R.H. (1986a). *J. Cardiovasc. Pharmacol.* **8**, 798.

Buckingham, R.E., Clapham, J.C., Coldwell, M.C., Hamilton, T.C. and Howlett, D.R. (1986b). *Br. J. Pharmacol.* **87**, 78P.

Buckingham, R.E., Hamilton, T.C., Howlett, D.R., Mootoo, S. and Wilson, C. (1989). *Br. J. Pharmacol.* **97**, 57.

Buckle, D.R., Arch, J.R.S., Fenwick, A.E., Houge-Frydrych, C.S.V., Pinto, I.L., Smith, D.G., Taylor, S.G. and Tedder, J.M. (1990). *J. Med. Chem.* **33**, 3028.

Buckle, D.R., Arch, J.R.S., Foster, K.A., Houge-Frydrych, C.S.V., Pinto, I.L., Smith, D.G., Taylor, J.F., Taylor, S.G., Tedder, J.M. and Webster, R.A.B. (1991). *J. Med. Chem.* **34**, 919.

Burrell, G. and Stemp, G. (1990). *Eur. Pat. Appl.* 399 834.

Burrell, G., Cassidy, F., Evans, J.M., Lightowler, D. and Stemp, G. (1990). *J. Med. Chem.* **33**, 3023.

Cain, C.R. and Metzler, V. (1985). *Naunyn Schmiedebergs Arch. Pharmacol.* **329**, R53.

Coldwell, M.C. and Howlett, D.R. (1987). *Biochem. Pharmacol.* **36**, 3663.

Evans, J.M. and Stemp, G. (1988). *Synth. Commun.* **18**, 1111.

Evans, J.M., Fake, C.S., Hamilton, T.C., Poyser, R.H. and Watts, E.A. (1983). *J. Med. Chem.* **26**, 1582.

Evans, J.M., Fake, C.S., Hamilton, T.C., Poyser, R.H. and Showell, G.A. (1984). *J. Med. Chem.* **27**, 1127.

Evans, J.M., Hadley, M.S. and Stemp, G., (1992). *In* 'Potassium Channel Modulators: Pharmacological, Molecular and Clinical Aspects' (T.C. Hamilton and A.H. Weston, eds), Blackwell Scientific, Oxford.

Foster, C.D., Fujii, K., Kingdon, J. and Brading, A.F. (1989a). *Br. J. Pharmacol.* **97**, 281.

Foster, C.D., Speakman, M.J., Fujii, K. and Brading, A.F. (1989b). *Br. J. Urol.* **63**, 284.

Gill, T.S., Davies, B.E., Allen, G.D. and Greb, W.H. (1988). *Br. J. Clin. Pharmacol.* **25**, 669P.

Hamilton, T.C., Weir, S.W. and Weston, A.H. (1986). *Br. J. Pharmacol.* **88**, 103.

Harfenist, M. and Thom, E. (1972). *J. Org. Chem.* **37**, 841.

Orlek, B.S. and Stemp, G. (1991). *Tetrahedron Lett.* **32**, 4045.

Stemp, G. (1990). *Eur. Pat. Appl.* 375 449.

Taylor, S.G., Bumstead, J., Morris, J.E.J., Shaw, D.J. and Taylor, J.F. (1988). *Br. J. Pharmacol.* **95**, 795P.

Vanden Burg, M.J., Woodward, S.M.A., Stewart-Long, P., Tasker, T., Pilgrim, A.J., Dews, I.M. and Fairhurst, G. (1987). *J. Hypertension* **5** (Suppl. 5), S193.

Williams, A.J., Lee, T.H., Cochrane, G.M., Hopkirk, A., Vyse, T., Chiew, F., Lavender, E., Richards, D.H., Owen, S., Stone, P., Church, S. and Woodcock, A.A. (1990). *Lancet* **336**, 334.

Further Reading

Cook, N.S. (1990). 'Potassium Channels, Structure, Function and Therapeutic Potential', Ellis Horwood, Chichester.

Evans, J.M. and Stemp, G. (1991). *Chem. Brit.* **27**, 439.

Evans, J.M. and Longman, S.D. (1991). *In* 'Annual Reports in Medicinal Chemistry, Vol. 25' (J.A. Bristol, ed.), Academic Press, San Diego.

Robertson, D.W. and Steinberg, M.I. (1990). *J. Med. Chem.* **33**, 1529.

–9–

Angiotensin-Converting Enzyme (ACE) Inhibitors and the Design of Cilazapril

S. REDSHAW

Roche Products Ltd,
Welwyn Garden City,
Herts., U.K.

Dr Sally Redshaw graduated in chemistry from the University of Nottingham in 1975 and also completed her Ph.D. there, in 1978. She subsequently joined Roche Products Ltd and is currently employed as a team leader in the Antiviral Chemistry Department. Her major research interest is the inhibition of therapeutically important proteinases.

I. Introduction

Despite an enormous research effort, the fundamental cause of hypertension is still not known. What is known is that high blood pressure, whatever the underlying cause, leads to increased morbidity and mortality as a consequence of arterial disease. If untreated over long periods of time, hypertension increases the risk of myocardial infarction, cerebral haemorrhage and renal failure. Gradually the heart and vascular muscle becomes hypertrophic, or thickened, and atherosclerotic changes occur in the blood vessels. Reduction of blood pressure to normal levels has been shown to lessen the incidence of coronary heart disease, stroke and kidney failure, thereby providing justification for therapy, even if the hypertension itself is asymptomatic.

MEDICINAL CHEMISTRY 2nd Edition
ISBN 0-12-274120-X

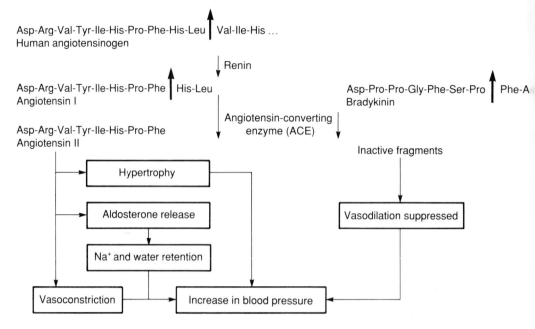

Fig. 9.1   The renin–angiotensin system.

II.   The Renin–Angiotensin System

In 1827, Bright at Guy's Hospital in London was amongst the first to recognize that renal disease was often accompanied by high blood pressure. Over a century of investigation subsequently unravelled the details of the renin–angiotensin system so that by the early 1960s the biochemical pathway was well understood. A diagrammatic representation of the renin–angiotensin system is shown in Fig. 9.1. The Leu-Val bond of a circulating globular protein known as angiotensinogen, or renin substrate, is hydrolysed specifically by the aspartic acid proteinase, renin, which is produced by the kidney. This releases the N-terminal decapeptide, angiotensin I, which has no known biological activity. Angiotensin-converting enzyme (ACE) then cleaves a two-amino acid fragment from the free C-terminus of this decapeptide to give the octapeptide, angiotensin II, which is responsible for the full pressor effect of the renin–angiotensin system. Besides acting directly via receptors on vascular smooth muscle to constrict the arteries and arterioles, angiotensin II also stimulates the adrenal cortex to release aldosterone which induces sodium and water retention, resulting in a further hypertensive effect through increased plasma volume. ACE has also been identified as kininase II (Yang and Erdos, 1967; Nakajima et al., 1973), the enzyme that degrades the vasodepressor peptide, bradykinin, to produce inactive fragments.

   Although the biochemical details of the renin–angiotensin system had been worked out by the early 1960s, its relevance to the control of blood pressure under either normal or pathological conditions was not well understood. Indeed, many researchers were of the opinion that the role of the renin–angiotensin system in blood pressure regulation was likely to be a very minor one under any circumstances, and it was only the

TABLE 9.1

In vitro Activities of Inhibitors of Angiotensin-Converting Enzyme Illustrating the Effects of Chain Length and Substitution on Inhibitory Potency

Compound	Structure	IC_{50} (μM)
1	< Glu-Trp-Pro-Arg-Pro-Gln-Ile-Pro-Pro (SQ20,881,Teprotide)	0.56
2	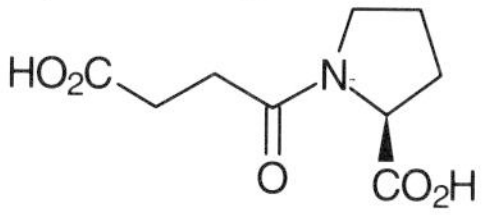 HO_2C ... N / CO_2H	630
3	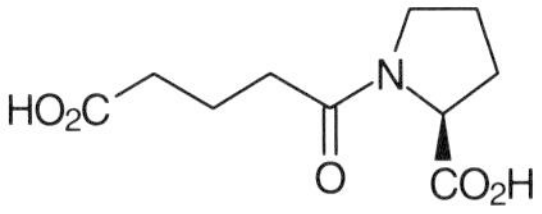 HO_2C ... N / CO_2H	70
4	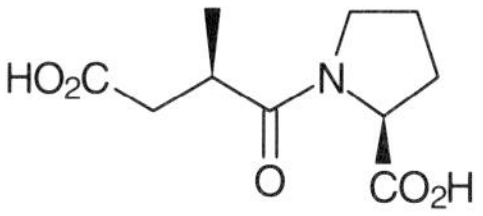 HO_2C ... N / CO_2H	52
5	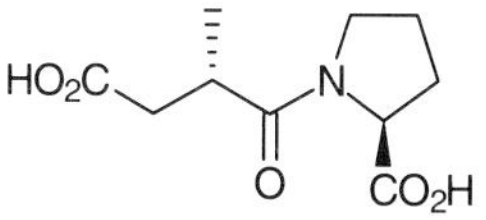 HO_2C ... N / CO_2H	1470
6	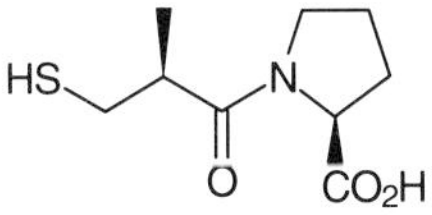 HS ... N / CO_2H (SQ14,225,Captopril)	0.023

development of specific means of blocking the system that allowed its full importance to be appreciated.

Angiotensin II antagonists such as saralasin ([1-*sarc*osine, 8-*ala*nine]angioten*sin*) were the first agents to be used to probe the renin–angiotensin system. Early results with saralasin (Brunner *et al.*, 1973) clearly showed the therapeutic potential of blocking the system, and a sustained effort has been directed towards identifying potent orally bio-available antagonists of angiotensin II. Much attention has also been focused on the design of inhibitors of the first enzyme of the cascade, renin. Many pharmaceutical companies have identified potent inhibitors of this highly specific enzyme, but the clinical development of these has been impeded so far by disappointing oral activity. In the late 1960s, the first peptidic inhibitors of the converting enzyme became available. These inhibitors were initially isolated from the venom of the Brazilian viper, *Bothrops jararaca* (Cushman and Ondetti, 1979) and one of them, teprotide [SQ20,811, compound (**1**), Table 9.1], was shown to have exciting potential as an antihypertensive agent (Engel *et al.*, 1972; Gavras *et al.*, 1974). The lack of oral bioavailability of these early peptidic ACE inhibitors again precluded their use as long-term therapy and prompted Cushman and Ondetti at the Squibb Institute to begin the search for an orally active compound. This search was to herald the dawn of a new era in medicinal chemistry: a largely empirical quest for new drugs was to become a true scientific discipline based on mechanistic theories and design.

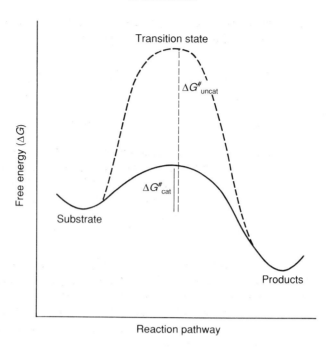

Fig. 9.2 Enzymes catalyse reactions by reducing ∆G. ———, Catalysed reaction; − − − −, uncatalysed
reaction.

III. Proteolytic Enzymes

Chemical reactions in biological systems rarely occur in the absence of a catalyst. This essential role is performed by proteins known as enzymes, a term coined in 1878 by Friedrich Wilhelm Kühne from the Greek *en*, in, and *zyme*, leaven. Enzymes have truly awesome catalytic abilities; reactions are accelerated by factors of at least a million-fold. Since enzymes are catalysts, they do not alter the equilibrium of a chemical reaction; the attainment of equilibrium is accelerated, but the position of that equilibrium is not shifted. A chemical reaction, $A \rightarrow B$, goes through a transition state which has a higher energy than that of A or B. The binding of a substrate by an enzyme is not equivalent to chemical activation, and in fact, results in stabilization of the substrate. The cardinal property of an enzyme is not, then, its ability to bind the substrate, but rather its ability to discriminate between the substrate and the transition state, binding the latter more tightly, and reducing the difference in energy that limits the rate of the reaction. Enzymes accelerate reactions, then, by decreasing ΔG, the activation barrier. The combination of enzyme and substrate creates a new reaction pathway, the transition state of which is of lower energy than that for the reaction taking place in the absence of enzyme (Fig. 9.2).

The making and breaking of chemical bonds by the enzyme is preceded by the formation of an enzyme–substrate (ES) complex. The substrate is bound to a specific region of the enzyme which is termed the active site. This region of the enzyme encompasses those residues that directly participate in the making and breaking of bonds, and which are known as the catalytic groups. The specificity of binding to the enzyme depends on the precisely defined arrangement of the residues forming the active site.

The chemical reaction carried out by proteolytic enzymes is the hydrolysis of the amide bonds linking amino acids in peptides and proteins:

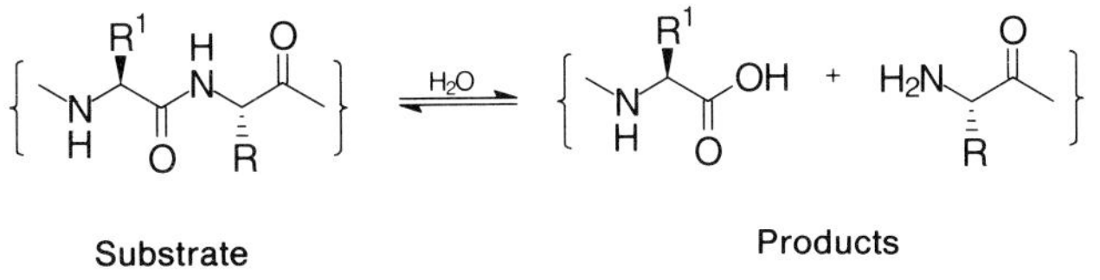

Substrate   Products

Proteinases differ enormously in how fastidious they are in their choice of reagent or substrate. At one end of the spectrum, the subtilisins, which are present in certain bacteria, are so undiscriminating that they can be used as ingredients in the so-called 'biological detergents' to digest all manner of protein residues, while at the other extreme, thrombin, one of the enzymes involved in blood clotting, will hydrolyse only peptide bonds between an arginine residue and a glycine residue in its target protein.

A number of devices have evolved to enable hydrolysis of the substrate to take place and proteolytic enzymes are classified into four categories depending upon their catalytic mechanism. The serine and cysteine proteinases are named after the nucleophilic residues in the active sites which form an initial acyl-enzyme intermediate. This intermediate is then hydrolysed by an activated water molecule to release the enzyme and products. The pathway for a cysteine proteinase is illustrated below:

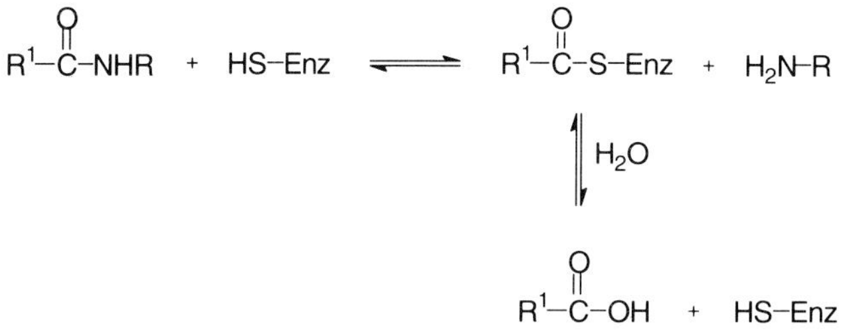

Examples of cysteine proteinases include papain and cathepsin H. Important members of the serine proteinase family are the mammalian enzymes trypsin, chymotrypsin and elastase, as well as the bacterial enzyme, subtilisin.

It is generally held that the enzymes comprising the other two classes, the aspartic proteinases and the metalloproteinases, do not form an initial acyl-enzyme intermediate as described above. The preferred mechanism is for specific residues in the active sites of these enzymes to act as general bases and in this way to 'pep up' the nucleophilicity of a bound water molecule which then attacks the scissile amide bond directly. Important aspartic proteinases include renin and also the proteinase encoded by the human immunodeficiency virus (HIV) which is essential for viral replication (Kohl et al., 1988; Peng et al., 1989). A commercially important aspartic proteinase is chymosin which is used extensively in the cheese-making industry. Metalloproteinases include ACE and collagenase as well as some of the β-lactamases, which, although not strictly speaking proteinases, are important in the development of resistance to antibiotics by bacteria.

IV. ACE and the Zinc Metalloproteinases

ACE (EC 3.4.15.1) is a glycoprotein which is widely distributed in mammals. It is primarily a membrane-bound enzyme of vascular endothelial cells, the pulmonary endothelial cells being a particularly important source.

The difficulties inherent in working with a large (1.3–1.6 kDa) membrane-bound and heavily glycosylated protein have made structure elucidation a Herculean task; the amino acid sequence was not determined until 1988 (Soubrier *et al.*), long after the first inhibitors had been marketed! Even today, the structural information available about this protein is not useful for inhibitor design and the most relevant factors remain a knowledge of the enzyme's substrate specificity together with an understanding of the catalytic mechanism.

The presence of an essential zinc ion in ACE encourages comparison with other zinc metalloproteinases. Carboxypeptidase A (Quiocho and Lipscomb, 1971), carboxypeptidase B (Schmid and Herriot, 1976) and thermolysin (Matthews *et al.*, 1974; Kester and Matthews, 1977) are zinc-dependent enzymes, the structures of which have been determined using X-ray crystallography. The binding of substrates and inhibitors to these enzymes has also been studied crystallographically, and this structural information, together with extensive mechanistic studies (Hartsuck and Lipscomb, 1971), has allowed an accurate picture of the active sites of these enzymes to be developed. A schematic representation of the catalytic site of carboxypeptidase A is shown in Fig. 9.3.

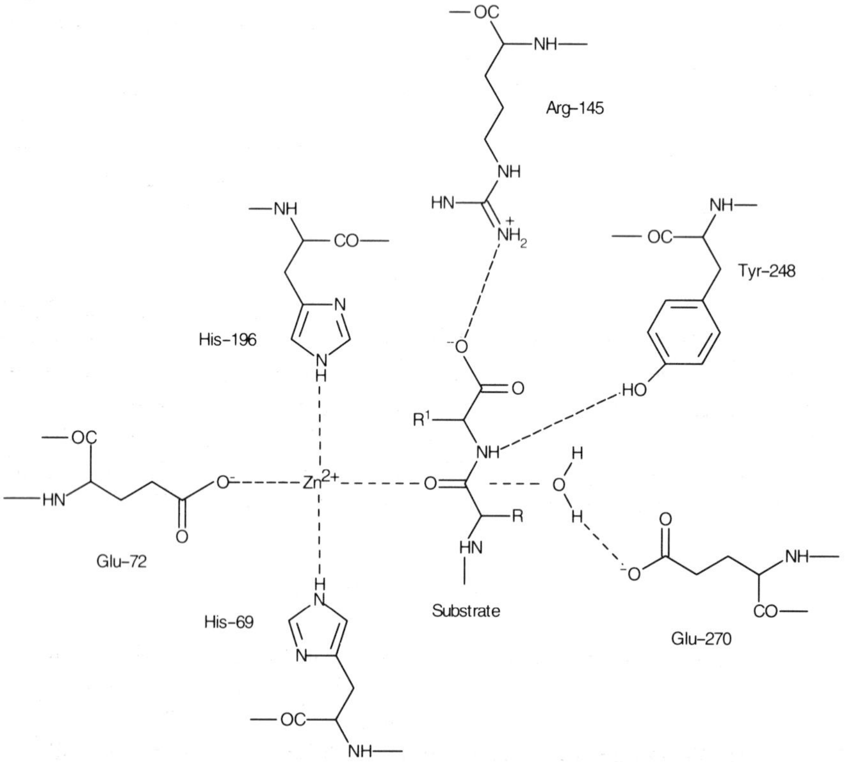

Fig. 9.3 Schematic representation of the active site of carboxypeptidase A showing a bound substrate.

Crystallographic studies on enzyme–substrate complexes show an electrostatic interaction between the positively charged guanidino function of Arg-145 and the negatively charged free carboxy group of a bound substrate.

An adjacent hydrophobic pocket (not shown) is responsible for the substrate specificity of carboxypeptidase A. The side chains of Glu-72, His-69 and His-196 form ligands to the zinc ion, and the scissile amide becomes polarized by forming a fourth ligand to the zinc, displacing a water molecule from the native enzyme as it does so. The mechanism of hydrolysis of the amide bond has been the subject of some controversy: direct attack of Glu-270 on the amide to form an anhydride-like intermediate, action of Glu-270 as a general base to activate a water molecule, and activation of a water molecule by the zinc ion have all been proposed. Studies using $H_2{}^{18}O$ (Breslow and Wernick, 1976) support the general base mechanism: Tyr-248 donates a proton to the released amino function, the resulting phenolate anion is positioned to accept a proton from the same water molecule which attacks the amide bond.

V. Assays for ACE

A reliable, and preferably high-throughput, assay is crucial to any enzyme inhibitor programme. Although a whole cell system or animal model can be used if the target enzyme is unavailable, these methods are less than ideal since they introduce complicating factors such as bioavailability, metabolism and cell penetration. These additional factors can have pronounced effects on activity, and as they are not easily taken into account, can generate very misleading structure–activity relationships (SARs).

Purified preparations of ACE are available and various assay systems have been devised to measure the potency of inhibitors. The most widely used of these involves spectrophotometric determination of the hippuric acid released from the N-terminal-protected tripeptide substrate hippuryl-L-histidyl-L-leucine [Fig. 9.4(a)], either alone (Cushman and Cheung, 1971), or as the complex formed with cyanuric chloride (Hayakari et al., 1978).

A schematic representation of the assay procedure using the cyanuric chloride complex method is shown in Fig. 9.4(b). The experiments are repeated at a number of different inhibitor concentrations and a graph is then plotted of the enzyme activity (determined by the amount of hippurylhistidine produced in the reaction) against the concentration of inhibitor present. This produces a sigmoidal curve as shown in Fig. 9.4(c).

The activity of an inhibitor is then usually quoted as an IC_{50} value which can be determined directly from the graph. The IC_{50} value thus represents the concentration of a given inhibitor that will reduce the activity of the enzyme by 50%.

VI. The Design of ACE Inhibitors

A. Using a Two-dimensional Model of the Enzyme Active Site

1. Captopril

In 1972, Byers and Wolfenden discovered a potent inhibitor of carboxypeptidase A. The activity of this compound, L-benzylsuccinic acid, was rationalized on the basis of its

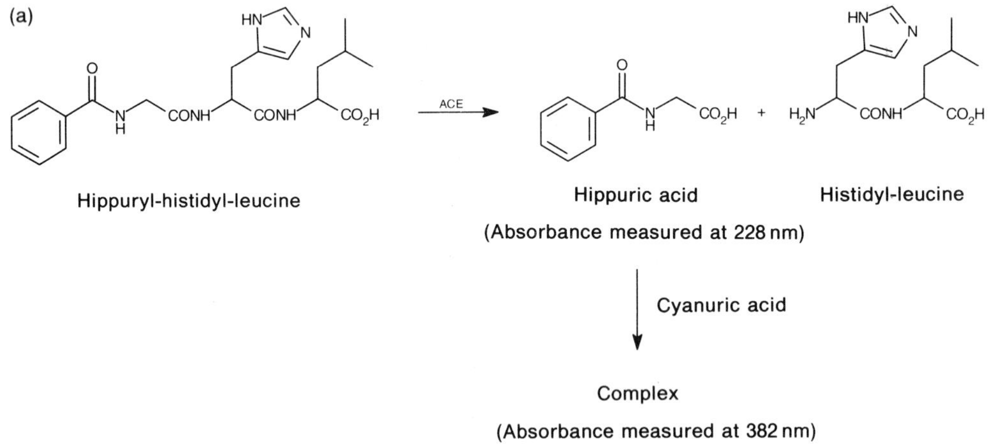

(a)

Hippuryl-histidyl-leucine   Hippuric acid   Histidyl-leucine

(Absorbance measured at 228 nm)

Cyanuric acid

Complex

(Absorbance measured at 382 nm)

Fig. 9.4   (a) Biochemistry of the ACE assay; (b) the ACE assay; (c) Plot of enzyme activity against inhibitor concentration. (b) and (c) on facing page.

resemblance to the substrates for the reverse enzymic reaction, i.e. to the products of the peptide hydrolysis. The benzyl group was considered to occupy the S_1' subsite, with the adjacent carboxylate anion forming an electrostatic interaction with the enzymic Arg-145 residue. The second carboxylate group could then act as a ligand to the active-site zinc ion (Fig. 9.5).

The first potent inhibitors of ACE were identified as a result of studies on the bradykinin-potentiating properties of peptides isolated from the venom of *Bothrops jararaca*. Knowledge gained from these studies about the chemical and enzymic properties of ACE, together with insights drawn from Wolfenden's recently published work on the biproduct inhibitors of carboxypeptidase A, allowed Cushman and Ondetti at the Squibb Institute to develop a two-dimensional model of the active site of ACE (Ondetti *et al.*, 1977).

Since ACE is a dipeptidylcarboxypeptidase, Ondetti and his colleagues reasoned that the distance between the cationic binding site and the zinc ion in ACE must be greater than in carboxypeptidase A, by approximately the length of one amino acid residue. It then followed that a succinyl derivative of an amino acid, rather than a simple succinic acid, should form the prototype for inhibitors of ACE (Fig. 9.6).

To test this hypothesis, they initially prepared succinyl-*S*-proline (2) choosing proline as the *C*-terminal amino acid since this residue occurs at the free *C*-terminus of all the naturally occurring peptidic inhibitors.

Although not exceptionally potent (IC$_{50}$ 630 μM), this compound provided support for the original hypothesis, as it was a specific inhibitor of the enzyme. After exploring the effect of different structural modifications, including length and substitution of the acyl group, 2-*R*-methylsuccinyl-*S*-proline (4) was identified as a significantly more active inhibitor (IC$_{50}$ 52 μM). The diastereoisomer (5) with the *S*-configuration at the methyl group was found to be much less active.

The Squibb group then reasoned that if the interaction of the carboxy group of such compounds with the zinc ion in the enzyme active site was crucial for inhibitory potency, then replacement of the carboxylate by another group capable of acting as a ligand to

(b)

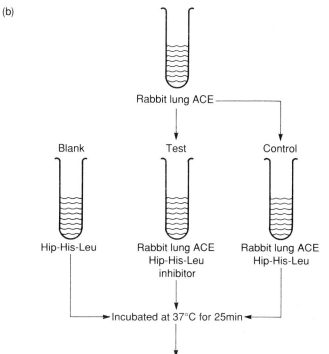

Reactions stopped by addition of ice-cold 200mM-phosphate buffer (pH 8.3). Rabbit lung ACE added to blank assay. Then 3% solution of cyanuric chloride in dioxan added and reactions incubated at 20°C for 15 min. Tubes centrifuged.

Absorbance measured at 382nm.

(c)

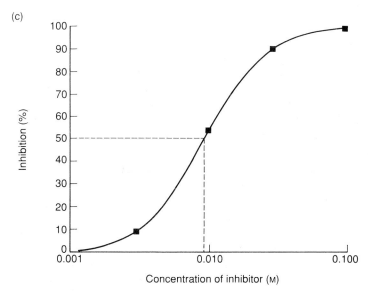

Fig. 9.4 Continued.

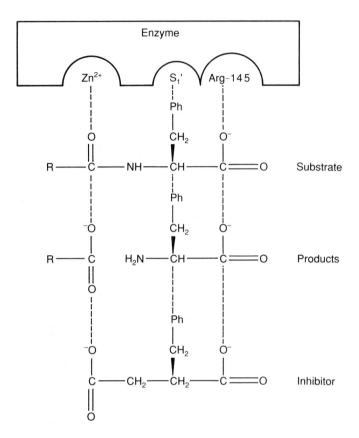

Fig. 9.5 Schematic representation of the binding of substrate, product and inhibitor molecules to the active site of carboxypeptidase A.

the zinc ion should result in a compound of equal or greater potency. Replacement of the carboxy group by nitrogen-containing functionalities (amines, amides or guanidines) gave no improvement in potency, but replacement by a thiol group led to a dramatic enhancement of activity. The most potent inhibitor of this series (**6**, SQ14,225) with an IC_{50} value of 23 nM was, on a molar basis, more potent than the previously described nonapeptide, teprotide (**1**). More importantly, this small non-peptidic inhibitor showed good oral bioavailability, and as *captopril* was destined to become the first commercially available ACE inhibitor.

2. Enalapril

The design of enalapril also grew out of Wolfenden's biproduct inhibitor hypothesis, but Patchett and colleagues (1980) at Merck used a different approach to improve the potency of glutarylproline (**3**), choosing to elaborate the molecule in such a way as to provide additional interactions with putative sites on the enzyme. In order to visualize some of the possibilities, they considered 2-methylglutarylproline as a biproduct inhibitor of the hydrolysis of *N*-acyl-Phe-Ala-Pro (Fig. 9.7).

It was hypothesized that incorporation of an NH group, a hydrophobic side chain or an RCONH group into the prototype inhibitor, 2-methylglutarylproline, should increase

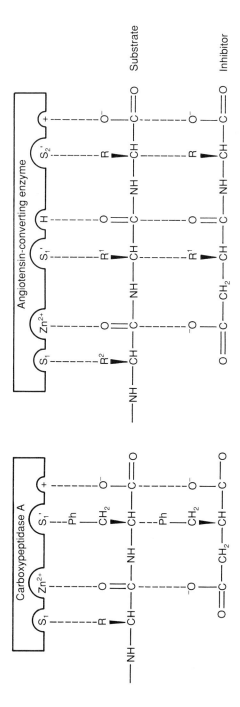

Fig. 9.6 Schematic representation of the binding of substrates and inhibitors to the active site of carboxypeptidase A, and to the hypothetical active site of angiotensin-converting enzyme.

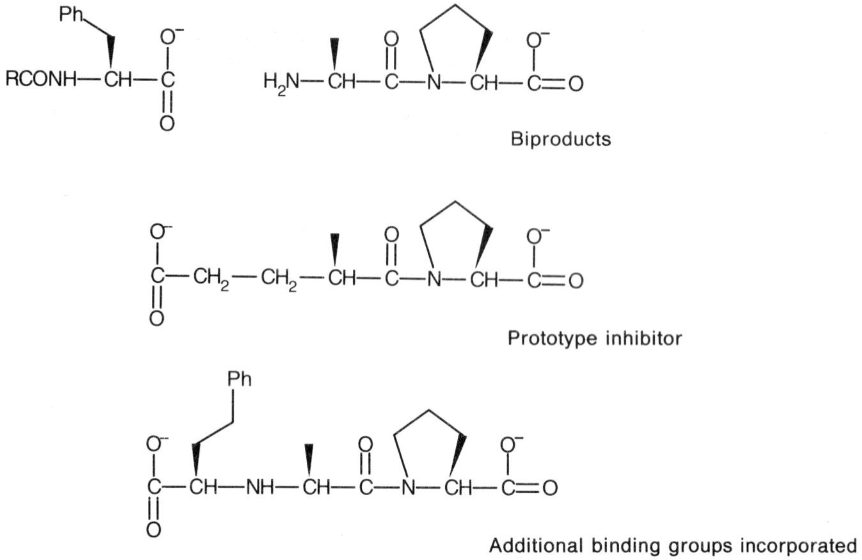

Fig. 9.7   Schematic representation of 2-methylglutarylproline as a biproduct inhibitor of the enzymic cleavage of N-acyl-Phe-Ala-Pro.

its resemblance to the products of the enzymic hydrolysis, and thus enhance inhibitory potency. Introduction of an NH group did not significantly improve activity (compound **8** compared with compound **7**), but incorporation of a hydrophobic substituent next to the N-terminal carboxy group (a manoeuvre suggested by the known preference of ACE for substrates with a hydrophobic amino acid in the antipenultimate position) led to a dramatic increase in potency (compounds **9–11**) (Table 9.2). The most potent compound (**11**), designated MK422, was that in which a phenethyl group was used as the hydrophobic side chain. The diastereisomer (**12**) with the R-configuration at the centre bearing the carboxylate group is a much less potent inhibitor.

MK422 is not well absorbed orally in laboratory animals or in man, but esterification of the N-carboxyalkyl group provided a simple solution to this problem. The ethyl ester, MK421, *enalapril*, is, as expected, a much less potent inhibitor of ACE, but *in vivo* the action of esterases rapidly converts the prodrug into the active diacid. It is fortunate that satisfactory oral absorption can be achieved with the monoester since esterification of the proline carboxy group strongly promotes the irreversible formation of the diketopiperazine (**13**).

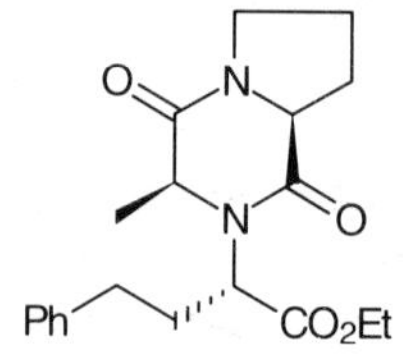

(**13**)

TABLE 9.2
In vitro Activities of Inhibitors of Angiotensin-Converting Enzyme Illustrating the Effects of Additional Binding Groups on Inhibitory Potency

Compound	Structure	IC_{50} (μM)
7	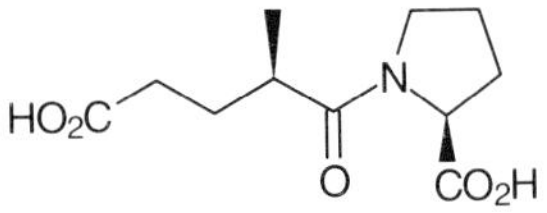 HO₂C ... N (pyrrolidine), O, CO₂H	4.9
8	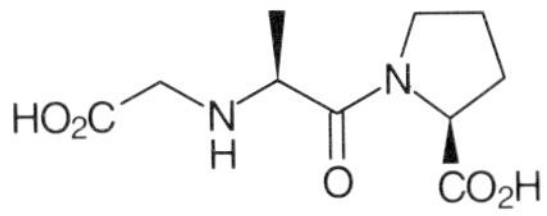 HO₂C–N(H) ... N (pyrrolidine), O, CO₂H	2.4
9	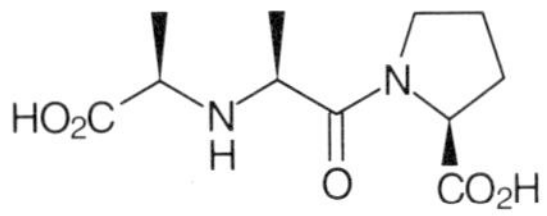 HO₂C ... N(H) ... N (pyrrolidine), O, CO₂H	0.09
10	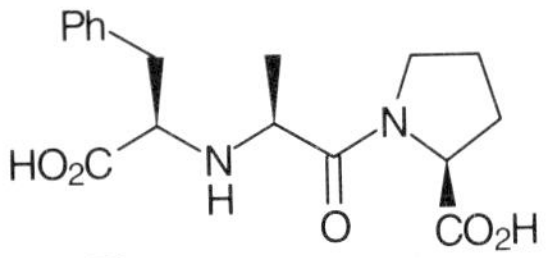 Ph, HO₂C ... N(H) ... N (pyrrolidine), O, CO₂H	0.04
11	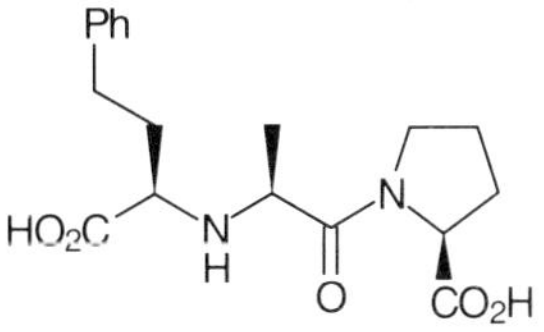 Ph, HO₂C ... N(H) ... N (pyrrolidine), O, CO₂H	0.001
12	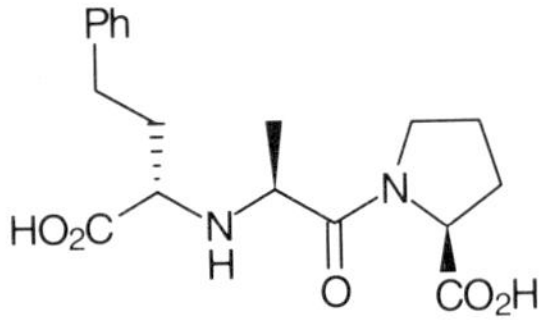 Ph, HO₂C ... N(H) ... N (pyrrolidine), O, CO₂H	0.8

B. Using a Three-dimensional Model of the Enzyme Active Site

1. *Thiol-containing Inhibitors*

Design approaches using a two-dimensional schematic representation of the enzyme active site have proved to be very effective, and, as in the case of captopril and enalapril, have led to potent enzyme inhibitors. Several groups, including our own at Roche, believed that it should be possible to achieve further improvements in potency if the bioactive conformation of these inhibitors could be used in a more rigid molecule. The importance of the three key binding groups, i.e. the carboxy group, the amide group and the zinc ligand, was well recognized, but information about their relative spatial orientation in the bound state was not accessible directly, since no X-ray crystallographic data were available, for either the native enzyme or enzyme–inhibitor complexes.

We addressed this problem by attaching the three important binding groups to a rigid template which could be varied systematically in order to change the relative spatial orientation of these groups (Hassall *et al.*, 1984). We began by imposing the constraint that the amide bond of captopril should remain in the *trans* conformation in the

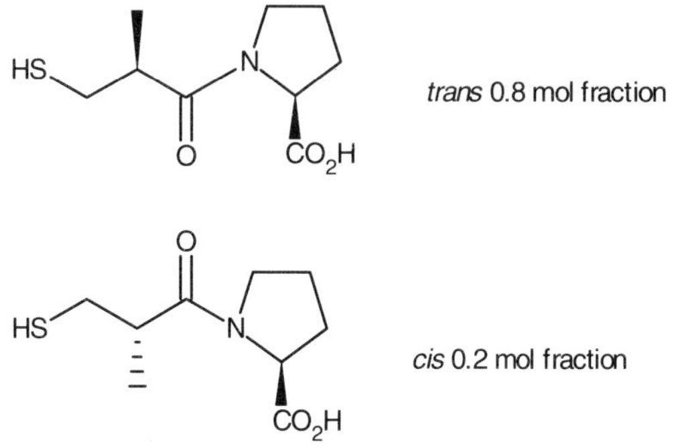

trans 0.8 mol fraction

cis 0.2 mol fraction

Fig. 9.8   *Trans* and *cis* conformers of captopril.

enzyme–inhibitor complex. This seemed to be a reasonable supposition since the *trans* form is clearly favoured both in solution and in the crystal (Fig. 9.8).

As our template, we chose a bicyclic system which is notionally formed by bridging the proline ring δ-carbon on to the alanylmethyl group [Fig. 9.9(a)]. In practice, we preferred to replace proline by hexahydropyridazine-3-carboxylic acid [Fig. 9.9(b)], since this avoids the inconvenience of an additional chiral centre and also provides the basis of a relatively straightforward route to a range of bicyclic compounds (Table 9.3).

We initially chose compounds which would place the thiol group in very different positions to allow us to map a large volume of space quickly.

In order to learn more about the bioactive conformation of captopril, we then needed to compare the positions of the three key binding groups (the thiol group, the amide group and the carboxy group) in each of our conformationally restricted inhibitors with the full range of spatial orientations possible for captopril. Keeping the amide bond of captopril in what we believed was the bioactive *trans* conformation, it was possible, using molecular graphics, to effect rotations about the CH$_2$-CH(Me) and CH(Me)–C(O) bonds (Fig. 9.10).

This generated the locus for the captopril S-atom depicted in Fig. 9.11(a). Using conventional molecular mechanics calculations, it was then possible to exclude high-energy (> 50 kcal/mol) conformations, leaving the mesh plot shown in Fig. 9.11(b) to represent the most likely positions of the S-atom relative to the amide and carboxy groups.

By comparing the restricted S-atom locus of each of our conformationally restricted inhibitors with the captopril S-atom locus, we were able to associate biological activities

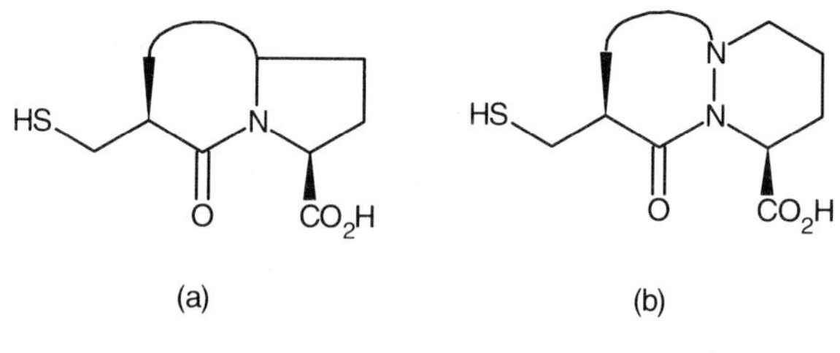

(a)

(b)

Fig. 9.9   (a) *Trans* conformer of captopril showing link required to form bicyclic derivative; and (b) preferred bicyclic derivative.

TABLE 9.3

In vitro Activities of Conformationally Restricted Thiol-containing Inhibitors of Angiotensin-Converting enzyme

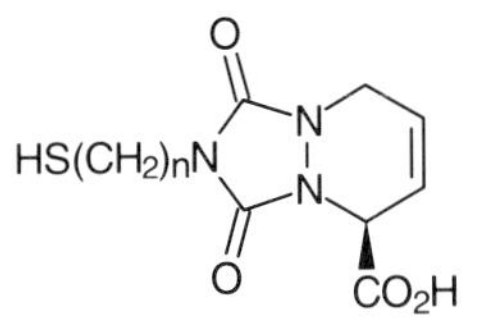

Compound	n	IC$_{50}$ (μM)	Compound	R^1	R^2	IC$_{50}$ (μM)
14	1	0.7	17	HSCH$_2$CH$_2$	H$_3$C	0.1
15	2	0.5	18	H$_3$C	HSCH$_2$CH$_2$	43
16	3	6.0	19	HSCH$_2$	H$_3$C	1.0

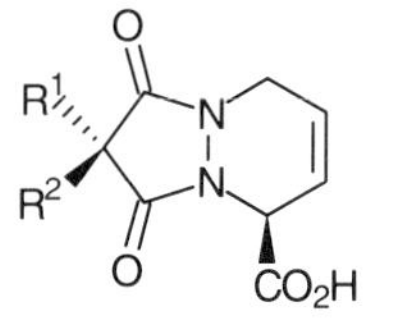

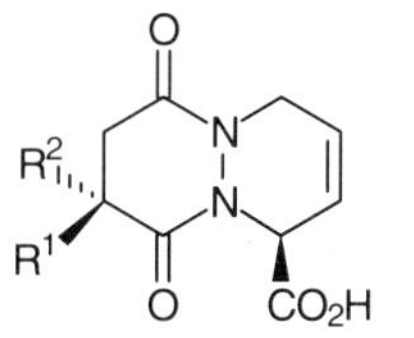

Compound	R^1	R^2	IC$_{50}$ (μM)
20	HSCH$_2$	H	0.1
21	H	HSCH$_2$	0.04

with the different regions of this locus, and hence identify the probable bioactive conformations of captopril.

Figure 9.12(a) represents the superposition of the mesh plot for compound (**21**) on that for captopril. The carboxy and amide groups of the two molecules have been aligned, and conformationally feasible positions for the thiol groups then traced. Figure 9.12(b) shows a scale representation of the portion of the mesh trace common to compound **21** and captopril. This process can be repeated for the individual conformationally restricted inhibitors (compounds **14–21**), producing in each case a restricted locus which corresponds to only part of the whole potential locus for the sulphur atom of captopril. If it is assumed that the enzyme inhibitory activity of the conformationally restricted inhibitors is directly dependent upon their ability to achieve the best fit in the active site, the mesh plot for the sulphur atom can be related to the biological activity (Γ) of each compound (Fig. 9.13).

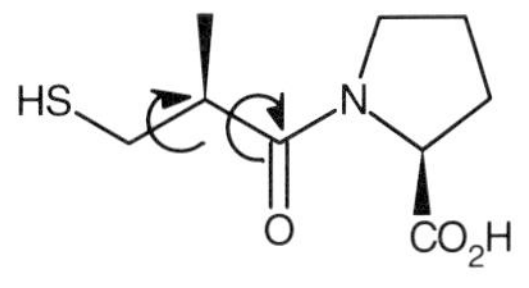

Fig. 9.10 Rotation about the CH$_2$-CH(Me) and CH(Me)-CO bonds of captopril allows generation of the mesh plot depicted in Figure 9.11(a).

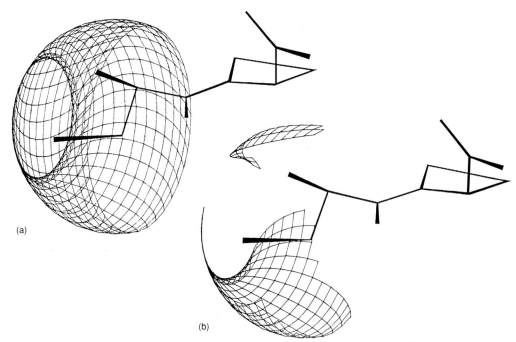

Fig. 9.11 (a) The surface mesh (locus) produced for the sulphur atom of captopril by allowing free rotation about the CH_2-CH(Me) and CH(Me)-CO bonds; and (b) the reduced locus after rejecting conformations > 50 kcal/mol in total energy.

A consideration of our initial thiol-containing compounds ascribed highest potency to two regions of the locus, Γ_4.

Although compounds **20** and **21** are reasonably potent enzyme inhibitors (IC_{50} 100 and 40 nM), they are less potent than captopril (IC_{50} 23 nM), indicating that the three key binding groups in these rigid molecules are not fixed in precisely the right orientation for optimal binding to ACE. This is probably because the torsion angle ψ (Fig. 9.14) is not optimal in these compounds which thus mimic relatively high-energy conformations of alanylproline. In these conformers, the methyl group of the alanine residue almost eclipses the δ-carbon atom of the proline ring. The energy profile for alanylproline does in fact show two rather broad energy minima, both of which correspond to ψ values outside the range calculated for compounds **20** and **21**.

2. Cilazapril

Having used our conformationally restricted thiol-containing inhibitors to map likely positions of the active site zinc ion relative to the hydrogen bond donor and cationic binding sites, we then further refined and extended our model to include other zinc ligands (Attwood *et al.*, 1986). We prepared a series of *N*-carboxyalkyl-bicyclic lactams in which the torsion angle, ψ, was varied systematically (Table 9.4).

The most potent ACE inhibitors (**24** and **28**) were those incorporating a seven-membered lactam with a torsion angle, ψ, of 164–166°. Interestingly, this value of approximately 165°, which appears to be optimal for binding to ACE, is different from that adopted by *N*-[1-carboxy-3-phenylpropyl]-L-leucyl-L-tryptophan ($\psi = 316°$) as

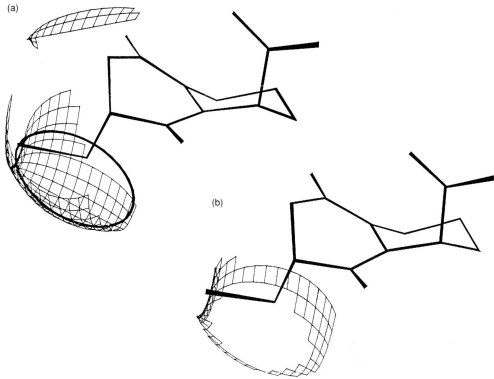

Fig. 9.12 (a) Superposition of captopril and bicyclic compound (21); and (b) the common locus of the sulphur atoms in compound (21) and captopril.

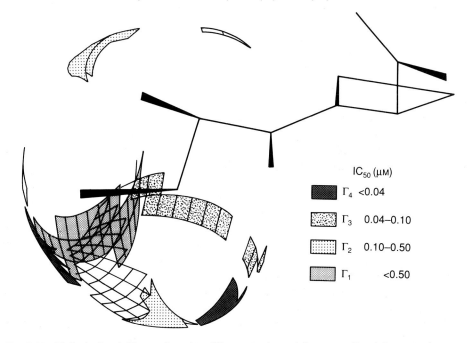

Fig. 9.13 Biological activities assigned to different regions of the captopril sulphur atom locus.

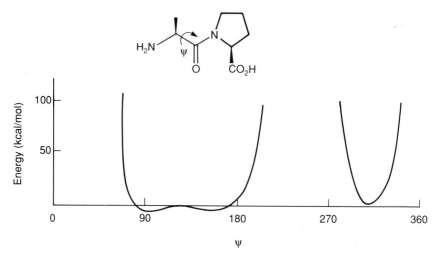

Fig. 9.14   The ECEPP energy profile of the H-Ala-Pro-OH fragment.

determined by X-ray crystallography in the complex with thermolysin (Monzingo and Matthews, 1984). This suggests that ACE binds inhibitors (and presumably substrates) in one low-energy conformation whilst the related metalloproteinase, thermolysin, binds its substrates in an alternative low-energy conformation. The proposed enzyme-bound conformation of compound **24** (cilazaprilat) is depicted in Fig. 9.15.

In common with other non-thiol-containing ACE inhibitors, the parent diacid, cilazaprilat is poorly absorbed orally. Conversion to the monoethyl ester affords the prodrug, cilazapril (**30**). Cilazapril shows excellent oral bioavailability and, once absorbed, the

TABLE 9.4

In vitro Activities of Conformationally Restricted N-carboxyalkyl Inhibitors of Angiotensin-Converting Enzyme Showing Variation with Torsion Angle, Ψ

		$Y = H_2$				$Y = O$	
Compound	n	ψ	IC$_{50}$ (nM)	Compound	n	ψ	IC$_{50}$ (nM)
22	0	254.9°	8000	**26**	0	237.6°	
23	1	245.6°	28	**27**	1	196.6°	20
24	2	163.9°	1.6	**28**	2	165.3°	4
25	3	142.1°*	4.5	**29**	3	146.4°*	15

Ψ values obtained from crystal structures of model bicyclic compounds.
*Values calculated by Modified Neglect of Diatomic Overlays (MNDO) (Dewar and Thiel, 1977).
The Ψ value $\sim$ 165° seems optimal for ACE. The value 316° was found for the thermolysin/*N*-(1-carboxy-3-phenylpropyl)-L-leucyl-L-tryptophan complex. These values are within the ranges of the two energy minima for alanylproline.

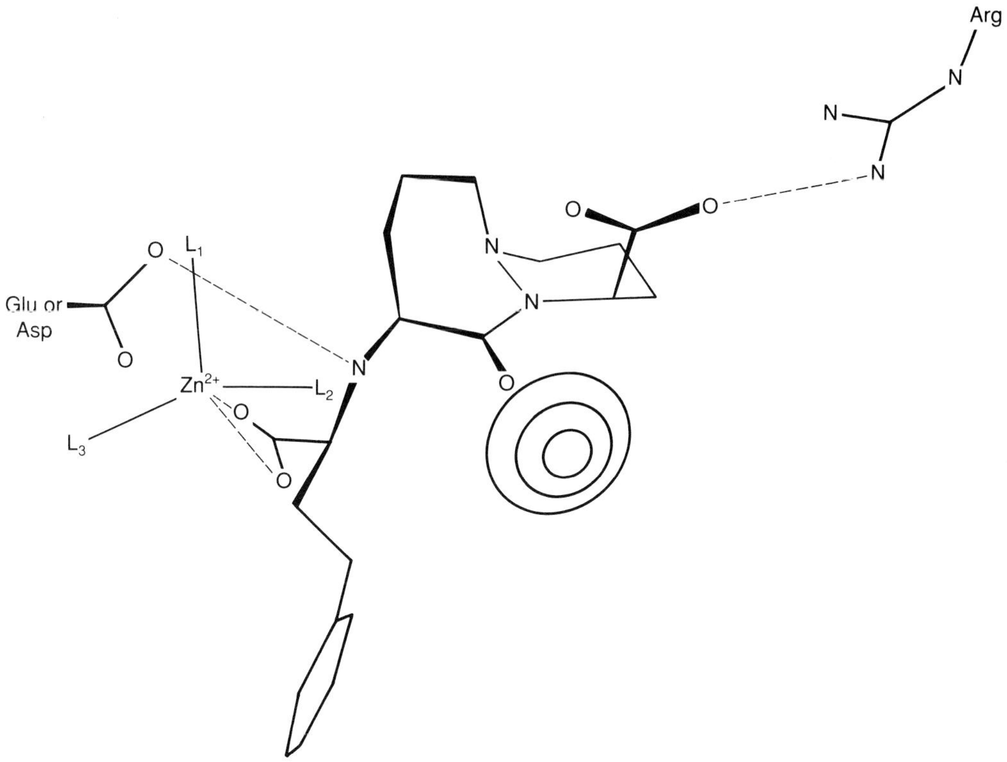

Fig. 9.15 The proposed orientation of cilazaprilat for binding to the catalytic site of ACE.

ethyl ester is rapidly hydrolysed by plasma esterases to liberate the parent diacid, cilazaprilat.

3. Synthesis of Cilazapril

The synthesis of cilazapril is set out in Scheme 9.1. The *t*-butyl ester of (N-1) protected hexahydropyridazine-3(*S*)-carboxylic acid (**31**) was acylated at N-2 using the protected glutamic acid derivative (**32**) to give compound **33**. The protecting groups were removed simultaneously by hydrogenolysis and the resulting acid was cyclized via the acid chloride to give the protected dipeptide mimetic **34**. We were gratified to find that reduction of compound **34** with borane/tetrahydrofuran (H₃B/THF) was highly regio-selective, possibly because the second amide is in a rather crowded environment. De-protection of intermediate **35** by hydrazinolysis then afforded the key dipeptide mimetic (**36**). Two methods of attaching the carboxyalkyl side chain were investigated. In our hands, reductive alkylation using keto acids or esters gave poor yields with little evidence of asymmetric induction at the newly formed chiral centre. Initial studies using racemic ethyl 2-bromo-4-phenylbutanoate afforded, in good yield, the required product (**37**) as a 1 : 1 mixture of epimers. Much to our chagrin, the same 1 : 1 mixture of epimers was obtained following alkylation with the hard-earned chiral bromo ester (**38**). The racemization was presumed to stem from nucleophilic exchange of bromide ion, and was avoided by the use of trifluoromethanesulphonate (triflate) as the leaving group. Our

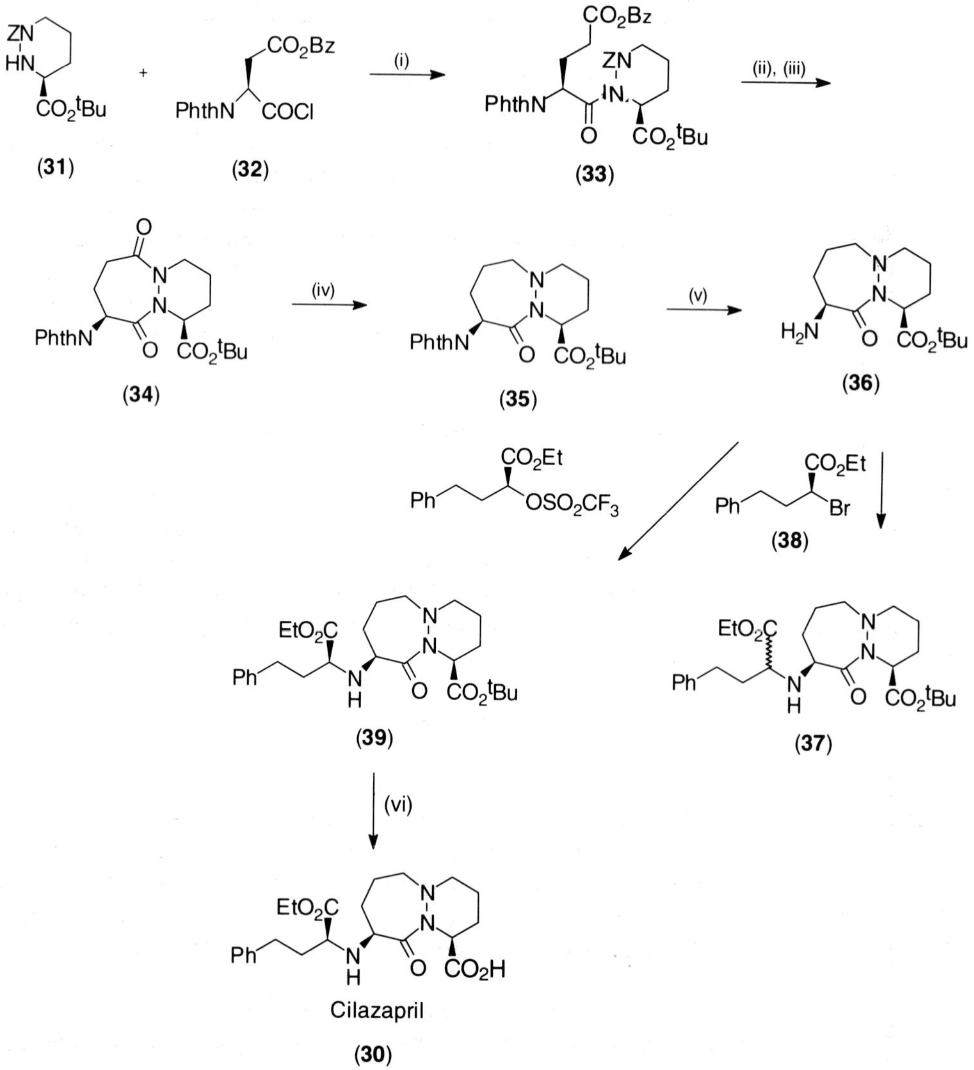

Scheme 9.1 Synthesis of cilazapril.

(i) Toluene, aqueous $NaHCO_3$; (ii) H_2, Pd/C, EtOH; (iii) $SOCl_2$, CH_2Cl_2; (iv) H_3B/THF; (v) $N_2H_4 \cdot H_2O$, EtOH; (vi) HCl/CH_2Cl_2.

experience was subsequently confirmed by the findings of Effenberger *et al.* (1983) who used triflate esters to prepare a range of chiral *N*-substituted α-amino carboxylic esters. Deprotection of the butyl ester (**39**) using anhydrous hydrogen chloride in dichloromethane then gave cilazapril (**30**).

VII. Biological Properties of ACE Inhibitors

A. Pharmacokinetics

In contrast with the position with enzyme inhibition, it is still barely possible to apply design principles to overcome pharmacokinetic problems, and serendipity remains the major operator. Fortune has favoured the medicinal chemist in the field of ACE inhibitors since captopril is well absorbed orally, and those inhibitors that use a carboxy group as the zinc ligand can be rendered orally bioavailable by the simple expedient of monoesterification. When cilazapril is administered orally to rats, approximately 98% of the dose is absorbed (Attwood *et al.*, 1984). The action of esterases then releases the parent diacid, cilazaprilat.

B. Efficacy in Animal Models of Hypertension

Cilazapril, and other ACE inhibitors, have long been established to lower blood pressure in a range of hypertensive animal models. Cilazapril lowers established high blood pressure in spontaneously hypertensive rats (SHRs) (Natoff *et al.*, 1985), and prevents the development of hypertension in young SHRs (Hefti *et al.*, 1986). In this animal model, cilazapril lowers blood pressure largely by a decrease in total peripheral resistance with little or no effect on heart rate or cardiac output.

C. Clinical Results

1. Hypertension

Numerous clinical trials have clearly demonstrated the efficacy of ACE inhibitors as monotherapy in mild, moderate and even severe hypertension. Although early theories suggested that angiotensin II might be important only in relatively rare forms of hypertension, e.g. renal hypertension, it is now known that ACE inhibitors, in conjunction with diuretics if necessary, will lower blood pressure in over 80% of patients.

2. Cardiac and Vascular Hypertrophy

High blood pressure damages the heart by causing thickening (hypertrophy) of the heart muscle (myocardium) and also of the walls of the blood vessels supplying the heart. These changes reduce the ability of the heart to increase its capacity beyond its normal work load (coronary reserve), and can also lead to an inadequate blood supply to the heart muscle itself (myocardial ischaemia). These pathological changes predispose the patient to heart failure, so the aim of any long-term therapy should be to reverse ventricular hypertrophy and improve coronary vascular reserve (Motz and Strauer, 1990). Cilazapril has been shown to prevent the development of cardiac and vascular hypertrophy and can also improve coronary reserve once hypertension is established (Clozel and Hefti, 1988).

3. Reperfusion Arrhythmias

Contrary to apparent logic, restoration of blood flow to the ischaemic heart is more likely to cause fibrillation and death than the period of ischaemia itself. Reperfusion can occur spontaneously or as a result of medical intervention and frequently leads to life-threatening arrhythmias. Studies have shown that ACE inhibitors protect against reperfusion arrhythmias (Van Gilst *et al.*, 1984; Lad and Manning, 1987), whereas angiotensin I and angiotensin II potentiate these arrhythmias (Van Gilst *et al.*, 1984).

4. Congestive Heart Failure

Congestive heart failure occurs more frequently with increasing age and is often linked to disease of the heart muscle caused by inadequate blood supply (ischaemic cardiomyopathy) or to hypertension. In order to maintain blood flow to vital organs, a number of compensatory mechanisms are triggered. Initially the loss of contractility is countered by cardiac dilation, ventricular hypertrophy and increased catecholamine production. As the condition progresses, decreased cardiac output and increased filling pressures result in activation of the sympathetic nervous and renin–angiotensin systems which in turn causes vasoconstriction and increased blood volume. This then sets up a vicious circle in which the failing heart has to work against increasing blood pressure. Recently interest has focused on the use of ACE inhibitors in congestive heart failure since a fall in systemic vascular resistance should reduce the excessive pre- and after-loads seen in this disease. ACE inhibition has been shown to be beneficial in congestive heart failure and it is becoming evident that early treatment may even prevent the development of overt heart failure.

5. Inhibition of Neo-intimal Thickening

In atherosclerosis, the blood vessels are narrowed by deposits of lipid and collagen which later become calcified. Atherosclerotic vascular stenosis, vascular surgery and percutaneous transluminal coronary angioplasty (PTCA) (a technique which attempts to clear extensively blocked arteries) are all accompanied by proliferation of the underlying smooth muscle cells (Ross, 1986) which further narrows the blood vessel. The walls of the large arteries and veins contain local angiotensin systems which seem to be involved in the thickening of the blood vessel walls seen in hypertension since ACE inhibitors such as cilazapril have been shown (Clozel *et al.*, 1989) to reverse this thickening.

In animal experiments, Powell and colleagues (1989) have shown that cilazapril similarly prevents this smooth muscle cell proliferation after vascular injury. The outcome of clinical studies in PTCA patients will determine whether these impressive results are reproduced in man.

VIII. Conclusions

Although early ideas about hypertension suggested that ACE inhibitors would be useful only in certain rather rare circumstances, this has proved to be an oversimplification, and ACE inhibitors have become an important class of drugs for controlling the commonly encountered forms of hypertension. Moreover, ACE inhibitors will also probably prove beneficial to patients suffering from myocardial infarction, congestive heart failure and possibly atherosclerosis.

References

Attwood, M.R., Francis, R.J., Hassall, C.H., Kröhn, A., Lawton, G., Natoff, I.L., Nixon, J.S., Redshaw, S. and Thomas, W.A. (1984). *FEBS Lett.* **165**, 201.

Attwood, M.R., Hassall, C.H., Kröhn, A., Lawton, G. and Redshaw, S. (1986). *J. Chem. Soc. Perkin Trans.* I, 1011.

Breslow R. and Wernick, D. (1976). *J. Am. Chem. Soc.* **98**, 259.

Brunner, H.R., Gavras, H., Laragh, J.H. and Keenan, R. (1973). *Lancet* **2**, 1045.

Byers, L.D. and Wolfenden, R. (1972). *J. Biol. Chem.* **247**, 606.

Clozel, J-P., Kuhn, H. and Hefti, F. (1989). *J. Hypertension* **7**, 267.

Cushman, D.W. and Cheung, H.S. (1971). *Biochem. Pharmacol.* **20**, 1637.

Cushman, D.W. and Ondetti, M.A. (1979). *In* 'Progress in Medicinal Chemistry' (G.P. Ellis and G.B. West, eds), Vol. 17, p. 41, Elsevier/North Holland, Amsterdam.

Dewar, J.S. and Thiel, W.J. (1977). *J. Am. Chem. Soc.* **99**, 4899.

Effenberger, F., Burkard, U. and Winfahrt, J. (1983). *Angew. Chem. Int. Ed. Engl.* **22**, 65.

Engel, S.L., Schaeffer, T.R., Gold, B.I. and Rubin, B. (1972). *Proc. Soc. Exp. Biol. Med.* **140**, 240.

Gavras, H., Brunner, H.R., Laragh, J.H., Sealey, J.E., Gavras, I. and Vukovich, R.A. (1974). *New Engl. J. Med.* **291**, 817.

Hartsuck, J.A. and Lipscomb, W.N. (1971). *in* 'The Enzymes' (P.D. Boyer, ed.), Vol. 3, p. 1, Academic Press, New York.

Hassall, C.H., Kröhn, A., Moody, C.J. and Thomas, W.A. (1984). *J. Chem. Soc. Perkin Trans.* I, 155.

Hayakari, M., Kondo, Y. and Izumi, H. (1978). *Anal. Biochem.* **84**, 361.

Hefti, F., Fischli, W. and Gerold, J. (1986). *J. Cardiovasc. Pharmacol.* **8**, 641.

Kester, W.R. and Matthews, B.W. (1977). *Biochemistry* **16**, 2506.

Kohl, N.E., Emini, E.A., Schleif, W.A., Davis, L.J., Heimbach, J.C., Dixon, R.A.F., Scolnick, E.M. and Sigal, I.S. (1988). *Proc. Natl. Acad. Sci. U.S.A.* **85**, 4686.

Lad, N. and Manning, A.S. (1987). *Br. J. Pharmacol.* **91**, 390P.

Matthews, B.W., Weaver, L.H. and Kester, W.R. (1974). *J. Biol. Chem.* **249**, 8030.

Monzingo, A.F. and Matthews, B.W. (1984). *Biochemistry* **23**, 5724.

Motz, W.H. and Strauer, B.E. (1990). *Am. J. Cardiol.* **65**, 60G.

Nakajima, T., Oshima, G., Yeh, H.S.J., Igic, R. and Erdos, E.G. (1973). *Biochim. Biophys. Acta* **315**, 430.

Natoff, I.L., Nixon, J.S., Francis, R.J., Klevans, L.R., Brewster, M., Budd, J.M., Patel, A.T., Wenger, J. and Worth, E. (1985). *J. Cardiovasc. Pharmacol.* **7**, 569.

Ondetti, M.A., Rubin, B. and Cushman, D.W. (1977). *Science* **196**, 441.

Patchett, A.A., Harris, E., Tristram, E.W., Wyvratt, M.J., Wu, M.T., Taub, D., Peterson, E.R., Ikeler, T.J., ten Broecke, J., Payne, L.G., Ondeyka, D.L., Thorsett, E.D., Greenlee, W.J., Lohr, N.S., Hoffsommer, R.D., Joshua, H., Ruyle, W.V., Rothrock, J.W., Aster, S.D., Maycock, A.L., Robinson, F.M. and Hirschmann, R. (1980). *Nature (London)* **288**, 280.

Peng, C., Ho, B.K., Chang, T.W. and Chang, N.T. (1989). *J. Virol.* **63**, 2550.

Powell, J.S., Clozel, J-P., Müller, R.K.M., Kuhn, H., Hefti, F., Hosang, M. and Baumgartner, H.R. (1989). *Science* **245**, 186.

Quiocho, F.A. and Lipscomb, W.N. (1971). *Adv. Protein Chem.* **25**, 1.

Ross, R. (1986). *New Engl. J. Med.* **314**, 488.

Schmid, M.E. and Herriot, J.R. (1976). *J. Mol. Biol.* **103**, 175.

Soubrier, F., Alhenc-Gelas, F., Hubert, C., Allegrini, J., John, M., Tregear, G. and Corvol, P. (1988). *Proc. Natl. Acad. Sci. U.S.A.* **85**, 9386.

Van Gilst, W.H., DeGraeff, P.A., Kingma, J.H., Wesseling, H. and Delangen, C.D.J. (1984). *Eur. J. Pharmacol.* **100**, 113.

Yang, H.Y.T. and Erdos, E.G. (1967). *Nature (London)* **215**, 1402.

$-10-$

Beta Blockers

B.G. MAIN
Chemistry Department,
ICI Pharmaceuticals p.l.c.,
Macclesfield, U.K.

H. TUCKER
Chemistry Department,
ICI Pharmaceuticals p.l.c.,
Macclesfield, U.K.

Dr Tucker received B.Sc. and Ph.D. degrees at the University College of Wales, Aberystwyth. After further study at the Technische Hochschule Darmstadt and at Bedford College, London he joined ICI Pharmaceuticals where he is now a Senior Scientist in the arthritis project. Research interests include the design of enzyme inhibitors and non-steroidal antihormonal agents.

Mr Main obtained his GRSC at Kingston College, and was elected FRSC in 1989. After periods with Glaxo and Parke-Davis Research he joined ICI Pharmaceuticals in 1967 where he is currently a Senior Scientist in the cardiovascular research group. Mr Main's primary areas of research have been opioids, prostaglandins and several aspects of heart failure treatment.

MEDICINAL CHEMISTRY 2nd Edition
ISBN 0-12-274120-X

I. Introduction

In this chapter we trace the development of compounds which selectively block β_1-receptors; these compounds have found many uses in the clinic including the treatment of angina and high blood pressure. In addition we show the origin of compounds which selectively block β_2-receptors; these latter compounds will allow the physiological importance of these receptors to be delineated.

II. Nervous Control of the Heart and Circulation

The nervous system which coordinates many of the body's functions is divided into two branches, the somatic system, which controls the activity of the skeletal (voluntary) muscles, and the autonomic system, which is concerned with the (involuntary) maintenance of a stable internal environment, and which is subdivided into the parasympathetic and sympathetic nervous systems (Fig. 10.1). In the heart, stimulation of the parasympathetic nerves (vagi) leads to a reduction in heart rate while stimulation of the sympathetic nerves increases the rate and force of contraction of the heart. The response of the heart to a given situation is governed by a fine balance between these two systems. Neural impulses along the parasympathetic or sympathetic nerves achieve their effect by the release from the nerve endings of the neurotransmitters acetylcholine (**1**) and noradrenaline (**2**), which bind to their respective receptors, eliciting biochemical reactions which ultimately express themselves in a physiological response. In addition, in response to stress the adrenal glands secrete into the bloodstream a hormone, adrenaline (**3**),

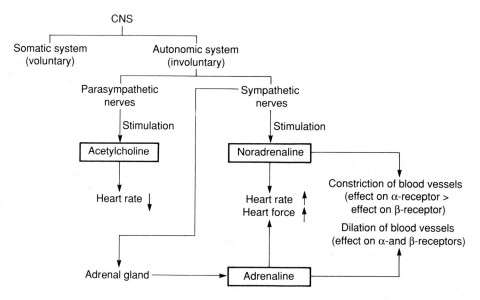

Fig. 10.1   Factors contributing to the control of heart rate and blood vessels.

(1) Acetylcholine

(2) Noradrenaline

(3) Adrenaline

(4) Isoprenaline

which is responsible for the 'flight or fight' syndrome by which the body prepares to protect itself or run away when threatened. Adrenaline stimulates the heart and dilates some blood vessels to supply the muscles with more blood; noradrenaline has the same cardiac actions but constricts the blood vessels (see Table 1.3).

In a search for bronchodilator agents, Konzett (1940) synthesized isoprenaline (**4**), the N-isopropyl analogue of noradrenaline, which has found widespread clinical use both as a bronchodilator and cardiac stimulant. Unlike noradrenaline and adrenaline, which are produced in the body, isoprenaline is a purely synthetic material.

In 1948 Ahlquist reported the effects of a number of sympathomimetic amines (agonists) including adrenaline, noradrenaline and isoprenaline on various tissues innervated by the sympathetic nervous system and concluded that there were two types of adrenergic (sympathetic) receptors which he termed α and β. Stimulation of the α-receptor mainly causes a contractile response while β-receptor responses are largely relaxant. The major exception to this is the cardiac β-receptor, stimulation of which increases the rate and force of contraction of the heart. From this, it follows that noradrenaline exhibits mainly α-stimulant activity with some (important) β-agonism, adrenaline has equipotent stimulant activity on α- and β-receptors and isoprenaline is a pure β-agonist.

In 1967, Lands et al. reported the effects of 15 sympathomimetic amines on a series of tissues and showed by statistical analysis of the results that there were two different populations of β-receptor which he designated as β_1 and β_2, the cardiac β-receptors being β_1 while those in the bronchial smooth muscle and in blood vessels are β_2 (see Table 11.2). More recently it has been shown that this is an oversimplification with most tissues containing both β_1- and β_2-receptors, the overall response of the tissue depending on the proportions of receptor present (Nahorski et al., 1981); thus cardiac β-receptors, for example, are mainly β_1 and bronchial β-receptors mainly β_2.

In the last few years, evidence has been accumulating that there is, in fact, a third β-receptor, the 'atypical' or β_3-receptor (Wilson et al., 1988; Strosberg et al., 1989). This appears to be responsible for the adrenergic control of the metabolic rate of an animal, activating 'brown fat' to burn up excess food intake (adipose tissue thermogenesis) (Muzzin et al., 1988).

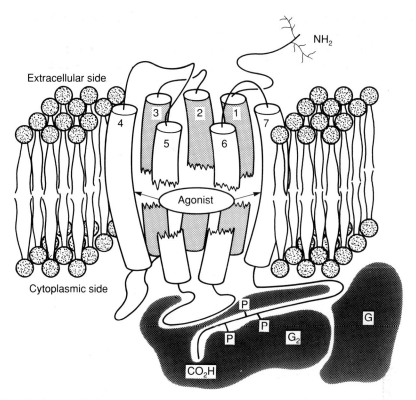

Fig. 10.2 Structure of β-receptor showing seven membrane-spanning hydrophobic α-helices. [Reproduced, with permission, from Taylor, C.W. (1990), *Biochem. J.* **272**, 1.]

III. Nature of the β-Receptor

The β-receptor is one of a family of receptors which mediate their actions through G-proteins (guanine nucleotide-binding regulatory proteins). Other classes of receptor in this family are the α-adrenergic and serotonin (5-hydroxytryptamine) receptors and rhodopsin (see Fig. 6.7). The binding of a β-agonist to its receptor leads to the activation of a G-protein which in turn stimulates the enzyme adenylate cyclase to convert ATP into cyclic AMP. Cyclic AMP, the so-called second messenger, initiates a complex biochemical cascade which ultimately leads to the physiological response.

Recent work on the structure of the β-receptor and related receptors has shown that these receptors consist of seven membrane-spanning hydrophobic α-helices which contain the binding site for agonists and antagonists and hydrophilic loops which are not required for ligand binding (Fraser and Venter, 1990) (Fig. 10.2). Site-directed mutagenesis studies (Chapter 6) have identified the specific amino acid residues which bind to the agonist. In the case of the catecholamines, an aspartic acid residue (Asp-113) in the third hydrophobic domain interacts with the amino group of the catecholamine; moreover two serine residues (Ser-204 and Ser-207) in the fifth hydrophobic helix interact with the agonist catechol hydroxy groups (Dixon *et al.*, 1991).

It is now established that the β_1- and β_2-receptors, though similar, are subtly different in structure and are encoded by separate genes.

In the case of the human β-receptor, the β_1-receptor is composed of 477 amino acid residues and the β_2-receptor 413. The two subtypes share 54% overall homology but the level of homology is much higher in the membrane-spanning region.

IV. β-Blockade and Cardiac Disease

Ahlquist's hypothesis of α- and β-receptors was ignored for the first 10 years largely because no agent was known that would selectively block β-receptors in the same way that, for example, ergotamine blocks the α-receptors. In 1958, Mills of the Eli Lilly Company prepared the dichloro analogue of isoprenaline, commonly known as DCI (5).

(5) Dichloroisoprenaline (DCI)

(6) Pronethalol

DCI was shown to inhibit the relaxation of bronchial smooth muscle elicited by isoprenaline (Powell and Slater, 1958) and also the cardiac actions of isoprenaline (Moran and Perkins, 1958), that is, it was the first β-adrenoreceptor blocking agent, now more commonly referred to as a β-blocker. In addition to its β-blocking actions, DCI also had a stimulant action, that is, it was a partial agonist, and at that time no clinical application was perceived for such a compound.

The true clinical potential of β-adrenoreceptor blocking agents was first recognized by Black at ICI when in 1959 he formulated a theory for a possible treatment for coronary artery disease and, in particular, angina. One of the body's responses to physical and mental stress is stimulation of the sympathetic nervous system causing (via β-receptors) an increase in the heart rate and force of contraction. These are energy-consuming processes and the heart muscle itself requires a greater supply of oxygen. If an adequate supply of blood cannot be maintained because of coronary arterial disease then the resulting oxygen deficiency in the cardiac muscle manifests itself in the intense pain of angina. Attempts to increase the oxygen supply to the heart by vasodilator drugs had not been successful, so Black proposed that reduction of the heart's demand for oxygen by blocking the effects of sympathetic nerve stimulation should be an effective means of treating angina. In particular, Black wanted a specific β-blocker without the stimulant properties of DCI. Within 18 months he had set up screening tests for this type of activity and his colleague Stephenson had synthesized pronethalol (6), a compound in which the two chlorine atoms of DCI (5) had been replaced by a fused benzene ring. This agent effectively blocked cardiac (and other) β-receptors and only had marginal partial agonist activity (Black and Stephenson, 1962).

V. Early β-Blockers

Pronethalol became the first clinically available β-blocker and was effective not only in the treatment of angina but also controlled certain kinds of cardiac arrhythmia and reduced elevated blood pressure. The discovery of these properties led to a systematic investigation around the pronethalol structure both within ICI and at other pharmaceutical companies. The structure–activity relationships of these arylethanolamines have been discussed elsewhere (Phillips, 1980) and will not be considered further. In the main, the ethanolamines are associated with varying degrees of partial agonism, with the notable exception of sotalol [(7), Table 10.1] and have achieved their greatest success as β-stimulants (see Chapter 11). Sotalol itself, however, is used clinically as a β-blocker.

Pronethalol was withdrawn from the clinic in 1963 due to concern over toxic symptoms (thymic tumours) in animals, but, during the development period of pronethalol, chemical work to optimize β-antagonist activity had continued. Besides studying the effects of aryl ring and amine substitutions in the molecule, the researchers modified the ethanolamine chain itself by, *inter alia*, the introduction of linking groups between the aryl ring and ethanolamine chain (Crowther, 1990). Of the many linking groups tried, oxymethylene proved to be the best, with the first analogue prepared, the α-naphthyloxypropanolamine propranolol (8), being 10–20 times more potent than pronethalol (Crowther and Smith, 1968). Smith, one of the chemists involved, has related that he used α-naphthol as a model compound since at the time he could not find the bottle of β-naphthol. Ten days later the β-naphthyloxy derivative (9), which is the direct analogue of pronethalol, was synthesized and found to be only slightly more potent than pronethalol. It is interesting to note that some years before the discovery of DCI, BDH workers (Petrov *et al.*, 1956) published papers on the local anaesthetic properties of a series of aryloxypropanolamines, which included the *n*-propylamino analogue of propranolol and compounds 10 and 11 which were later found to be β-adrenergic agonists. 'Where observation is concerned, chance favours only the prepared mind' (Louis Pasteur).

(8) Propranolol   (9)

(10) R = Me
(11) R = H

TABLE 10.1
Early β-Blockers in Clinical Use

Structure	Name (manufacturer)	Relative potency[a]	isa[b]	msa[c]
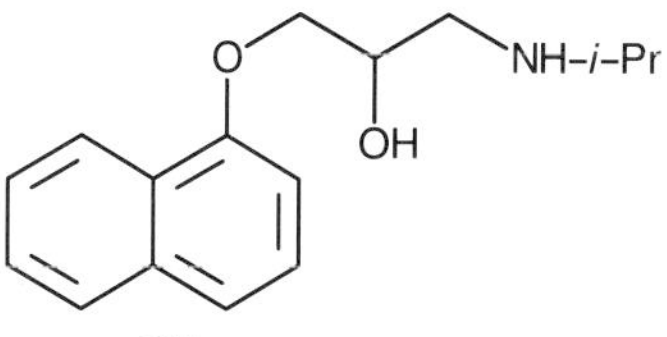	(8) Propranolol (ICI)	1	−	+
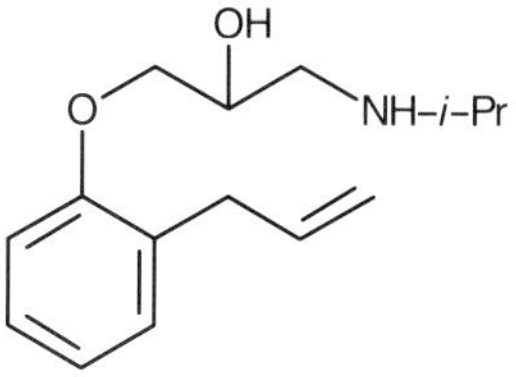	(12) Alprenolol (Hassle)	1	+	+
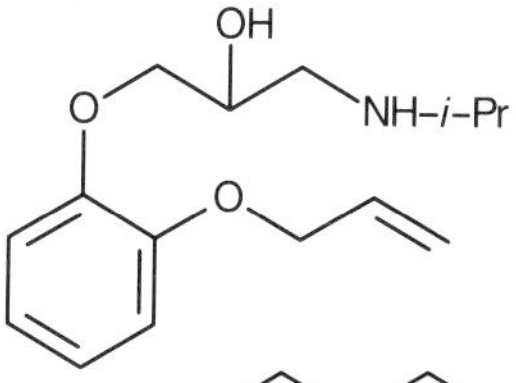	(13) Oxprenolol (Ciba-Geigy)	0.3	+	+
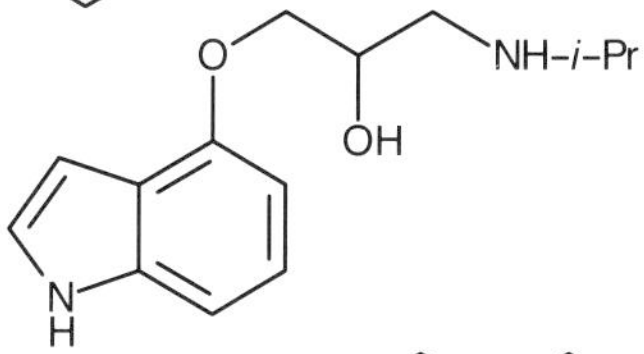	(14) Pindolol (Sandoz)	30	+ +	+
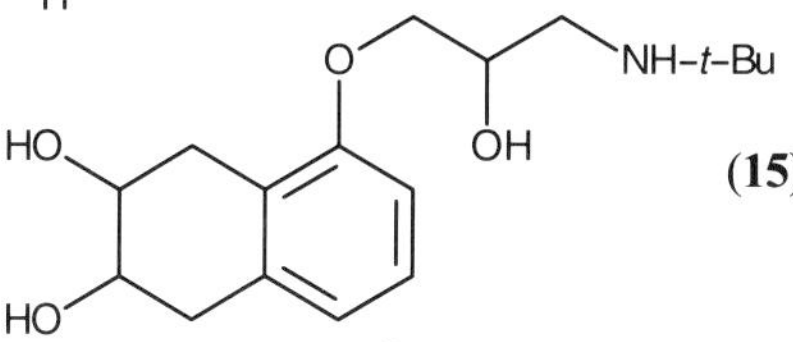	(15) Nadolol (Squibb)	1	−	?
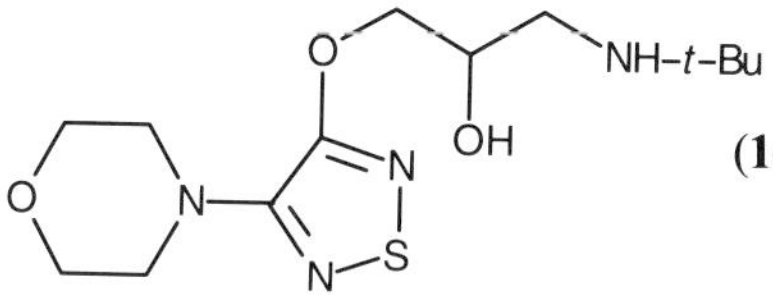	(16) Timolol (Merck)	4	−	?
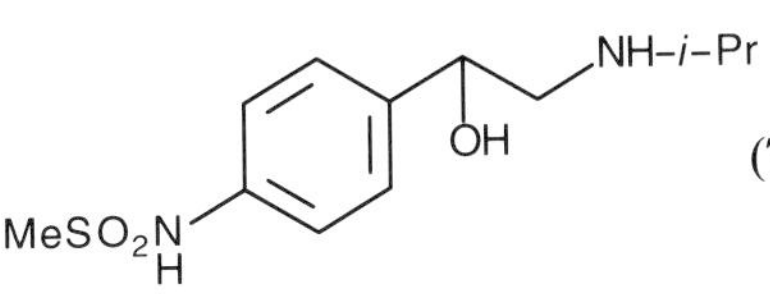	(7) Sotalol (Mead Johnson)	0.08	−	−

[a] Ratio of ED_{50} values in the test described by Smith (1978).
[b] isa, intrinsic sympathomimetic activity.
[c] msa, membrane stabilizing activity.

Many companies at this time began working in the aryloxypropanolamine area, and a number of successful drugs have emerged. A selection of these is listed in Table 10.1. Propranolol has become the reference compound, and established the clinical concept of β-blockade for the treatment of angina. In addition it pioneered the clinical use of β-blockers in many other disease states.

It is appropriate at this stage to summarize the pharmacology, clinical properties and toxicological problems common to all the drugs of this early type.

VI.   Pharmacology of β-Blockers

All β-blockers have two properties in common; namely, they are competitive antagonists, and they are specific for β-receptors. For any β-blocker the following parameters are used in defining its profile: potency, partial agonism and membrane stabilizing activity (msa).

A.   Potency

Potency is usually expressed, *in vivo*, as an ED_{50}, or *in vitro*, as a pA_2. ED_{50} is the dose of antagonist required to reduce the response to an agonist by 50%, and pA_2 has been defined by Nogrady, 1985. Because many different species and tissue types have been used throughout the world to evaluate β-blockers, wherever possible potencies relative to propranolol are used in this text.

B.   Partial Agonism

Partial agonism has generally been termed 'intrinsic sympathomimetic activity' (isa), and is a measure of the ability of a compound to stimulate β-receptors directly. The majority of β-blockers have isa, and the usual value quoted is the absolute increase in heart rate observed when a compound is given at a fixed (high) dose of 2.5 mg/kg i.v. to rats whose natural catecholamine stores have been depleted by pretreatment with syrosingopine (a reserpine derivative).

In this test propranolol causes no increase in heart rate, pronethalol causes a 70 beat per minute (bpm) rise, and isoprenaline causes the maximum rise of 200 bpm (from about 300 to 500). There is still much speculation about the clinical relevance of low (up to ~ 100 bpm) amounts of isa.

C.   Membrane Activity

In 1964 it was shown that pronethalol had local anaesthetic activity, more potent than procaine, and it was suggested that this might be the mode of action of β-blockers in angina and arrhythmias. Subsequent investigations showed, however, that this activity, which is believed to be caused by electrical stabilization of cell membranes, was not present at normal clinical doses.

Hellenbrecht *et al.* (1974) showed that, for a series of related β-blockers, this membrane stabilizing activity (msa) was directly related to the partition coefficient of the compound, highly lipophilic compounds such as propranolol being very effective local anaesthetics.

VII.   Structure–Activity Relationships of Early β-Blockers

Considering the fundamental structure (17) for arylethanolamines (X = direct link) and aryloxypropanolamines (X = OCH_2), one can identify two functions which are essential for β-blockade, namely the β-aminoethanol chain and the aromatic ring. Experience with the arylethanolamine series had shown that optimum β-antagonist activity resided with branched alkylamine substituents, the best of which were isopropyl and t-butyl. This was also found to be true for the aryloxypropanolamines and in all the early work on β-blockers the isopropyl- or t-butyl-amino substituents were retained. The insertion of additional groups between the amine and carbinol group, removal or alkylation of the hydroxy group or acylation of the amine all led to a substantial or total loss of β-blocking activity. With the exception of X = OCH_2, all other linking groups tried have given compounds with little or no activity (X = CH=CH, SCH_2, —CH_2CH_2—) although modest activity was obtained when X = $NHCH_2$.

R—[benzene ring]—(X)—CH(OH)—CH_2—NHR'

(17)

Ar(X)—CH(H)(OH)—CH_2—NHR

Biologically active
enantiomer of 17

Methyl substitution at the α, β or γ carbon atom resulted in a reduction in β-blocking activity which was least for α-substitution. The β-antagonist activity resides with one enantiomer, the (R)-ethanolamine and the (S)-aryloxypropanolamine, both of which have the same absolute configuration. Most commercially available β-blockers are sold as the racemates.

With some minor exceptions, all β-blockers have an aromatic ring which need not necessarily be benzenoid but can be heterocyclic [e.g. timolol (16)], or benzheterocyclic [e.g. pindolol (14)] (Table 10.1). The nature and position of the substituents in the aromatic ring have a crucial effect on the potency and overall pharmacological profile of the molecule. For the series of positional isomers listed in Table 10.2 it is clear that the order of β-antagonist potency is ortho > meta > para. In addition, the ortho position can accommodate large substituents and retain β-antagonist activity whereas the para position is very sensitive to group size. This is clearly demonstrated by the isomers of the phenoxy analogue. Polysubstitution of the aromatic ring yielded compounds with a variety of potencies, except for 2,6-disubstitution which resulted in an almost complete loss of β-antagonist activity.

The size of the ortho substituent plays an important part in governing the levels of isa exhibited by the molecule. The larger the ortho substituent, the less is the isa observed. In an attempt to explain this phenomenon, the effect of the ortho substituent on the conformation of the ethanolamine chain was computed and related to the isa observed for a series of aryloxypropanolamines (Richards et al., 1975). In another approach (Main, 1982), Taft's steric factor for the ortho substituent was plotted against isa. Both approaches gave a straight line relationship.

A graphic example of the influence of the aromatic ring substituent on the pharmacological profile is provided by the hydroxy substituent. The aryloxypropanolamine analogue of isoprenaline (19) has been shown (Kaiser et al., 1977) to be a potent β-agonist. The corresponding 3,5-dihydroxy analogue (20) is a partial agonist, as is the meta-hydroxy (21) and para-hydroxy analogue (22) [the laevorotatory isomer of which,

TABLE 10.2
Comparison of the Potencies of *ortho-*, *meta-* and *para-*
Substituted Phenoxypropanolamines (18)[a]

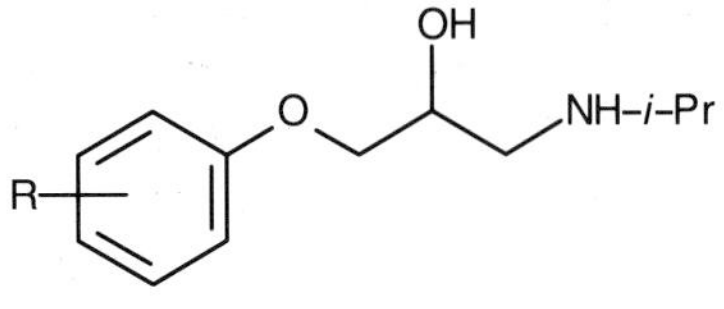

(18)

R	ortho	meta	para
Cl	5	0.52	0.085
NO$_2$	2	0.39	0.028
CH$_3$	1.71	0.66	0.09
OCH$_3$	1.71	0.91	0.54
OC$_6$H$_5$	2	0.3	0.001

[a] Potency values related to propranolol = 1.

prenalterol (Carlsson *et al.*, 1977), has been under development as a cardiac stimulant for the treatment of heart failure]. The *ortho* analogue, however, is a β-blocker. The interesting heterocyclic derivative tazolol (23) is also a partial β_1-agonist, which contrasts markedly with timolol (16) which is devoid of isa.

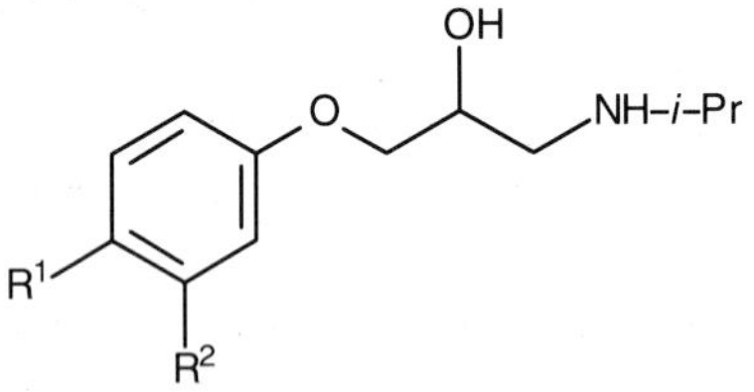

(19) R^1 = R^2 = OH
(22) R^1 = OH; R^2 = H

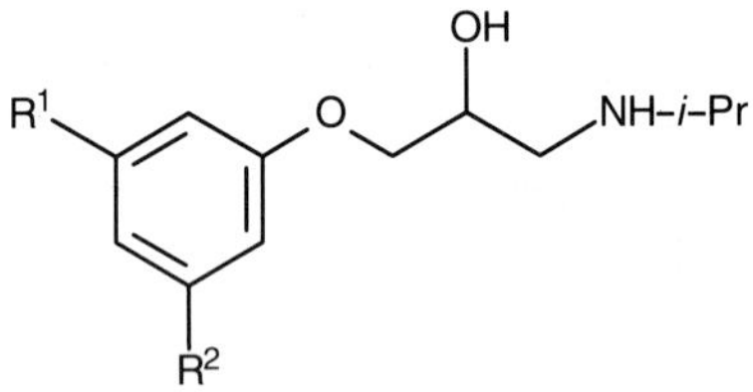

(20) R^1 = R^2 = OH
(21) R^1 = OH; R^2 = H

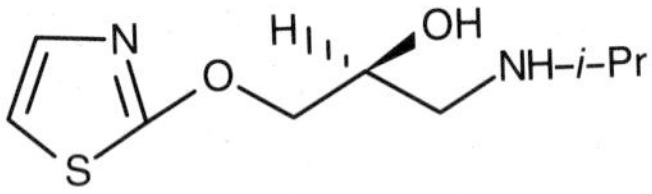

(23) Tazolol

VIII.   Cardioselective β-Blockade

Propranolol (8) is a lipophilic compound, that is, it has a high partition coefficient favouring transfer from aqueous to non-polar (lipid) media. This is known to facilitate transfer across the blood–brain barrier and could, perhaps, be part of the reason for the CNS side effects of certain β-blockers. It was felt that these effects could be reduced by making more hydrophilic compounds, and in 1964 the first reports of sotalol (7) prompted Crowther *et al.* (1971) to carry out the synthesis of the oxypropanolamine analogue (24). The synthesis of this compound was not trivial, and while this was underway Smith prepared the corresponding acetamido compound from the readily available *para*-acetamidophenol. This compound, practolol (**25**), was not as potent as propranolol, and in addition it had some isa. The sulphonamide (**24**), prepared very soon afterwards, had similar properties. It was later found that practolol blocked cardiac β-receptors ($β_1$) without blocking vascular receptors ($β_2$), that is, it was cardioselective. It was urgently tested against bronchial β-receptors ($β_2$) and shown to be without effect here also. This was of great significance because giving a non-selective β-blocker to asthmatics could trigger an asthmatic attack; $β_1$-selective blockade should be much safer in these patients. The sulphonamide was similarly selective.

Selectivity is usually defined by the ratio

$$\frac{ED_{50} \text{ vasodilation}}{ED_{50} \text{ heart rate}}$$

for antagonism of isoprenaline responses in animals, and by ratios of K_B values for $β_1$ and $β_2$ responses *in vitro*. All cardioselective β-blockers have a value for this ratio greater than 1, but it must be emphasized that the compounds are cardioselective not cardio-specific, that is, if enough drug is given (greater than the cardiac ED_{50} × this selectivity ratio), $β_2$ effects will be seen.

Practolol has a log P of 0.79 (i.e. a partition coefficient of 6:1 in favour of octanol versus water) compared with propranolol 3.65 (i.e. partition coefficient of 4500:1). It achieved the goal of less CNS side effects, and could be given with greater (though not absolute) safety to asthmatics. It also caused a smaller reduction of cardiac output than propranolol (due to its isa). Practolol, the first cardioselective β-blocker, was launched in 1970 and was used in the treatment of angina, hypertension and also immediately post-infarction. Unfortunately, after several years of clinical use practolol-related toxicity was recognized in a small percentage of patients receiving the drug. The skin rashes, eye problems (sometimes leading to blindness) and a severe form of peritonitis (fatal in some cases) led to the withdrawal of practolol from the market. The reasons for this toxicity are still neither known nor predictable from animal studies, and appear to be unique to practolol.

IX. Structure–Activity Relationships of Cardioselective β-Blockers

The discovery of practolol resulted in widespread chemical activity aimed at introducing other amidic groups into aryloxypropanolamines to achieve a practolol-like profile of activity. The structure–activity relationships described earlier still hold for this type of compound, with one exception: *para*-amidic substituents give β-blockers which are more

TABLE 10.3
Isomers of Amide-substituted β-Blockers

(18)

R	Potency[a]	Selectivity	isa[b]
o-NHCOCH₃	0.075	−	+
m-NHCOCH₃	0.13	−	?
p-NHCOCH₃ (practolol)	0.36	+	+

[a]Propranolol = 1.
[b]isa, intrinsic sympathomimetic activity.

potent than their *ortho* counterparts. Only these *para* isomers are cardioselective (Smith, 1978) (Table 10.3) and all compounds of this type, like practolol, have isa.

The next development came when Hull and co-workers (Barrett *et al.*, 1973) introduced a methylene group between the amide function and the aromatic ring. The first compound prepared, atenolol [(27), Table 10.4], was as potent as propranolol, cardioselective, but most surprisingly devoid of isa. It was found subsequently that this was a general phenomenon, and the related p-CH₂NHCOR and p-CH₂NHCONHR analogues were likewise cardioselective and without isa. Independently, Carlsson *et al.* (1973) prepared the *para*-methoxy ethyl compound, metoprolol (28), and showed that this had the same profile; betaxolol (29) is very similar.

The cardioselective β-blockers in clinical use are shown in Table 10.4. Compounds of all these series are considerably more hydrophilic than propranolol, and one school of thought attributed cardioselectivity to interaction of the drug with a relatively hydrophilic β_1-receptor rather than with a more lipophilic β_2-receptor. However, Somerville *et al.* (1979) showed clearly that, in a series of isomeric *meta* and *para* compounds, selectivity resided with the *para* isomer only, although the log P values were identical in each case (Table 10.5).

It is thought that cardioselectivity arises from an additional interaction of a *para* substituent, possibly via hydrogen bond formation, with a complementary site on the β_1- but not β_2-receptor.

X. β_2-Selective Blockade

In β-antagonists and -agonists of the ethanolamine series introduction of an alkyl group on the carbon atom α to the amine resulted in an overall reduction in potency. The introduction of this alkyl group gives rise to two isomers, the *erythro* and *threo*, and in general the *erythro* isomers are more potent than the *threo*. The α-methyl analogue of dichloroisoprenaline (30) was found to block the vascular (β_2) receptors more effectively than the cardiac (β_1) receptors. Other substituted ethanolamines found to be β_2-selective were butoxamine (31) and H35/25 (32). However, they were not very potent and

TABLE 10.4
Cardioselective β-Blockers in Clinical Use

Structure	Name (manufacturer)	Relative potency[a]	isa[b]
	(25) Practolol (ICI)	0.25	+
	(26) Acebutolol (May & Baker)	0.4	+
	(27) Atenolol (ICI)	0.8	−
	(28) Metoprolol (Hassle)	0.8	−
	(29) Betaxolol (Synthelabo-Searle)	1.0	−

[a]Propranolol = 1.
[b]isa, intrinsic sympathomimetic activity.

TABLE 10.5
Cardioselectivity and Hydrophilicity

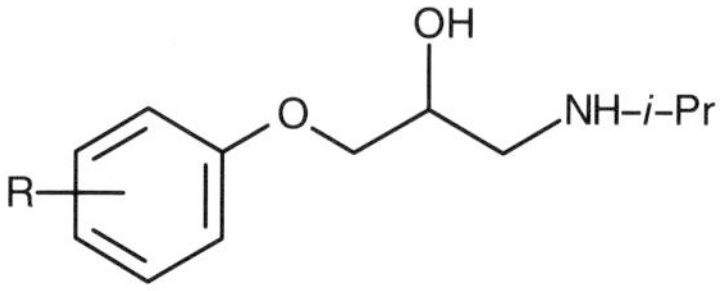

R	log P	K_D (heart, β_1)	K_D (corpus luteum, β_2)	β_2/β_1
p-NHCOCH$_3$	0.79	5.4×10^{-7}	7.2×10^{-5}	133
m-NHCOCH$_3$	0.74	8.2×10^{-7}	2.4×10^{-6}	2.9
p-CH$_2$CONH$_2$	0.23	3.7×10^{-7}	1.5×10^{-5}	40
m-CH$_2$CONH$_2$	0.22	5.9×10^{-7}	3.2×10^{-7}	0.5
p-CH$_2$NHCOCH$_3$	0.32	2.8×10^{-7}	4.6×10^{-6}	17
m-CH$_2$NHCOCH$_3$	0.34	8.0×10^{-7}	1.2×10^{-6}	1.5

butoxamine exhibited non-specific cardiodepressant activity at doses slightly higher than those necessary for β_2-blockade, and H35/25 had marked β_2-agonist activity.

In the aryloxypropanolamines too, α-methyl substitution results in an approximately 10-fold decrease in β-antagonist activity, and in the case of α-methylpropranolol (**33**) the *erythro* isomer is more potent than the *threo* by a factor of 2. However, like propranolol itself, both *erythro* and *threo* α-methyl derivatives of propranolol are β_2-selective.

During a more comprehensive study of the effects of α-methyl substitution on β-receptor-blocking potency, a potent β_2-selective antagonist ICI 118551 (**34**) was discovered. This compound exhibits a β_2-selectivity (over β_1) *in vivo* of greater than 250:1. The related *threo* isomer is also a selective β_2-antagonist but is less potent and selective than the *erythro* isomer. The β_2-antagonist activity of ICI 118551 was confirmed in clinical trials but the compound was withdrawn from development following the discovery of

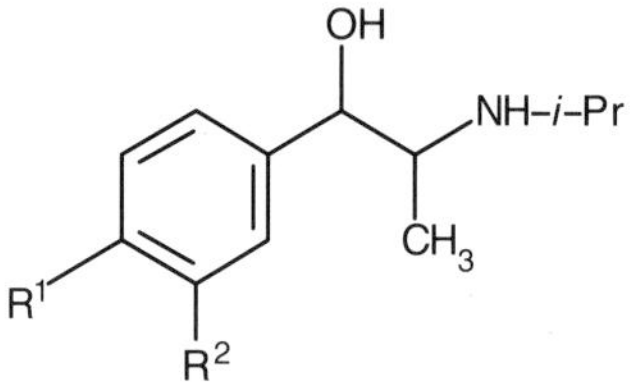

(**30**) R^1 = R^2 = Cl
(**32**) R^1 = CH$_3$; R^2 = H; H35/25

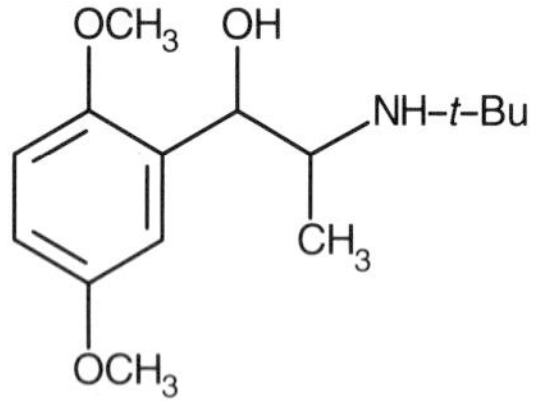

(**31**) Butoxamine

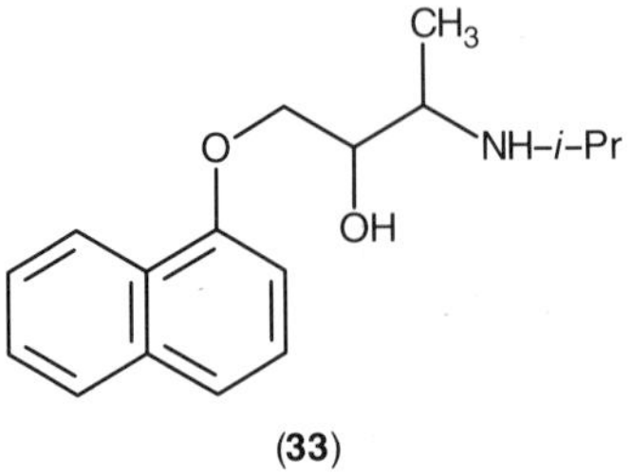

(**33**)

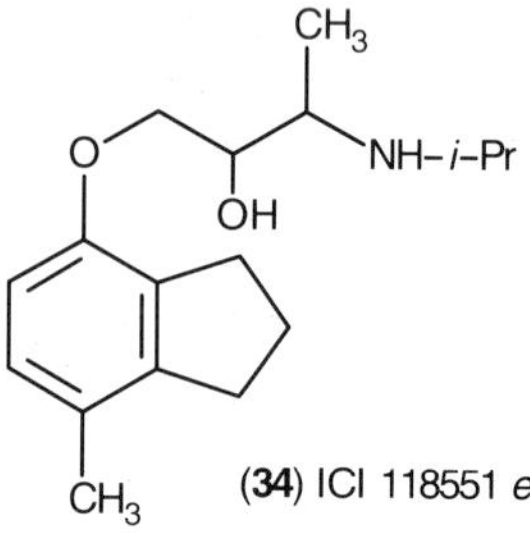

(**34**) ICI 118551 *erythro*

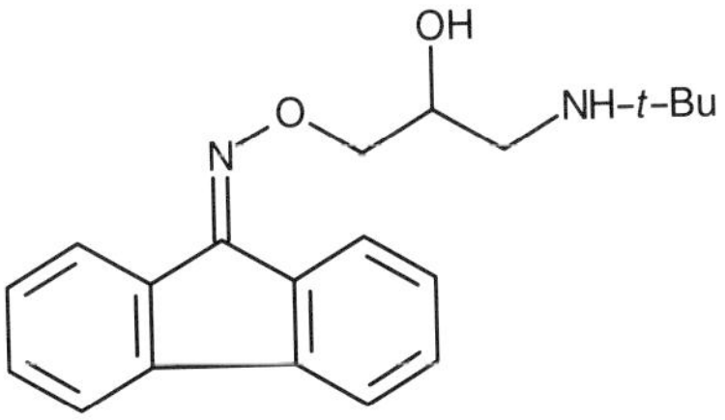

(35) IPS339

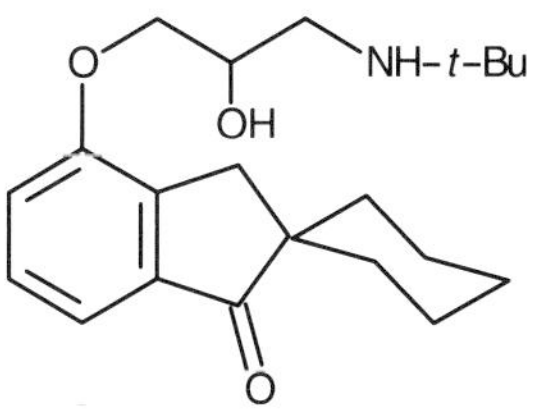

(36) Spirendolol

unacceptable toxicity in animal studies. It still finds widespread use, however, as a pharmacological and biochemical tool.

Among other β_2-selective antagonists reported are the fluorene oximinopropanolamine IPS339 (**35**) and spirendolol (**36**).

XI. 'Third Generation' β-Blockers

Early work on β-blockers had established that optimum potency resided with the isopropyl- or t-butylamino substituents, as compounds listed in Tables 10.1 and 10.4 demonstrate. Hoefle *et al.* (1975), however, reported, on the basis of *in vitro* experiments, that compounds (e.g. **37**) containing the 3,4-dimethoxyphenethylamino group were cardioselective. Augstein *et al.* (1973) had reported previously that tolamolol (**38**), which

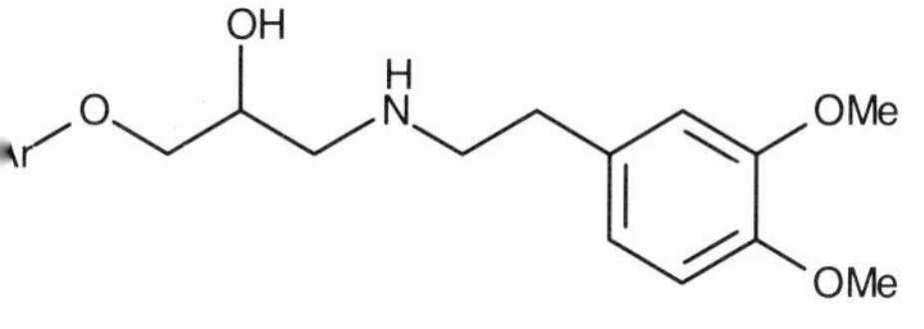

(37)

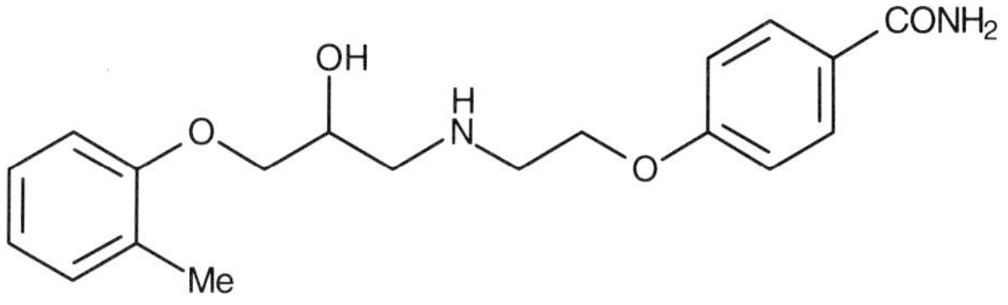

(38) Tolamolol

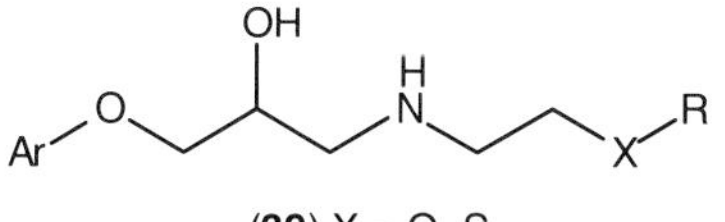

(39) X = O, S
R = aryl, alkyl

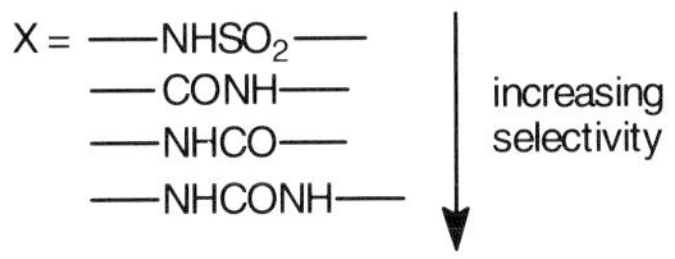

$$X = \begin{array}{l} \text{—NHSO}_2\text{—} \\ \text{—CONH—} \\ \text{—NHCO—} \\ \text{—NHCONH—} \end{array}$$

increasing selectivity

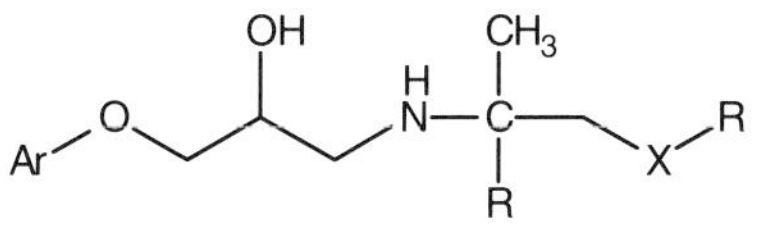

(40)

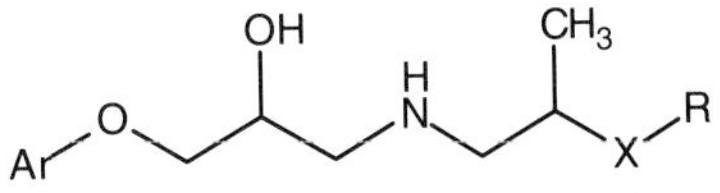

(41)

R = H or CH₃

contains an amide-substituted phenoxyethylamino group, was cardioselective, and suggested that this might be due to hydrogen bonding of the amide to the receptor. Smith
and Tucker (1977) showed that compounds containing non-amidic phenoxy- and thio-
phenoxy-substituted amine groups were also cardioselective, as were simple alkoxy- and
alkylthioethylamine analogues (39) (Tucker and Coope, 1978). This led them to conclude
that there was an additional interaction at this part of the molecule with the β_1-receptor.

Smith and Large (1982) followed up this work with an extensive series of compounds
in which the (thio) ether group, X in (39), was replaced by a variety of amidic alternatives. These compounds proved to be both potent and cardioselective and strongly
suggested a specific drug–receptor interaction at the amide site. The group R may be
varied widely without losing activity, but the length of the alkylene chain is critical; thus
extension to —NH(CH$_2$)$_3$ —X—R caused a reduction in potency, and branching of the
chain, as in compounds 40, tends to lessen cardioselectivity. (Note: For 40, R = H also
gives rise to inconvenient diastereoisomers.) Branching adjacent to the amide, as in
compound 41, also reduced activity. It is important to note that, in this type of
compound, amide substitution of the aryl ring (i.e. practolol-like compounds) either is
unnecessary for cardioselectivity or actually decreases potency.

(42) epanolol; R^1 = CN, R^2 = H, X = NHCOCH$_2$Ph (ICI)

(43) primidolol; R^1 = Me, R^2 = H, X =   (Pfizer)

(44) xamoterol; R^1 = H, R^2 = OH, X =   (ICI)

Several cardioselective agents have been selected from this class of compound for
clinical study, for example, epanolol (42) and primidolol (43).

Further work at ICI led to the discovery of xamoterol (44) (Barlow et al., 1981), a very
selective β_1-partial agonist. This compound, in addition to being a β-blocker itself,
exhibits agonism to the extent of about 45% of the maximum, so providing cardiac
stimulation at rest but β-blockade during strenuous exercise. Because of this stabilizing
effect, it is used in the treatment of mild heart failure (German and Austrian Study
Group, 1988).

XII.   Chemistry of β-Blockers

β-Blockers of the ethanolamine series are prepared by the routes described in Chapter

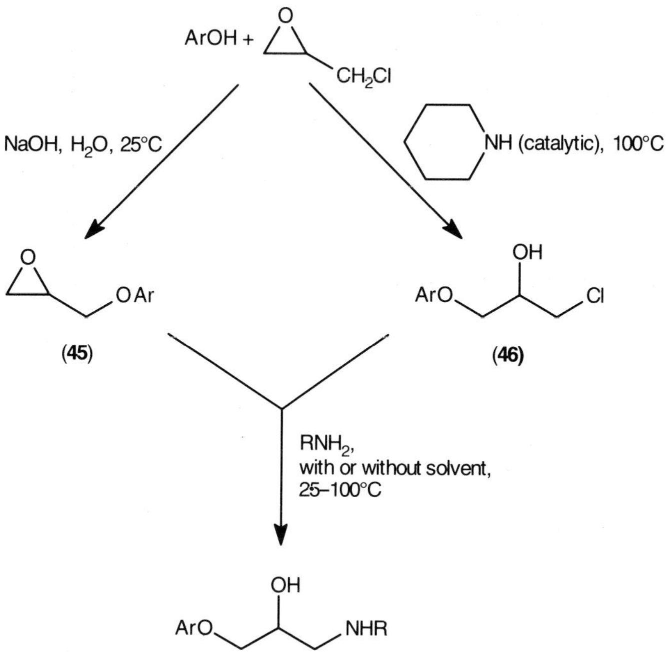

Scheme 10.1   Synthesis of aryloxypropanolamines.

11 for the synthesis of arylethanolamine β-agonists. β-Blockers of the aryloxypropanolamine series are commonly prepared by base-promoted reaction of the corresponding phenol with epichlorohydrin which, depending on the reaction conditions employed, can give either the epoxide (45) or chlorohydrin (46) or mixtures of both which are then reacted with the appropriate amine (Scheme 10.1). The production of side products such as the bis-ether (47) and the tertiary amine (48) can be minimized by using an excess of epichlorohydrin and amine respectively.

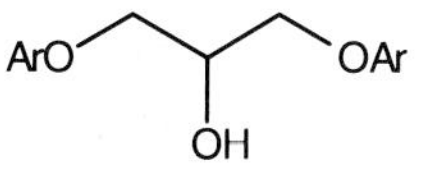

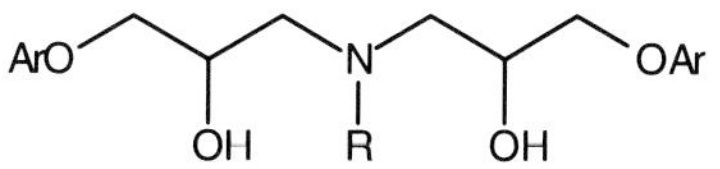

The β-antagonist activity of the aryloxypropanolamines resides with one enantiomer which was shown in the case of propranolol to have the (S) absolute configuration by relating (+)-propranolol to (+)-lactic acid (Smith and Dukes, 1971). The normal resolution techniques using optically active acids were found to be time-consuming and unpredictable, and much effort has been devoted to devising a general asymmetric synthetic route which would furnish both enantiomers reliably. In the first of these approaches, a Pfizer group (Danilewicz and Kemp, 1973) prepared the (R) (less active) enantiomer of practolol from D-mannitol using the route outlined in Scheme 10.2. Success of this route was critically dependent on cleanly monotosylating the primary hydroxy group of diol (49). The synthesis of the more active (S) enantiomer using an analogous route was not attractive because L-mannitol is not readily available.

More recently S glycerol acetonide (50) has become cheap and readily available due to

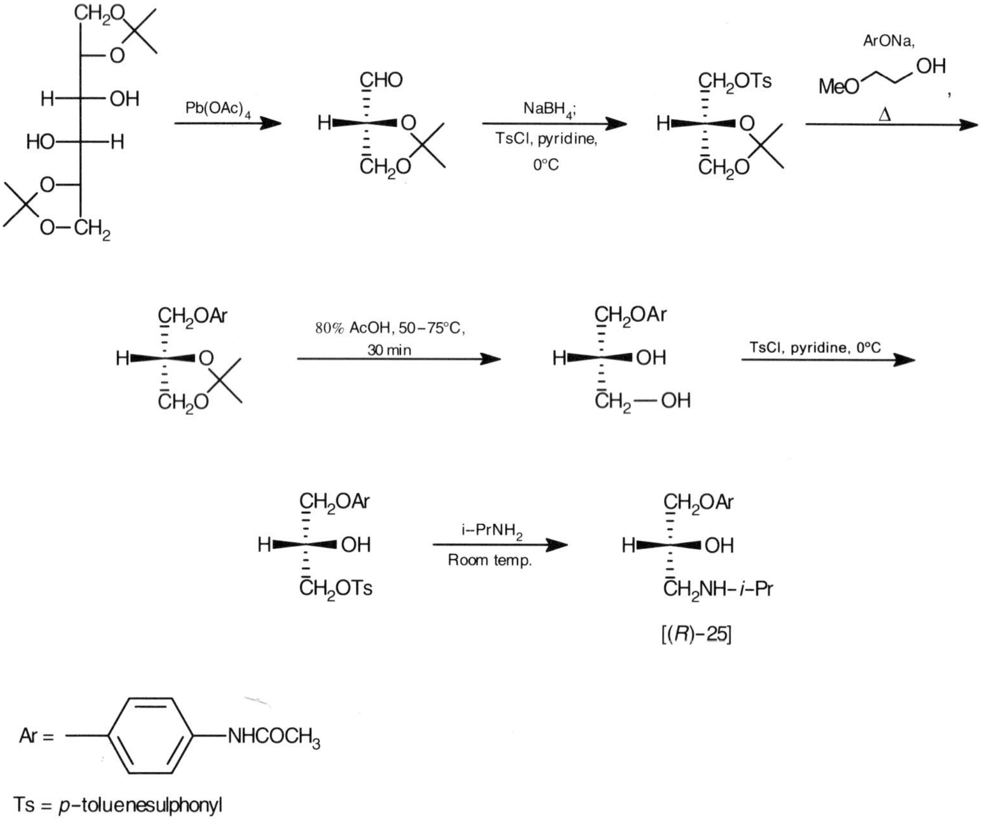

Scheme 10.2   Preparation of (R)-practolol.

$Ar = $ —(benzene ring)— $NHCOCH_3$

Ts = p-toluenesulphonyl

the discovery of a selective enzymic oxidation process which oxidizes only one isomer from racemic glycerol acetonide. This allows access to the more active S-enantiomer of the β-blocker (Scheme 10.3). The R-enantiomer may be made from the other enantiomer of glycerol acetonide which is also commercially available, though more expensive.

The diol intermediate (51) may be cyclized cleanly using either triphenylphosphine/ diethylazodicarboxylate (DEAD) (Takano et al., 1983) or HBr/acetic acid followed by sodium methoxide (Golding et al., 1973), thus making laboratory-scale production of the two enantiomers a straightforward process.

Commercially, enantiomerically pure β-blockers are still made by traditional resolutions of racemic products. Intuitively it might be thought that chiral epichlorohydrin could be used to give optically pure products. This is not the case, however, as it can react with nucleophiles by two mechanisms, direct substitution at carbon 1 to give (52) and attack at carbon 3 with subsequent ring closure to give the opposite enantiomer (53) of the product. When X is Cl both pathways operate, approximately 90% by attack at C-3, giving partial racemization. With X as o-toluene p-sulphonyl, 95% of the attack occurs at C-1, and with X as $OSO_2 CF_3$ (trifluoromethanesulphonate; triflate) (McClure et al., 1979) or O-SO$_2$-m-nitrophenyl (Klunder et al., 1989) this increases to 100%. The nitrobenzenesulphonate is preferred as it is a stable solid which may be crystallized to high purity.

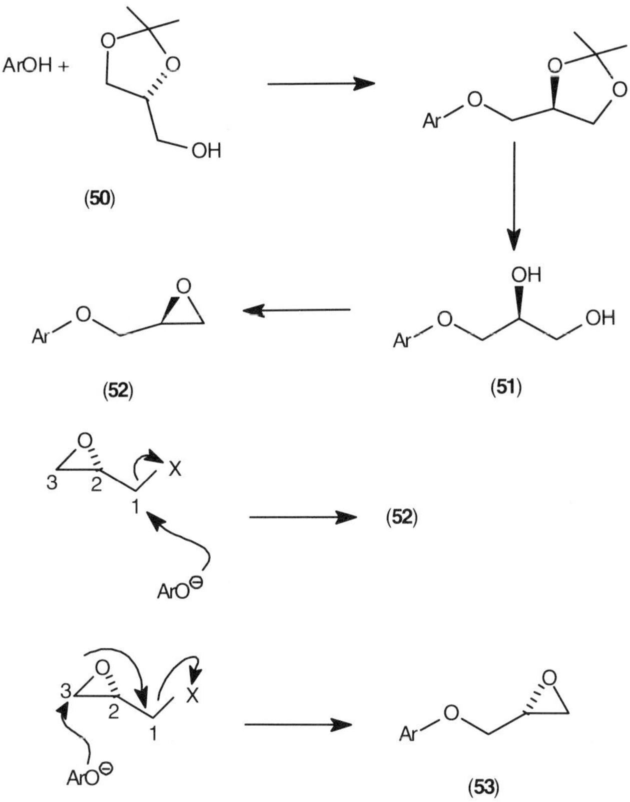

Scheme 10.3 Synthesis of (S)-practolol.

XIII. Clinical Effects of β-Blockers

Black's original postulate that β-blockers would alleviate angina and control catecholamine-elicited arrhythmias was rapidly justified. Early observations that these agents would lower blood pressure were very unexpected, and in fact hypertension is now the main area of use. β-Blockers are used in a number of disease states in which raised sympathetic amine levels are implicated. A discussion of these states follows.

Thyrotoxicosis. An enlargement of the thyroid gland (due to iodine deficiency) causing palpitations and tremor due to overproduction of catecholamines.

Phaeochromocytoma. A tumour of the adrenal gland causing overproduction of adrenaline. This condition is normally treated with β-blockers in combination with an α-blocker.

Tremor. Physiological tremor of the skeletal muscles may be due to the effects of an overactive sympathetic nervous system. It is readily treated by β-blockers (β_2-response). Perhaps related (though totally unexpected) is the ability of β-blockers to alleviate the trauma of alcohol and drug withdrawal, and also relieve the stress associated with examinations, race car driving, snooker and orchestral playing, public speaking, etc., all of which are due to an inappropriate level of sympathetic activity.

Other indications for β-blockers where there is less obvious adrenergic involvement are as follows.

Glaucoma. An eye disease characterized by an increase in intraocular pressure which can cause headache, visual disturbances and eventually blindness. It is effectively controlled by some β-blockers, e.g. timolol (**16**) (the compound which established this therapy) and more recently betaxolol (**29**). The exact mechanism by which β-blockers control this disease is not yet known.

Anxiety. Propranolol (**8**) is active in this condition, presumably because the peripheral (cardiac, tremor) symptoms are known to be mediated by the sympathetic nervous system. Whether propranolol has an action in the central nervous system as well is still a matter of debate.

Migraine. Propranolol is used in the prophylaxis of migraine. The mode of action is not known, and the type of receptor has not yet been identified in this important area.

Myocardial Infarction ('Heart Attack'). Trials have shown that some β-blockers are effective in preventing re-infarction and death in patients who have suffered heart attacks (Anon, 1986). The drugs have proved to be efficacious when administered within the first 12 h following the onset of chest pain, or given prophylactically following a heart attack (Yusuf *et al.*, 1985). β-Blockers may become the first line of treatment for the victims of heart attack, and compounds used in this indication include atenolol (**27**), propranolol (**8**), timolol (**16**) and acebutolol (**26**).

XIV.   Side Effects of β-Blockers

In common with all other classes of drugs, β-blockers have side effects, the most important of which are as follows.

Bronchoconstriction in Asthmatics. Asthmatic patients require continuous beta ($β_2$) stimulation to keep their bronchi dilated. β-Blockers, by preventing this bronchodilation, can cause (potentially fatal) bronchoconstriction in some sensitive patients.

Tiredness of the Limbs. This is due to the reduction in cardiac output and is common to all β-blockers.

Effects on the Central Nervous System. Among the side effects which have been observed with β-blockers are dizziness, vivid dreams and sedation. These are more common in lipophilic β-blockers, i.e. those that have a high partition coefficient [propranolol (**8**), pindolol (**14**) and oxprenolol (**13**) etc.].

Heart Failure/Bradycardia. All β-blockers produce a fall in resting heart rate (bradycardia) due to blockade of on-going sympathetic activity. Patients on the verge of heart failure are very dependent on the sympathetic nervous system, hence β-blockade may well push them into overt heart failure. Some authorities feel that drugs having isa will cause fewer problems as less sympathetic activity will be removed.

XV Conclusion

The concept of β-adrenergic blockade was pioneered in the early 1960s. From modest beginnings in the treatment of angina pectoris, the clinical utility of this class of compound has progressed steadily to the present, where diverse disease states such as hypertension, anxiety and migraine are routinely treated with β-blockers. β-Blockers have made a major impact on the treatment of cardiovascular disease since their discovery in the early sixties. Estimated world sales of β-blockers exceed £1000 million per annum, nearly two thousand patents have been filed, and approximately 20 major products are marketed. A figure of 100 000 might be a reasonable estimate of the number of compounds synthesized to get to the present position in this field of research.

References

Ahlquist, R.P. (1948). *Am. J. Physiol.* **153**, 586.

Anon. (1986). *Lancet* (July 12th), 79.

Augstein, J., Cox, D.A., Ham, A.L., Leeming, P.R. and Sanrey, M. (1973). *J. Med. Chem.* **16**, 1245.

Barlow, J.J., Main, B.G. and Snow, H.M. (1981). *J. Med. Chem.* **24**, 315.

Barrett, A.M., Carter, J., Hull, R. and Le Count, D.J. (1973). *Br. J. Pharmacol.* **48**, 340P.

Black, J.W. and Stephenson, J.S. (1962). *Lancet* **2**, 311.

Carlsson, E., Ablad, B. and Ek, L. (1973). *Life Sci.* **12**, 107.

Carlsson, E., Dahloff, C-G. and Hedberg, A. (1977). *Naunyn-Schmiedebergs Arch. Pharmacol.* **300**, 101.

Crowther, A.F. (1990). *Drug Design and Delivery* **6**, 149.

Crowther, A.F. and Smith, L.H. (1968). *J. Med. Chem.* **11**, 1009.

Crowther, A.F., Howe, R. and Smith, L.H. (1971). *J. Med. Chem.* **14**, 511.

Danilewicz, J.C. and Kemp, J.E.G. (1973). *J. Med. Chem.* **16**, 168.

Dixon, R.A.F., Candelore, M.R., Strader, C.D. and Tota, M.R. (1991). *TIPS* **12**, 4.

Fraser, C.M. and Venter, J.C. (1990). *Annu. Rev. Respir. Dis.* **141**, S22.

German and Austrian Study Group (1988). *Lancet* **1** (8584), 489.

Golding, B.T., Hall, D.R. and Sakritar, S. (1973). *J. Chem. Soc.*, 1214.

Hellenbrecht, D., Grobecker, H. and Muller, K-F. (1974). *Eur. J. Pharmacol.* **29**, 223.

Hoefle, M.L., Hastings, S.G., Meyer, R.F., Corey, R.M., Holmes, A. and Stratton, C.D. (1975). *J. Med. Chem.* **18**, 148.

Kaiser, C., Bower, W.D., Colella, D.F., Garvey, E., Jen, T. and Wardell, J.R. (1977). *J. Med. Chem.* **20**, 687.

Klunder, J.M., Onami, T. and Sharpless, K.B. (1989). *J. Org. Chem.* **54**, 1295.

Konzett, H. (1940). *Arch. Exp. Pathol. Pharmakol.* **197**, 27.

Lands, A.M., Arnold, J.P., Brown, T.G., Luduena, F.P. and McAuliff, J.P. (1967). *Nature (London)* **214**, 597.

Main, B.G. (1982). *J. Chem. Technol. Biotechnol.* **32**, 617.

McClure, D.E., Arison, B.H. and Baldwin, J.J. (1979). *J. Am. Chem. Soc.* **101**, 3666.

Moran, N.C. and Perkins, M.E. (1958). *J. Pharmacol. Exp. Ther.* **124**, 223.

Muzzin, P., Colomb, C., Giacobino, J-P., Venter, J.C. and Frazer, C.M. (1988). *J. Receptor Res.* **8**, 713.

Nahorski, S.R., Dickinson, K. and Richardson, A. (1981). *Mol. Pharmacol.* **19**, 194.

Nogrady, T. (1985). 'Medicinal Chemistry, A Biochemical Approach', Oxford University Press, p. 64.

Petrov, V., Stephenson, O. and Thomas, A.J. (1956). *J. Pharm. Pharmacol.* **8**, 666.

Phillips, D.K. (1980). *Handb. Exp. Pharmacol.* **54**, 3.

Powell, C.E. and Slater, I.H. (1958). *J. Pharmacol. Exp. Ther.* **122**, 480.

Richards, W.G., Clarkson, R. and Ganellin, C.R. (1975). *Philos. Trans. Roy. Soc. (London), Ser. B.* **272**, 75.

Smith, L.H. (1978). *J. Appl. Chem. Biotechnol.* **28**, 201.

Smith, L.H. and Dukes, M. (1971). *J. Med. Chem.* **14**, 326.

Smith, L.H. and Tucker, H. (1977). *J. Med. Chem.* **20**, 1653.

Smith, L.H. and Large, M.S. (1982). *J. Med. Chem.* **25**, 1286.

Somerville, A.R., Coleman, A.J. and Paterson, D.S. (1979). *Biochem. Pharmacol.* **28**, 1011.

Strosberg, A.D., Emorine, L.J., Marullo, S., Briend-Sutren, M-M., Patey, G., Tate, K. and Delavier-Klutchko, C. (1989). *Science* **245**, 1118.

Takano, S., Seya, K., Goto, E., Hirama, M. and Ogasawara, K. (1983). *Synthesis*, 116.

Tucker, H. and Coope, J.F. (1978). *J. Med. Chem.* **21**, 769.

Wilson, C., Wilson, S., Piercy, V., Sennitt, M.V. and Arch, J.R.S. (1988). *Eur. J. Pharmacol.* **100**, 309.

Yusuf, S., Peto, R., Lewis, J., Collins, R. and Sleight, P. (1985). *Prog. Cardiovasc. Dis.* **27**, 335.

Reviews

Overall

Main, B.G. (1990). *In* 'Comprehensive Medical Chemistry' (C. Hansch, P.G. Sammes, J.B. Taylor, eds), Vol. 3, p. 187, Pergamon Press, Oxford.

Pharmacological Aspects

Barrett, A.M. (1972). *In* 'Drug Design' (E.J. Ariens, ed.), Vol. III, p. 205,. Academic Press, New York, London.

Clinical Aspects

Weetman, D.F. (1977). *Drugs Today* **13**, 261.

Lees, G.M. *Br. Med. J.* **283**, 173.

Structure–Activity Relationships

See Phillips (1980).

Biochemical Aspects

Levitski, A. (1981). *In* 'Topics in Molecular Pharmacology' (A.S.V. Burger and G.C.K. Roberts, eds) vol. 1, p. 23. Elsevier, Amsterdam.

Triggle, D.J. (1981). *In* 'Burger's Medicinal Chemistry', 4th edn (M.E. Wolff, ed.), Part III, p. 225, Wiley, New York, 1981.

–11–

Salbutamol: A Selective β_2-Stimulant Bronchodilator

L.H.C. LUNTS

Glaxo Group Research Ltd,
Ware, U.K.

Dr Lawrence Lunts graduated from the University of Sheffield and obtained a Ph.D. in chemistry for work connected with the steroidal alkaloid conessine. In October 1955 he joined Allen and Hanburys Ltd as a research chemist and has worked in several areas including parasitology, antibacterials, antiinflammatories and cardiovascular. He remained with the company (which merged with Glaxo Group), becoming a Senior Research Leader, until retiring in June 1991.

I. Introduction and Biological Background

Bronchial asthma is a widespread disease from which about 5% of Western society will suffer at some time in their lives (Howarth and George, 1983). It affects about one in 25 children and kills at least 1500 people a year in the United Kingdom (Colmer and Gray, 1983). The aetiology of the disease is often of allergic origin but its causes are complicated and not completely understood.

Bronchial asthma is characterized by breathlessness and wheezing; in the early stages of the disease this is due to constriction of respiratory smooth muscle (bronchoconstriction) which is easily reversed by administration of a bronchodilator. In more severe cases, physical obstruction of the airways also occurs due to inflammation of the bronchial mucosal cells and by the formation of a viscous bronchial secretion. This condition is not reversible by bronchodilators, and glucocorticoid steroids are the only

MEDICINAL CHEMISTRY 2nd Edition
ISBN 0-12-274120-X

drugs that will reverse the inflammatory process and re-establish the response to bronchodilators.

Many natural mediators have been implicated in causing bronchoconstriction. Some are stored inside specialized cells, called mast cells, which, in allergic individuals, rupture in response to a signal at the cell surface and initiate an asthmatic attack.

Blocking the actions of the released mediators is a potential method for treatment but this requires knowledge of which mediators are the most significant and the availability of suitable antagonists. Although histamine is one of the substances released from the mast cell and causes bronchoconstriction, antihistamines (H_1-antagonists and H_2-antagonists) are of little value in the treatment of asthma.

Nature's own method of bronchodilation utilizes the hormone adrenaline (2) which is released from the adrenal gland and mediates its actions through stimulation of the enzyme adenylate cyclase. This enzyme converts adenosine triphosphate to cyclic adenosine 3′,5′-monophosphate (cyclic AMP) which is responsible for the relaxation of smooth muscle.

However, while adrenaline (2) is effective in relaxing bronchial muscle, it also causes unwanted stimulation of the heart and elevation of blood pressure.

(1) R = H   Noradrenaline
(2) R = Me   Adrenaline
(3) R = i-Pr Isoprenaline
(4) R = t-Bu

It was clearly desirable, therefore, to find a compound that would reverse the constriction of the airways without causing cardiac or other side effects. It should also be effective given by mouth or inhaled and have a long duration of action.

An important step forward was the classification of adrenoceptors into α and β subtypes by Ahlquist (1948). This was based on their sensitivities to noradrenaline and adrenaline and some close analogues of these natural stimulants. α-Adrenoceptors were defined as those very sensitive to adrenaline and insensitive to isoprenaline and β-receptors as those most sensitive to isoprenaline and least sensitive to noradrenaline (1) (Table 11.1).

Isoprenaline (3) largely replaced adrenaline as a bronchodilator because of its greater selectivity of action on β-receptors but it still had substantial disadvantages. It has marked cardiovascular side effects causing an increase in both the force and rate of contraction of the heart and could only be used by the inhaled route.

Lands and his colleagues (1967) were responsible for defining that further selectivity for bronchial smooth muscle was possible using techniques similar to those employed by Ahlquist. They found that catecholamines with a relatively large N-alkyl substituent (4) in the side chain were more active at some β-receptors than others. On the basis of the observed order of activity, they subclassified β-receptors into β_1 and β_2 types. Lands's work indicated that a selective β_2-stimulant given by inhalation or by mouth would be an excellent bronchodilator free from cardiovascular side effects (Table 11.2).

TABLE 11.1
Some Properties of α- and β-Adrenoceptor Stimulants

Tissue	α-Effect	β-Effect
Bronchus	NSE[a]	Relaxation
Uterus	NSE	Relaxation
Blood vessel	Constriction	Dilatation
Skeletal muscle	NSE	Decrease in duration of muscle twitch
Heart muscle	NSE	Increase in force and rate
Alimentary tract	Constriction of sphincters	Relaxation

[a] NSE, No significant effects.

Isoetharine (5), which was distinctly more active on respiratory smooth muscle than on the heart, proved to be a useful selectively acting bronchodilator in man, but it and

(5)
Isoetharine

the *t*-butyl substituted compound (4) were short-acting compounds, which limited their therapeutic value.

TABLE 11.2
Lands Classification and Distribution of β-Adrenoceptor Subtypes

Organ	Main receptor type or subtype
Heart	β_1
Lung (bronchial smooth muscle)	β_2
Blood vessels (vascular smooth muscle)	α, β_2
Uterus	β_2
Skeletal muscle	β_2

II. Biological Test Procedures

In our laboratories, primary screening for bronchodilator activity is carried out using the guinea-pig isolated tracheal smooth muscle preparation or in anaesthetized guinea-pigs. In these tests, the abilities of new compounds to inhibit contractions of smooth muscle elicited by a variety of chemical stimuli is determined.

That the mechanism of the bronchodilation is via β-receptors is shown by blocking its effect with a β-blocking agent. To ensure that the compound has little or no activity on the heart (β_1-receptor), its ability to cause an increase in the rate of contraction of the guinea-pig isolated right atrium preparation is determined. The ratio of $\beta_2 : \beta_1$ activities gives a measure of selectivity.

To quantify the results, the *in vitro* test potency is expressed as the EC_{50} value [the concentration of drug required to cause a 50% decrease in tracheal contractions (β_2) or a 50% increase in the heart rate (β_1)]. An alternative way of expressing potency is to calculate the equipotent dose ratio relative to isoprenaline.

Further bronchodilator tests may be carried out in anaesthetized cats and these can simultaneously determine other parameters such as the effect on blood pressure and heart rate. In man, the efficacy of bronchodilators is measured by abilities to (a) increase the volume of air that can be expired in one second (FEV_1, forced expiratory volume in 1 second) (b) increase in the peak expiratory flow rate (PEFR).

III. Structure–Activity Relationships and Metabolism of Catecholamines

Adrenaline (**2**) and noradrenaline (**1**) are the natural hormones of the sympathetic nervous system. They are substituted phenylethanolamines with a catechol (1,2-benzenediol) nucleus so are referred to as catecholamines. On the above classification, noradrenaline (**1**) acts mainly as an α-adrenergic stimulant whereas adrenaline (**2**) acts both as an α- and a β-stimulant. An increase in the bulk of the substituent on the nitrogen atom, from hydrogen (noradrenaline) to isopropyl (isoprenaline), results in a loss of potency of the compound as an α-stimulant and an increase in its potency as a β-stimulant. Isoprenaline (**3**) is a powerful β-stimulant devoid of α-agonist properties. This makes it a more satisfactory bronchodilator than adrenaline since it causes no vasoconstrictor (α) side effects.

As mentioned earlier, the two catecholamines **4** and **5** have a higher selectivity than isoprenaline (**3**) for β_2 than for β_1 sites and so produce less unwanted cardiac stimulant actions.

The isopropyl or *t*-butyl substituent of β-stimulant catecholamine and similar amines can also be replaced by certain aralkyl groups and potency is maintained or increased. Some examples are shown in Table 11.3.

Phenylethanolamines have an asymmetric centre at the carbon atom that bears the hydroxy group. The natural hormones have the (*R*) configuration, and biological activity in all phenylethanolamines resides in these enantiomers.

The shortcomings of the catecholamines (**3–5**) are that they have a short duration of action and are inactive when given by mouth. Catecholamines are readily taken up into tissues and inactivated by enzymes such as catechol *O*-methyltransferase (COMT). This results in methylation of the *meta*-hydroxy group to give an inactive ether:

TABLE 11.3
Bronchodilator Activity of Some Substituted Catecholamines

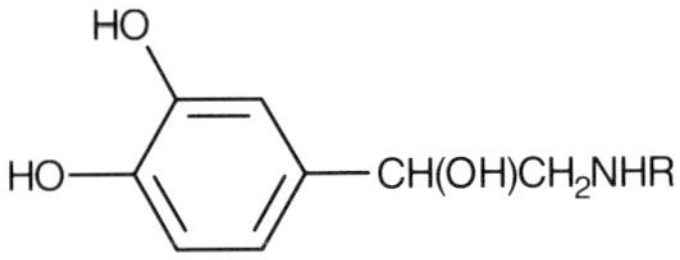

$$\text{HO} \underset{\text{HO}}{\bigcirc} \text{—CH(OH)CH}_2\text{NHR}$$

R	Activity (isoprenaline = 1)
i-Pr	1
t-Bu	1.5–2[a]
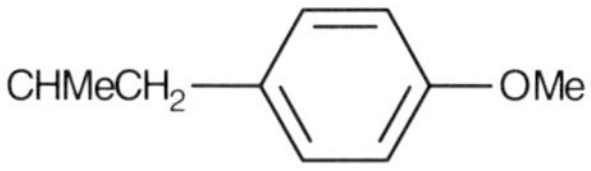 CHMeCH$_2$—(C$_6$H$_4$)—OMe	2[b]
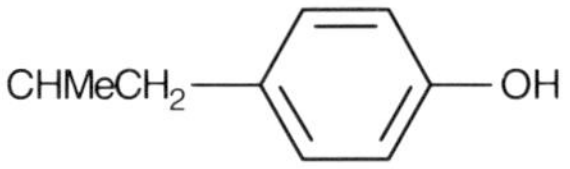 CHMeCH$_2$—(C$_6$H$_4$)—OH	8[c]

[a] Various sources (Moed et al., 1955; Lands et al., 1967).
[b] Perfused guinea-pig lung: constriction induced by histamine (Biel et al., 1954).
[c] Anaesthetized guinea-pig: constriction induced by acetylcholine (Moed et al., 1955).

IV. Chemical Approach to Finding an Alternative β-Stimulant

In order to obtain an improved bronchodilator we needed a compound that (a) was stable to metabolic inactivation and (b) was more selective for β-adrenoceptors in the lung than for those in the heart.

Our approach was to endeavour to replace the *meta*-hydroxy group of catecholamines with another group which in conjunction with the *para*-hydroxy group would possess the characteristics that were necessary for biological activity.

An analysis of the properties of this *meta*-hydroxy group identifies the following parameters that may be significant:

1. Size
2. Electronic effects on the aromatic system (resonance and inductive)
3. Capacity to form hydrogen bonds
4. Acidity
5. Ability to chelate with metals, particularly assisted by the *para*-hydroxy group
6. Capacity to form a redox system, again assisted by the *para*-hydroxy group.

Hydrogen-bonding and chelation powers were considered to be important properties because they had been invoked as the major binding features of the catecholamines at the receptor level. These properties together with acidity would be provided by a carboxylic acid group.

The synthesis of this salicylic acid derivative (11) is outlined in Scheme 11.1 (Collin et al., 1970). The ketoacid (7) was prepared by a Fries rearrangement of aspirin (6) and its

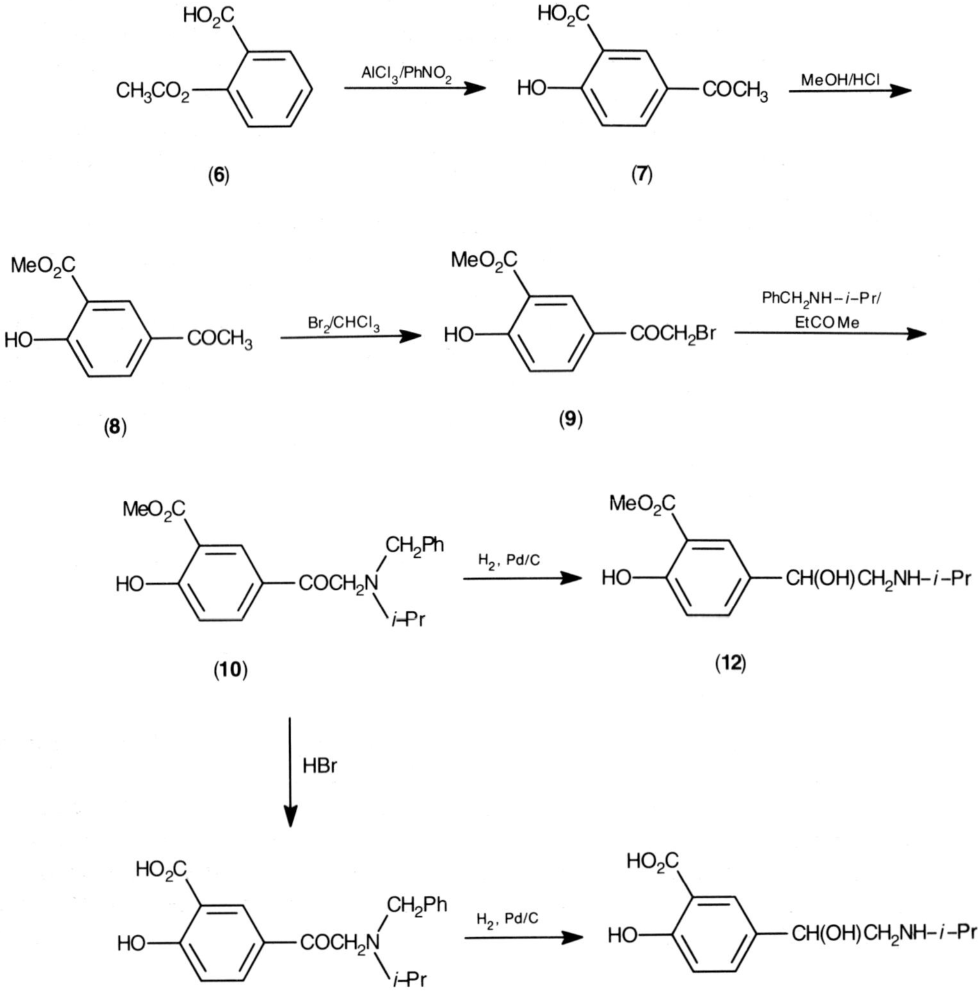

Scheme 11.1   Preparation of a carboxylic acid analogue of isoprenaline.

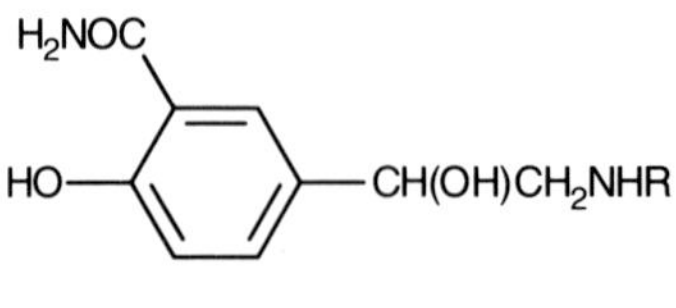

(13) R = i-Pr
(14) R = t-Bu

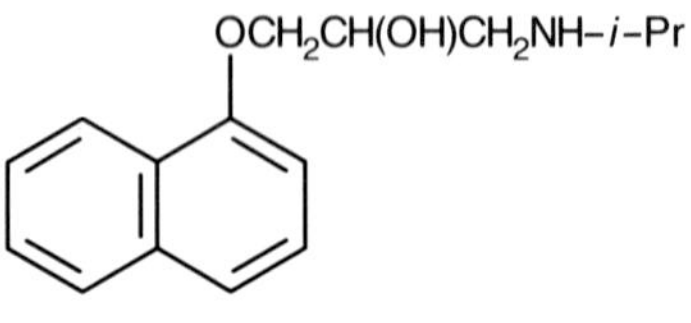

(15)

Propranolol

methyl ester (8) was converted via a bromoketone (9) into the aminoketone (10) which was demethylated with hydrobromic acid. Catalytic hydrogenation then removed the protecting benzyl group and reduced the ketone to give the salicylic acid analogue (11) of isoprenaline. Reduction before hydrolysis afforded the methyl ester (12).

If the bromoketone (9) was condensed with isopropylamine, a very poor yield of a secondary aminoketone was obtained.

The ester (12) reacted with ammonia to give a salicylamide derivative (13) (Clifton *et al.*, 1982). These three compounds 11, 12 and 13 proved not to be β-stimulants, but the ester (12) and particularly the amide (13) were β-adrenergic *antagonists*. In fact, the closely related *t*-butylamine (14) was about one-tenth as active as the standard β-blocker propranolol (15) (see Chapter 10). This result afforded some encouragement since it showed that the compounds had an affinity for the receptor, but had not the capacity to elicit a biological response.

A similar approach to finding an improved bronchodilator was carried out concomitantly by workers at Mead Johnson. They considered that the important property of the *meta* substituent was its acidity and that it would be mimicked by the methanesulphonamide group (pK_a 8.35) since it was of a comparable acidity to a phenol (pK_a 9.56); the bulky $MeSO_2$ moiety could be oriented in such a way that it would not interfere with binding to a receptor (Larsen and Lish, 1964). Their compound soterenol (16) was a long-acting, selective β-stimulant, but it does not appear to have found clinical application. The isomer (17) was inactive (Larsen *et al.*, 1967).

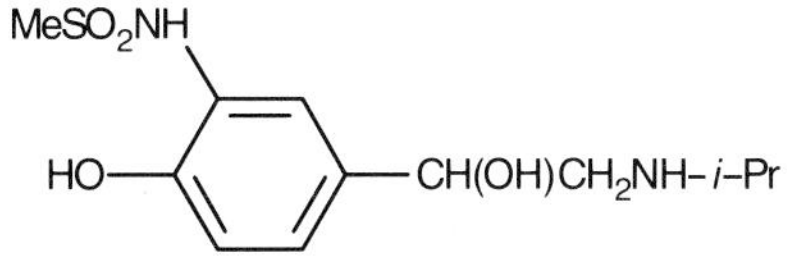

HOCH$_2$

MeSO$_2$NH

HO—⟨ ⟩—CH(OH)CH$_2$NH–i–Pr   MeSO$_2$NH—⟨ ⟩—CH(OH)CH$_2$NH–i–Pr

(16)   (17)

Soterenol

Another suitable candidate was a saligenin (2-hydroxybenzenemethanol) derivative since it retained a hydrogen-bonding and a chelating ability. Consequently the keto esters (18) were reduced by lithium aluminium hydride and debenzylated (Scheme 11.2). The products were indeed potent β-stimulants. The isopropyl compound (19) had almost half of the activity of isoprenaline as a bronchodilator in the guinea-pig (β_2) but with only one-thousandth of its activity on guinea-pig cardiac muscle (β_1). The *t*-butyl homologue (20) was even more selective [as was to be expected by comparison with the

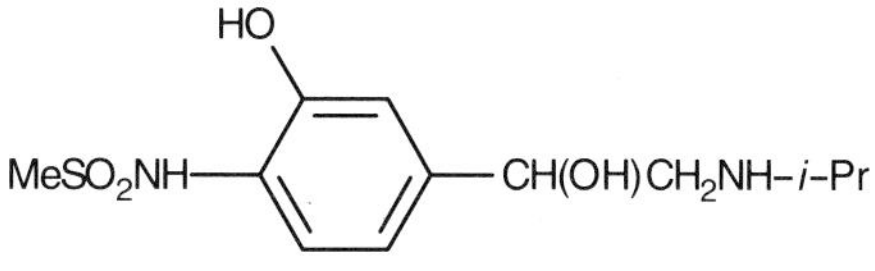

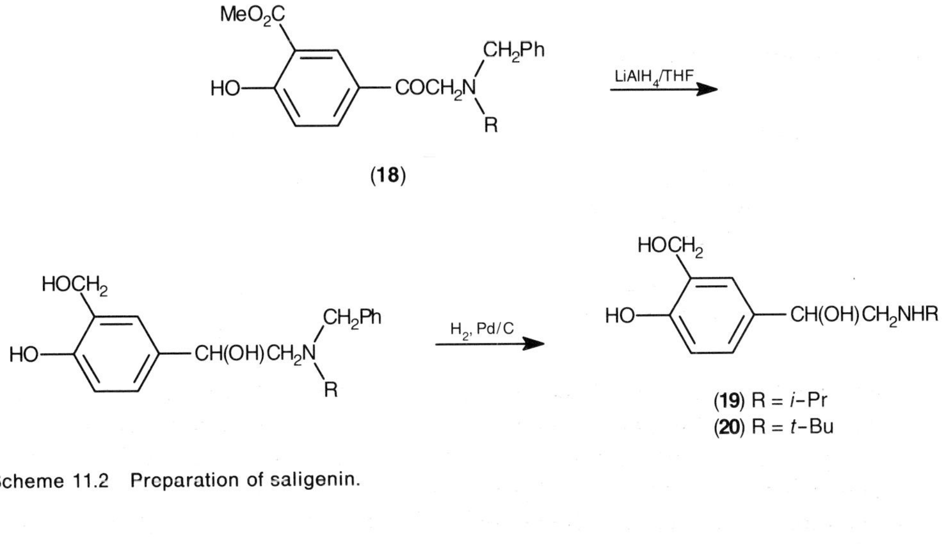

MeO$_2$C

HO—⟨ ⟩—COCH$_2$N(CH$_2$Ph)(R) $\xrightarrow{\text{LiAlH}_4/\text{THF}}$

(18)

HOCH$_2$

HO—⟨ ⟩—CH(OH)CH$_2$N(CH$_2$Ph)(R) $\xrightarrow{\text{H}_2,\ \text{Pd/C}}$

HOCH$_2$

HO—⟨ ⟩—CH(OH)CH$_2$NHR

(19) R = i–Pr
(20) R = t–Bu

Scheme 11.2 Preparation of saligenin.

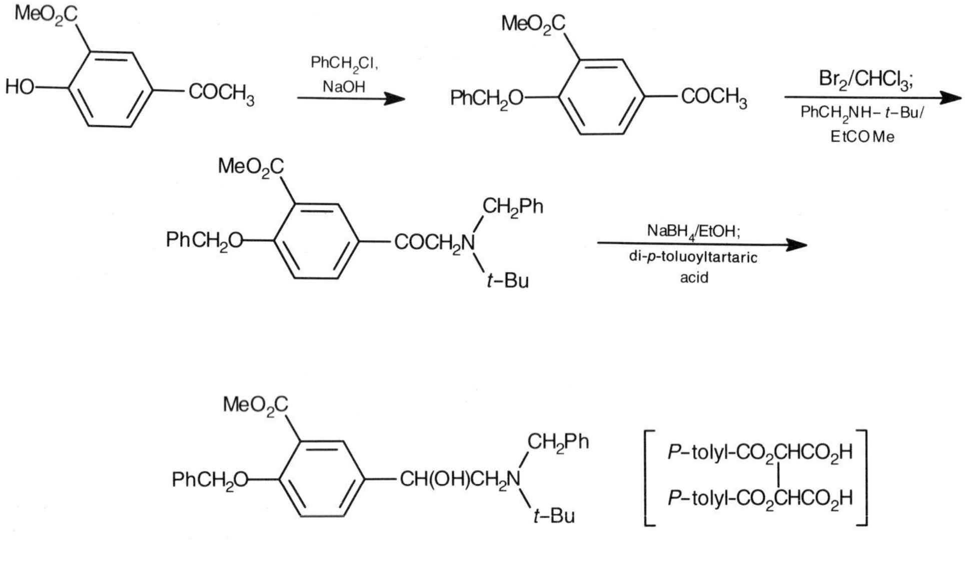

Scheme 11.13 Preparation of an intermediate ester di-*p*-toluoyl-tartrate.

catecholamine analogue (**4**)]. It was equipotent with isoprenaline on bronchial muscle but two thousand times less potent on the heart.

This compound, salbutamol, was developed as a bronchodilator drug.

A. Enantiomers of Salbutamol

A very convenient method to prepare the enantiomers of salbutamol was by resolution of an intermediate protected ester (Hartley and Middlemiss, 1971) (Scheme 11.3). Only one of the diastereoisomers of the di-*p*-toluoyltartrate salt (**21**) was insoluble in ethyl acetate. This was recrystallized twice and converted into the free base which after reduction with lithium aluminium hydride and catalytic debenzylation gave enantiomerically pure salbutamol. By using (−)-di-*p*-toluoyltartaric acid, (*R*)-(−)-salbutamol was obtained; the other enantiomer was formed similarly after resolving the original ester as its (+)-di-*p*-toluoyltartrate.

The absolute configuration of (*R*)-(−)-salbutamol (**20a**) was determined by comparison of its circular dichroism (CD) spectrum with that of (*R*)-(−)-octopamine (**22**). They both showed a negative Cotton effect at 276–280 nm.

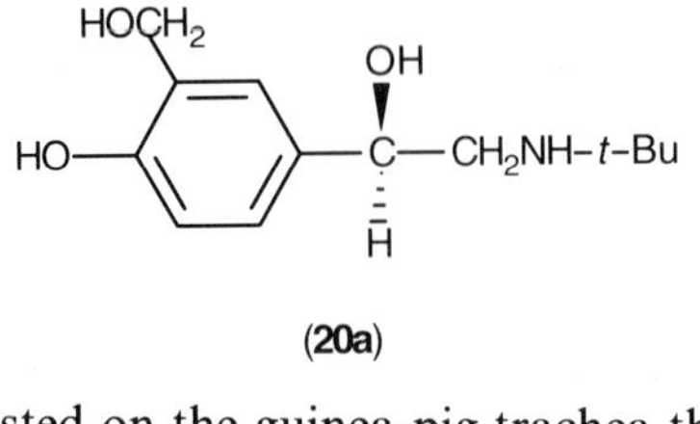

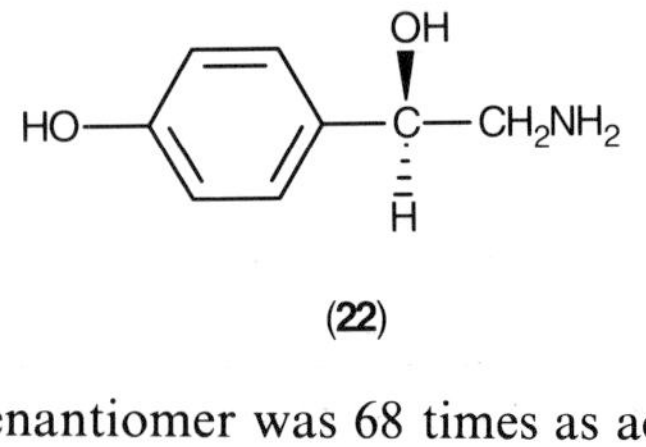

When tested on the guinea-pig trachea this enantiomer was 68 times as active as the (*S*)-(+)-isomer as a β_2-stimulant, as was to be expected by analogy with other phenylethanolamines which have been referred to already (Brittain *et al.*, 1973).

V. Structure–Activity Relationships between Salbutamol and Congeners

A. Other Saligenin Derivatives

Salbutamol analogues with substituents on nitrogen similar to those on active catecholamines described in Section III had high potency and also β_2-selectivity (Table 11.4). One of these, the 4-methoxy-α-methylphenethyl derivative (salmefamol) (27), was one and a half times as active as salbutamol. It is a clinically effective bronchodilator with a duration of action of up to 6 h.

Chemically these analogues are usually prepared by methods using reductive alkylation of the primary amine with an aldehyde or a ketone. There are several ways of accomplishing this reaction as illustrated in Scheme 11.4 for the preparation of salmefamol (27).

The intermediate esters could be prepared from a primary phenylethanolamine (25) which itself was derived by hydrogenolysis of a dibenzylamino precursor (24). It was more convenient to use the latter compound directly, when concomitant debenzylation and reductive alkylation can occur, so that the reaction may proceed via a tertiary benzylamine (28). The reaction sequence could be further shortened by starting from the dibenzylamino ketone (23).

Esters were then reduced by lithium aluminium hydride as before. Of course the reductive alkylations can be undertaken with saligenin precursors instead of esters.

TABLE 11.4

Bronchodilator and Cardiac Stimulating Activity of Some Substituted Saligenin Ethanolamines[a]

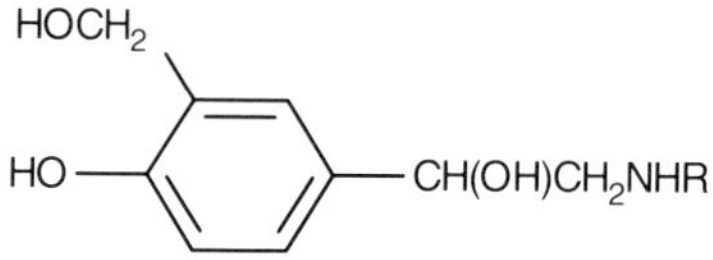

HOCH$_2$

HO ——⟨benzene ring⟩—— CH(OH)CH$_2$NHR

	Activity (isoprenaline = 1)	
R	Bronchodilator[b]	Cardiac stimulant[c]
i-Pr	0.4	0.001
t-Bu	1	0.005
Salbutamol (20)		
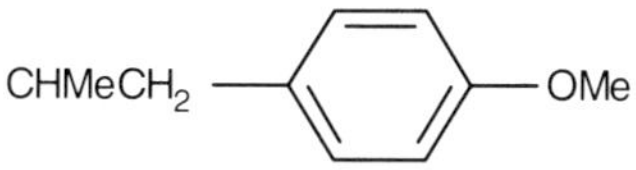 CHMeCH$_2$—⟨benzene ring⟩—OMe	1.5	0.00075
Salmefamol (27)		
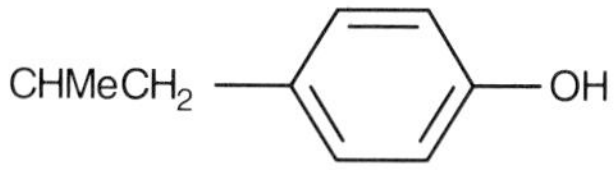 CHMeCH$_2$—⟨benzene ring⟩—OH	3	0.0005

[a] Adapted from Howarth and George (1983).
[b] Anaesthetized guinea-pig (Konzett–Rössler preparation).
[c] Guinea-pig isolated atria.

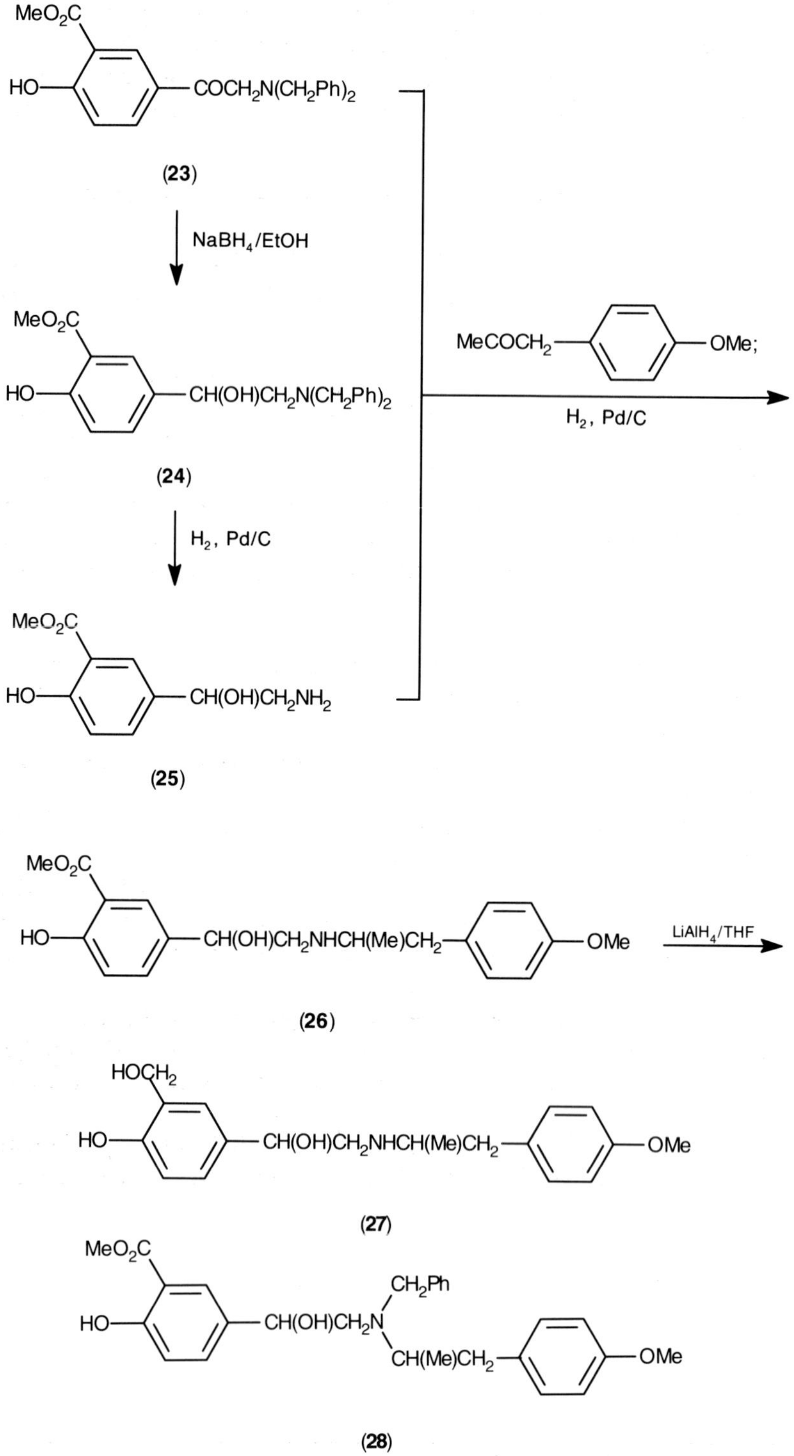

Scheme 11.4 Preparation of salmefamol.

The more active of these aralkyl phenylethanolamines were prepared from methyl ketones and thus contain another asymmetric carbon atom (which bears the methyl group). In these instances the products exist as a mixture of four diastereoisomers. Each isomer of salmefamol (27) has been tested, and β-stimulant activity resides mainly in the (R,R)-enantiomer.*

B. Variants of the Saligenin Moiety of Salbutamol

When the activity of salbutamol had been established it was necessary to prepare other analogues and derivatives to gain an insight into the influence of structure on activity.

Modification of the Hydroxymethyl Group

As the catecholamines and their saligenin analogues had similar β_2-stimulant activity it was likely that they interacted in a similar fashion at the β_2-adrenoceptor. In contrast, only the catecholamines had potency at β_1-adrenoceptors. To investigate the sensitivity of these interactions and effects on $\beta_1:\beta_2$ selectivity we introduced steric hindrance at the benzylic carbon atoms and extended the methylene group to move the alcohol group further from the ring. We also put a second hydroxy group into the *meta* side chain.

An aldehyde (29) was obtainable by oxidation of salbutamol with manganese dioxide and this or a protected derivative was a convenient source of a secondary alcohol (30), the hydroxypropyl compound (31) and a 1,3-diol (32) (Schemes, 11.5, 11.6, 11.7 and 11.8).

Grignard reactions on ester derivatives afforded tertiary alcohols (33) and (34) (Schemes 11.9 and 11.10).

The hydroxyethyl homologue (35) of salbutamol was built up from phenol, protected as a tetrahydropyranyl ether (Scheme 11.11).

The most interesting compound from this selection was the hydroxyethyl derivative (35) which was very potent and selective as a β_2-adrenergic stimulant. We found the hydroxypropyl homologue (31) to be much less active as were the other compounds described in these schemes (Brittain *et al.*, 1976).

Therefore it seems that the methylene of the hydroxymethyl group of salbutamol cannot be substituted without causing significant loss of potency. Even polyfunctional groups such as diols cannot replace the hydroxymethyl substituent. Only homologation to a two-carbon chain afforded a very active stimulant.

Neither the phenolic ether (36) nor the methyl phenol (37) had any β-stimulant activity (Colin *et al.*, 1970). Their syntheses are shown in Schemes 11.12 and 11.13.

---

*The first assignment of configuration refers to the carbon atom which bears the hydroxy group; the second is that of the carbon atom with the methyl group.

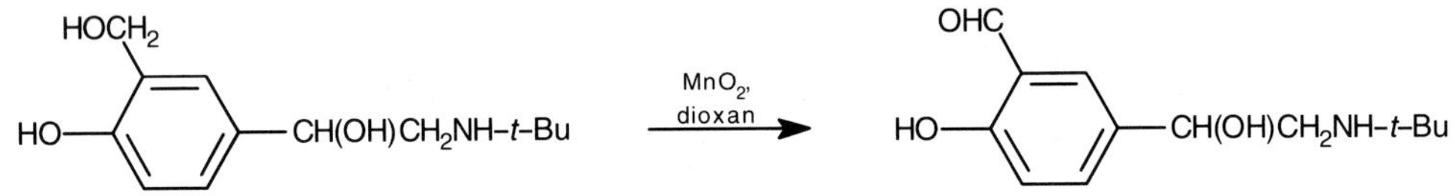

HOCH$_2$—[ring, HO]—CH(OH)CH$_2$NH–t–Bu $\xrightarrow[\text{dioxan}]{\text{MnO}_2,}$ OHC—[ring, HO]—CH(OH)CH$_2$NH–t–Bu

(29)

Scheme 11.5

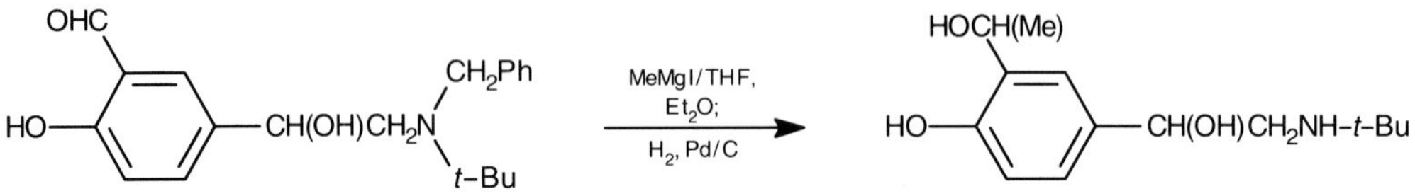

OHC—[ring, HO]—CH(OH)CH$_2$N(CH$_2$Ph)(t–Bu) $\xrightarrow[\text{H}_2,\,\text{Pd/C}]{\substack{\text{MeMgI/THF,} \\ \text{Et}_2\text{O};}}$ HOCH(Me)—[ring, HO]—CH(OH)CH$_2$NH–t–Bu

(30)

Scheme 11.6

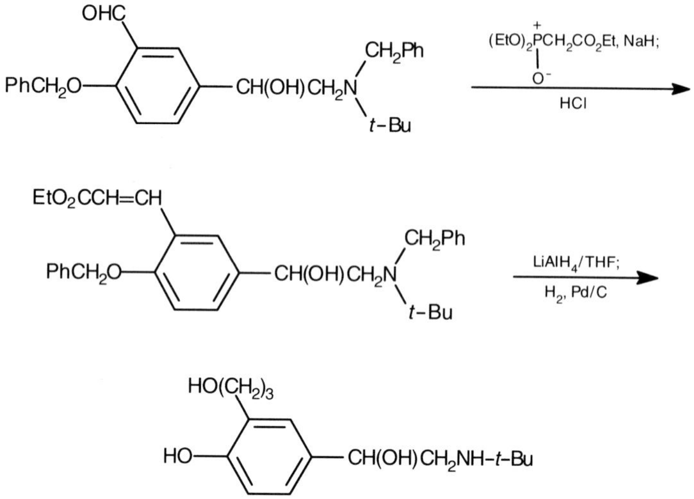

OHC—[ring, PhCH$_2$O]—CH(OH)CH$_2$N(CH$_2$Ph)(t–Bu) $\xrightarrow[\text{HCl}]{\overset{+}{(\text{EtO})_2\text{PCH}_2\text{CO}_2\text{Et, NaH};} \; \text{O}^-}$

EtO$_2$CCH=CH—[ring, PhCH$_2$O]—CH(OH)CH$_2$N(CH$_2$Ph)(t–Bu) $\xrightarrow[\text{H}_2,\,\text{Pd/C}]{\text{LiAlH}_4/\text{THF};}$

HO(CH$_2$)$_3$—[ring, HO]—CH(OH)CH$_2$NH–t–Bu

(31)

Scheme 11.7

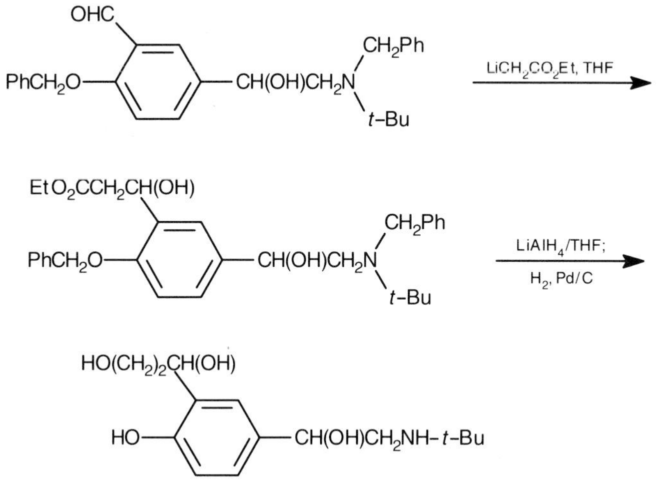

OHC

PhCH$_2$O—⟨ring⟩—CH(OH)CH$_2$N(CH$_2$Ph)(t-Bu) $\xrightarrow{\text{LiCH}_2\text{CO}_2\text{Et, THF}}$

EtO$_2$CCH$_2$CH(OH)

PhCH$_2$O—⟨ring⟩—CH(OH)CH$_2$N(CH$_2$Ph)(t-Bu) $\xrightarrow[\text{H}_2, \text{Pd/C}]{\text{LiAlH}_4/\text{THF};}$

HO(CH$_2$)$_2$CH(OH)

HO—⟨ring⟩—CH(OH)CH$_2$NH–t–Bu

(**32**)

Scheme 11.8

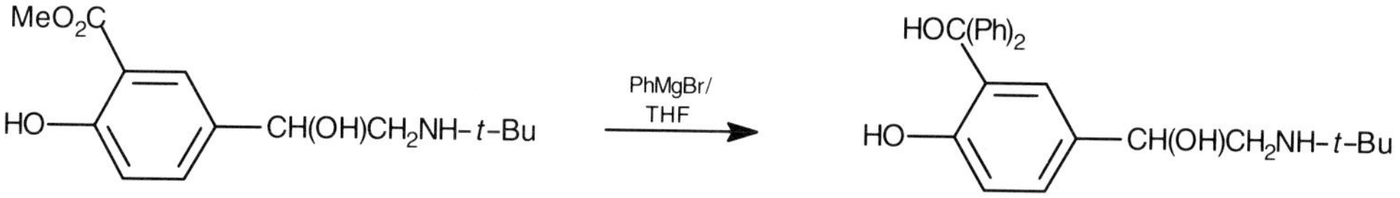

MeO$_2$C

HO—⟨ring⟩—CH(OH)CH$_2$NH–t–Bu $\xrightarrow[\text{THF}]{\text{PhMgBr/}}$ HOC(Ph)$_2$ HO—⟨ring⟩—CH(OH)CH$_2$NH–t–Bu

(**33**)

Scheme 11.9

MeO$_2$C

PhCH$_2$O—⟨ring⟩—CH(OH)CH$_2$N(CH$_2$Ph)(i-Pr) $\xrightarrow[\text{H}_2, \text{Pd/C}]{\text{MeMgBr/} \text{Et}_2\text{O};}$ HOC(Me)$_2$ HO—⟨ring⟩—CH(OH)CH$_2$NH–i–Pr

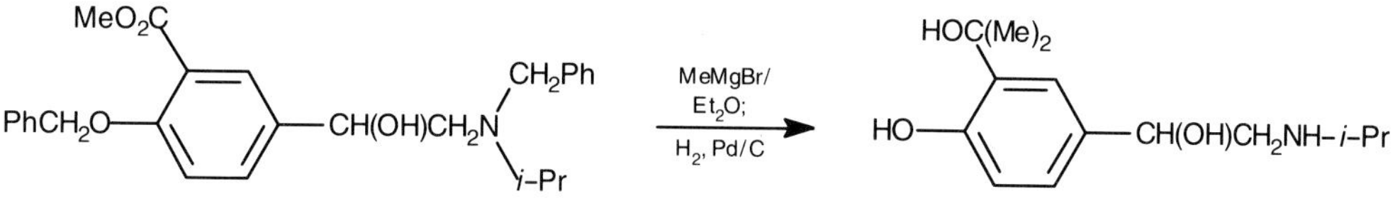

(**34**)

Scheme 11.10

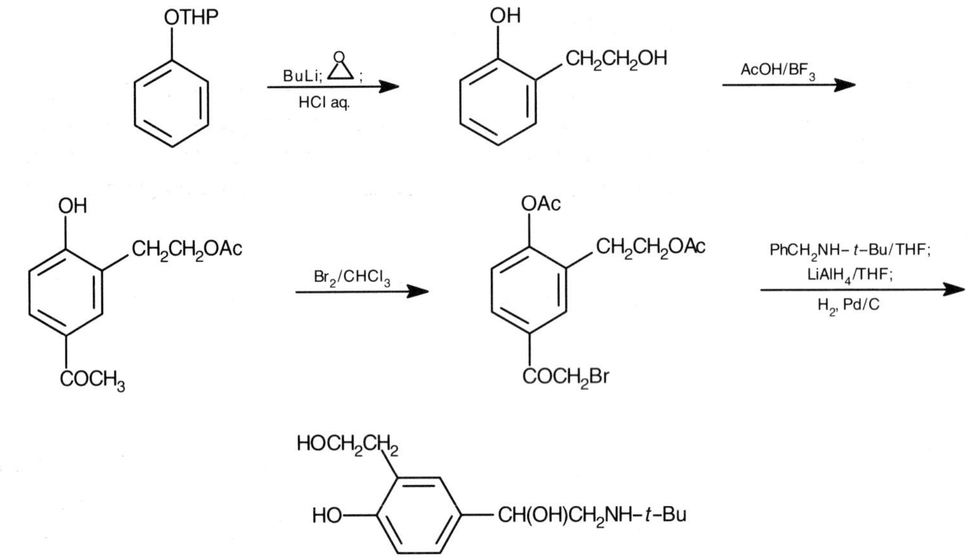

(35)

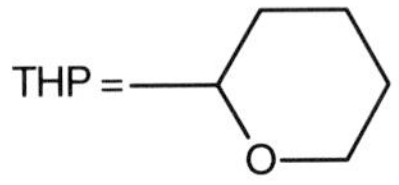

THP =

Scheme 11.11

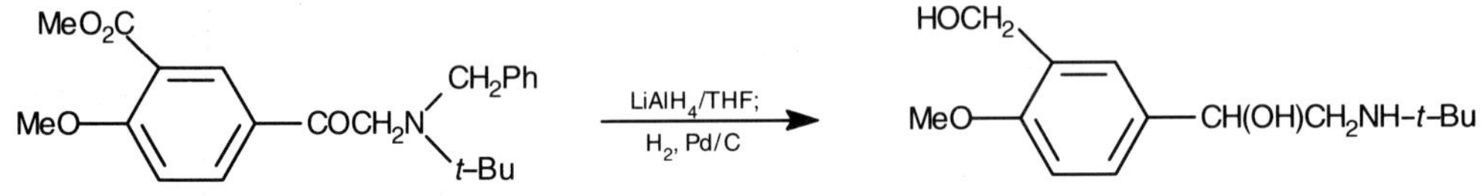

(36)

Scheme 11.12

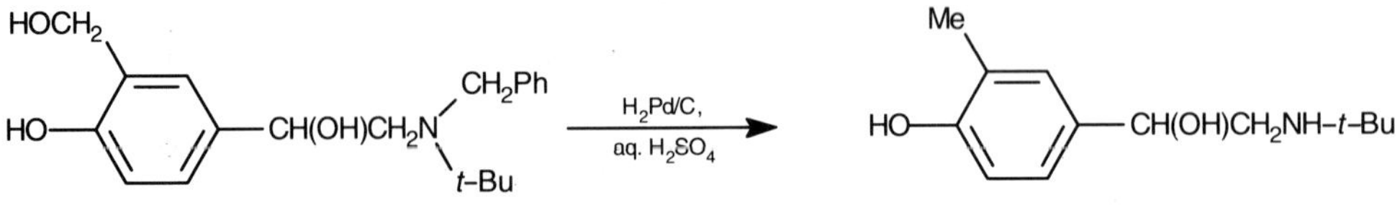

(37)

Scheme 11.13

C.   Salicylamine Derivatives

Reduction of derivatives of the amide mentioned in Section IV gave salicylamine derivatives (Scheme 11.14). This replacement of one of the hydroxy groups of salbutamol by an amino group afforded an inactive compound (**38**), but when it was converted into a variety of amides it yielded some very active and selective β_2-stimulants (Table 11.5).

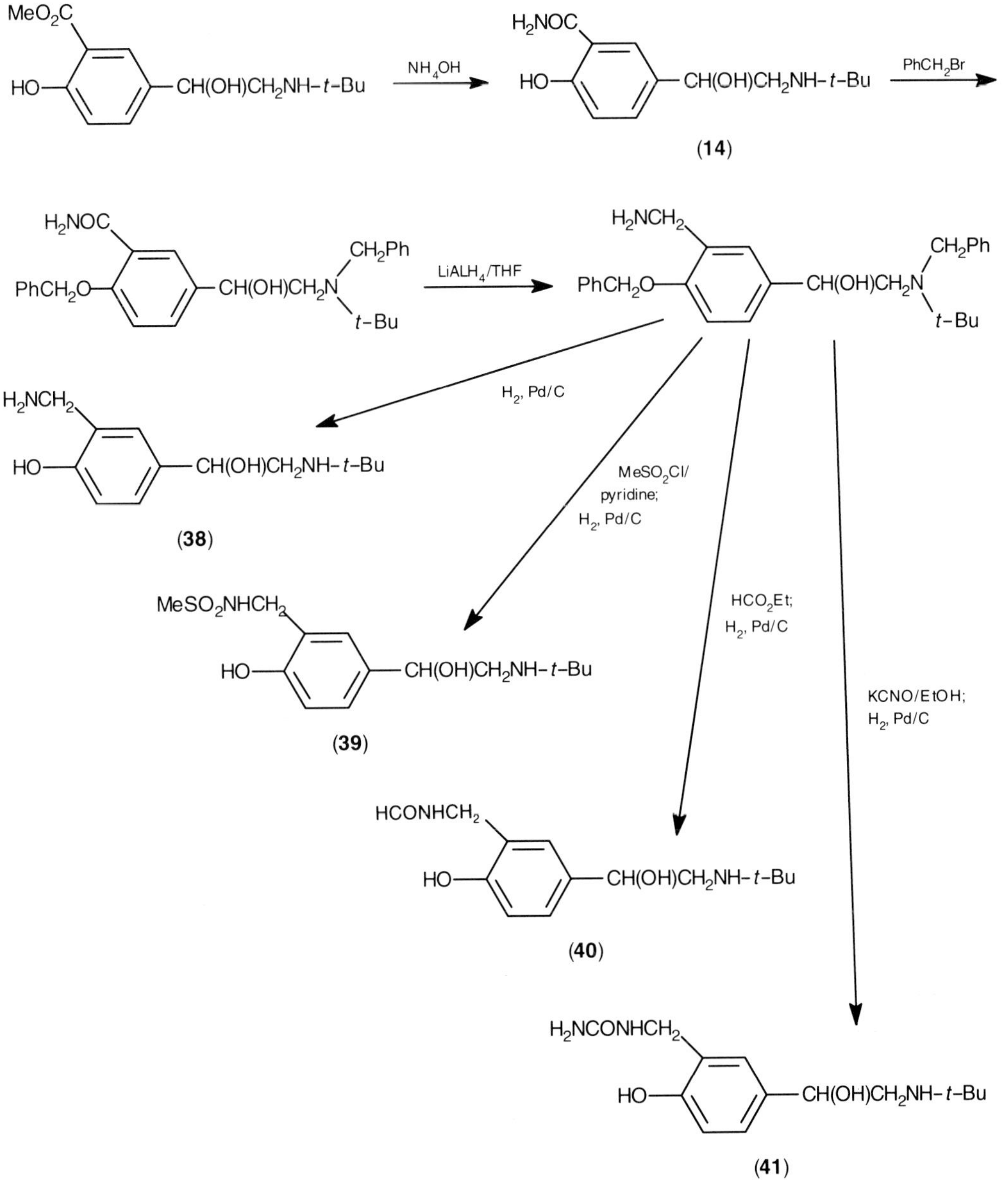

Scheme 11.14   Preparation of salicylamine analogues and derivatives.

TABLE 11.5

Bronchodilator and Cardiac Stimulating Activity of Some Salicylamine Derivatives[a]

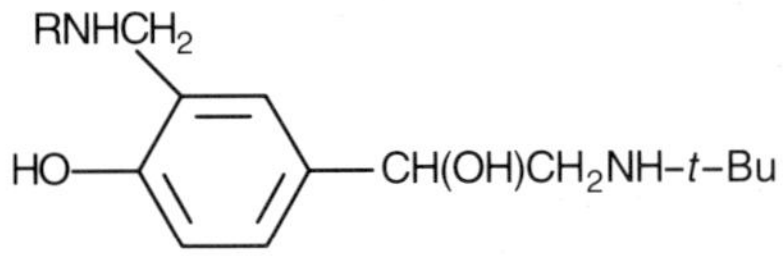

RNHCH$_2$

HO— —CH(OH)CH$_2$NH-t-Bu

	Activity (isoprenaline = 1)	
R	Bronchodilator[b]	Cardiac stimulant[c]
H	0	NT[d]
MeSO$_2$	0.1	0.001
HCO	0.05	0
NH$_2$CO	0.1	< 0.0001

[a] Adapted from Brittain *et al.* (1976).
[b] Anaesthetized guinea-pig (Dixon–Brodie preparation). Salbutamol = 0.1.
[c] Guinea-pig isolated atria (rate).
[d] NT, not tested.

D. Conclusions Regarding Structure–Activity Relationships

A methanesulphonamide (**39**), formamide (**40**) and urea (**41**) stand out with potencies of the order of salbutamol.

Their mobile hydrogen atoms were situated relative to the phenolic group in a manner similar to that of the alcoholic group of salbutamol.

The common features in the aromatic ring of the β-stimulants that we have obtained by chemical derivation from salicylic acid are

1. A phenolic hydroxy group in the 4-position relative to the ethanolamine side chain
2. A substituent in the 3-position which is capable of taking part in hydrogen-bonding and which does not have an electron-withdrawing effect on the phenyl ring
3. A steric arrangement such that the functional moiety on the 3-substituent is not restricted by nearby bulky groups, and is not removed from the aromatic ring by more than two carbon atoms.

Properties possessed by catechols but which do not seem to be relevant to β_2-stimulant activity are

1. Chelation *per se* since the aldehyde (**29**) and its oxime are not very active
2. Redox capacity
3. Acidity of the 3-substituent, since the hydroxymethyl group is not acidic, the methanesulphonamidomethyl substituent of **39** is acidic and the methanesulphonanilide group in soterenol (**16**) is more acidic than phenol* (Larsen *et al.*, 1967).

---

*The irrelevance of comparative acidities of the 3- and 4-substituents is confirmed by the inactivity of the 3-hydroxy-4-methanesulphonamide isomer (**17**).

VI.   Summary and Final Comments

This chapter has outlined the problems presented by the requirement for a bronchodilator therapy for asthma and how several selective β-stimulant bronchodilators were devised and synthesized.

The principle was to modify an active drug (isoprenaline) by replacement of the functional group that was responsible for its deficiencies to give an analogue that would retain the key physicochemical properties of the original. Simple chemical manipulation of an abundant starting material, aspirin, generated the antiasthmatic drug salbutamol. Effects of changes in structure on biological activity were applied to the preparation of further potent and selective β_2-adrenoceptor stimulants.

The success of this project has had profound beneficial effects for asthmatics who are now able to obtain rapid and sustained relief from their distressing symptoms without significant side effects.

Salbutamol* is efficacious in man in relieving attacks of asthma, active by inhalation and orally with minimal side effects and having a duration of about 4 h. It is not metabolized by COMT. In man it is converted into a phenolic sulphate.

Salbutamol was introduced as a bronchodilator in the United Kingdom in 1969 where it quickly became the most widely prescribed treatment for asthma. It is marketed throughout the world and is the market leader in 26 countries.

Bronchodilator and other properties of salbutamol have been reviewed extensively (Brittain *et al.*, 1976).

VII.   Progress beyond Salbutamol

Although the duration of action of salbutamol is adequate for many attacks there remained a need for a longer-acting bronchodilator to combat nocturnal asthma, that is bronchoconstriction that occurs at about 4 a.m. in many patients – the so called 'morning dip'. We decided to tackle this problem by modifying the molecule to produce a more lipophilic derivative on the assumption that it would bind more strongly at, or in the vicinity of, the receptor and so be available to act for a longer period before being removed.

As has been mentioned earlier (p. 217) the *t*-butyl group on the nitrogen atom can be replaced by some aralkyl moieties. We made many analogues with groups which enhanced lipophilicity and eventually found that the long-chain 4-phenylbutoxyhexyl substituent afforded a compound (**42**) which had a high potency (more than twice that of salbutamol) and a very extended duration of action in *in vitro* and *in vivo* tests in guinea-pigs.

$$\text{HOCH}_2$$
$$\text{HO}\quad\text{CH(OH)CH}_2\text{NH(CH}_2)_6\text{O(CH}_2)_4\text{Ph}$$

(**42**)

*Albuterol in the United States.

These results are borne out in man and the compound (as a salt with 1-hydroxy-2-naphthoic acid) (Salmeterol) when administered in 50 µg doses by inhalation affords protection for more than 12 h, and also against the second, late phase, reaction which is not usually controlled by other β-receptor agonists (Ullman and Svedmyr, 1988; Twentyman *et al.*, 1990).

References

Ahlquist, R.P. (1948). *Am. J. Physiol.* **153**, 586.

Biel, J.H., Schwarz, E.G., Sprengler, E.P., Leiser, H.A. and Friedman, H.L. (1954). *J. Am. Chem. Soc.* **76**, 3149.

Brittain, R.T., Farmer, J.B. and Marshall, R.J. (1973). *Br. J. Pharmacol.* **48**, 144.

Brittain, R.T., Dean, C.M. and Jack, D. (1976). *Pharmacol. Ther., Part B.* **2**, 423.

Clifton, J.E., Collins, I., Hallett, P., Hartley, D., Lunts, L.H.C. and Wicks, P.D. (1982). *J. Med. Chem.* **25**, 670.

Collin, D.T., Hartley, D., Jack, D., Lunts, L.H.C., Press, J.C., Ritchie, A.C. and Toon, P. (1970). *J. Med. Chem.* **13**, 674.

Colmer, L.J. and Gray, D.J.P. (1983). *Practitioner* **227**, 271.

Hartley, D. and Middlemiss, D. (1971). *J. Med. Chem.* **14**, 895.

Howarth, P.H. and George, G.F. (1983). *Adverse Drug React. Acute Poisoning Rev.* **2**, 25.

Lands, A.M., Arnold, A., McAuliff, J.P., Luduena, F.P. and Brown, T.G. (1967). *Nature (London)* **214**, 597.

Larsen, A.A. and Lish, P.M. (1964). *Nature (London)* **203**, 1283.

Larsen, A.A., Gould, W.A., Roth, H.R., Comer, W.T., Uloth, R.J., Dungan, K.W. and Lish, P.M. (1967). *J. Med. Chem.* **10**, 462.

Moed, H.D., van Dijk, J. and Niewind, H. (1955). *Recl. Trav. Chim. Pays-Bas Belg.* **74**, 919.

Twentyman, O.P., Finnerty, J.P., Harris, A., Palmer, J. and Holgate, S.T. (1990). *Lancet* **ii**, 1338.

Ullman, A. and Svedmyr, N. (1988). *Thorax* **43**, 674.

—12—

Discovery of Cimetidine, Ranitidine and Other H₂-Receptor Histamine Antagonists

C.R. GANELLIN
Department of Chemistry,
University College London,
London, U.K.

Professor C. Robin Ganellin, FRS read chemistry at Queen Mary College London and studied under M.J.S. Dewar for a Ph.D. in Organic Chemistry. He conducted postdoctoral research with A.C. Cope at MIT (1959) and subsequently joined Smith Kline and French Laboratories as a medicinal chemist, eventually becoming Vice-President for Research at Welwyn. In 1986 he was elected as a Fellow of the Royal Society and was appointed to the SK&F Chair of Medicinal Chemistry in the Chemistry Department of University College London. Co-discoverer of the H₂-receptor histamine antagonists and of the drug cimetidine, he has special interest in applying the principles of physical–organic chemistry to structure–activity analysis. He has been accorded wide recognition for his work and, in particular, has received awards from the Royal Society of Chemistry, American Chemical Society, La Societé de Chimie Thérapeutique, the Sociedad Espanola de Quimica Therapeutica, Society of Chemical Industry and Society for Drug Research, and in 1990 was elected into the USA National Inventors Hall of Fame.

MEDICINAL CHEMISTRY 2nd Edition
ISBN 0-12-274120-X

I. Background Rationale

A. Peptic Ulcer Disease

Duodenal and gastric ulcers (collectively known as peptic ulcers) affect large numbers of people who are otherwise relatively fit. Peptic ulceration is the most common disease of the gastrointestinal tract and it has been estimated that approximately 10–20% of the adult male population in Western countries will experience a peptic ulcer at some time in their lives. The disease produces considerable illness and pain, and results in great economic loss to the patients and their communities; it can even be fatal. In 1970, for example, in the United States there were some 3.5 million peptic ulcer sufferers and 8600 deaths were attributed to this disease.

Duodenal and gastric ulcers are localized erosions of the mucous membrane of the duodenum or stomach respectively which expose the underlying layers of the gut wall to the acid secretions of the stomach and to the proteolytic enzyme pepsin. What causes acute peptic ulcer is still not properly understood but for many years the main medical treatment was aimed at reducing acid production, based on the hope that neutralizing gastric acid would reduce its irritating effects and also reduce the efficacy of pepsin, and so allow ulcers to heal. The subsequent therapeutic use of cimetidine and ranitidine has clearly demonstrated the value of this medical approach.

The stomach contains many different types of highly specialized secretory cells controlled by the nervous system and hormones. For example, the parietal cells secrete hydrochloric acid. Prior to and during a meal, the volume of acid, pepsin and mucus secretion increases to as much as 10 times the basal secretion rate, and the pH may fall to 1–2.

B. Chemical Messengers

Secretion of gastric acid is initiated by the thought, sight, smell or taste of food and is mediated by the autonomic nervous system via the vagus nerves which provide parasympathetic innervation to the stomach and small intestine (Fig. 12.1); the neurotransmitter released by stimulation of the vagus is acetylcholine (Fig. 1.11).

Branches of the vagus, innervating the antral region of the stomach, stimulate the release of the peptide hormone gastrin from special gastrin-producing 'G' cells. The presence of food in the stomach further stimulates release of gastrin, which passes into the bloodstream and is carried to the parietal cells where it acts to stimulate them to secrete hydrochloric acid.

In addition to acetylcholine and gastrin, a third chemical secretagogue, histamine, was known to be involved (Fig. 12.2). As long ago as 1920, histamine was shown to stimulate gastric acid secretion when injected into the dog; subsequently it was found to be a natural constituent of the stomach lining and this led inevitably to the suggestion that it was involved in the physiological process of gastric secretion.

The relationship between the three secretagogues acetylcholine, gastrin and histamine has been a source of considerable controversy among physiologists for many years. When it was found (in the late 1940s and early 1950s) that the antihistamine drugs did not reduce acid secretion, the role of histamine was placed in serious question. By 1964, when gastrin had been isolated and sequenced at Liverpool University, most gastric

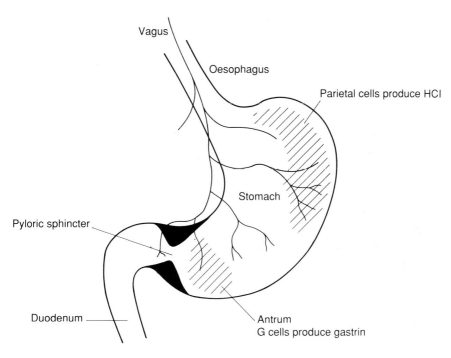

Fig. 12.1   Diagram of the stomach showing vagus nerve and position of G cells (producing gastrin) and parietal cells (producing HCl)

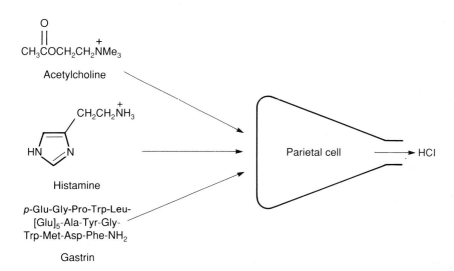

Fig. 12.2   Three chemical messengers stimulate the production of hydrochloric acid from the gastric parietal cell. The formulae show acetylcholine cation, histamine monocation (the most prevalent species at physiological pH of 7.4), and human gastrin I.

physiologists were convinced that gastrin played the key role in the physiological control of gastric acid and considered histamine to be unimportant.

C.   Controlling Gastric Acid

For many years the main medical treatment for peptic ulcers relied on the use of antacids to neutralize the gastric acid; these come in various forms, for example, magnesium trisilicate, aluminium hydroxide, sodium bicarbonate, calcium carbonate. All, however, when taken in sufficient quantities may cause unpleasant side effects. For real neutralization, very large quantities are required; for example, it has been calculated that 60 g of sodium bicarbonate daily would be required by intragastric drip to keep the pH of the stomach contents at 4.0 in patients with gastric ulceration.

Anticholinergic drugs (to block the acetylcholine transmission) can decrease gastric acid secretion, but their use in the treatment of peptic ulceration was limited by two factors: first, a dose high enough to decrease gastric acid secretion also blocks the action of acetylcholine at other receptor sites (Fig. 1.11) and so causes 'side effects' such as dryness of the mouth, urinary retention and blurred vision; secondly, some of these drugs are quaternary ammonium salts and are poorly absorbed on oral administration so that patients may fail to receive an effective dose. Subsequently, pirenzepine (**1**), a selective anticholinergic (acts at muscarinic M_1-receptors) was discovered and shown to be an effective treatment for peptic ulceration which, because of its selectivity, avoided many of the side effects associated with the use of anticholinergic drugs.

(1) Pirenzepine

The alternative to drug treatment is surgery. This aims to cut out part of the acid-secretory and gastrin-producing regions of the stomach (e.g. partial gastrectomy) or to selectively cut the branches of the vagal nerve (e.g. selective vagotomy) that supply the acid-secretory region. This is a difficult and sometimes dangerous operation.

D.   The Search for New Antiulcer Drugs

With this background to treatment it is not surprising that most major pharmaceutical companies set up research programmes aimed at discovering antiulcer agents. For many years the main approach was to induce ulcer formation in rats, or to stimulate production of gastric acid in rats, and to screen compounds for their ability to protect against ulcer formation or to inhibit acid production. The record of success was pretty poor. One of the main problems is that acid secretion is a very active metabolic process which is easy to inhibit with metabolic poisons or via actions which reduce the metabolism but are not selective. Thus many agents were found to be active in laboratory animals but very few survived initial human trials.

In the 1960s, with increasing understanding of the physiology of gastric acid secretion, several companies established research programmes to discover specific inhibitors of the action of the chemical messengers. In the United Kingdom, ICI Pharmaceuticals initiated a search for a gastrin antagonist, and Pfizer and Smith Kline & French (SK&F) independently sought a histamine antagonist. The idea that hormones and transmitter substances act at specific sites that we call receptors has been extremely valuable in the development of new drugs and is discussed in Chapter 7.

The two secretagogues, gastrin and histamine, therefore presented alternative targets for inhibiting acid production. If one of them has a controlling influence on acid secretion, then blocking it might succeed in controlling acidity, but, since blocking one site still leaves two others to act, it is by no means certain that such an approach will be successful. Furthermore, as with acetylcholine, gastrin and histamine have other actions, and an agent that blocks their effects on acid stimulation may also block other sites leading to unacceptable side effects (as with the anticholinergics). Thus, it can be appreciated that such an approach to drug discovery is highly speculative and one certainly cannot presume that it will be successful in providing a new therapy. An important part of this analysis is provided by the pharmacologist's view of drug receptors.

E. Histamine Receptors

Histamine [2-(imidazol-4-yl)ethylamine] appears to be a locally acting transmitter substance that has very specific actions. In 1910, Dale and Laidlaw, working at the Wellcome Physiological Research Laboratories in London, published the first of a series of papers describing the pharmacological effects of histamine; in particular, they noted the powerful effect of histamine in stimulating contractions of smooth muscle and its potent action in lowering blood pressure.

The early studies on histamine indicated a similarity in some of its effects with the symptoms that appear during inflammation, and with symptoms characteristic of shock produced by trauma or allergic reactions. It became widely assumed that histamine was a principal mediator of inflammation and shock and this stimulated a search by Bovet in Paris for substances capable of counteracting these apparent injurious effects (Bovet, 1950). His findings led to the development of the antihistamine drugs in the 1940s and to their introduction for the treatment of allergic conditions such as urticaria and hay fever.

(2a) Mepyramine   (2b) Diphenhydramine

Two typical antihistamine drugs, mepyramine (2a) [Neoantergan^R (Rhone-Poulenc)] and diphenhydramine (2b) [Benadryl^R (Parke-Davis)] were shown to be specific in antagonising histamine-stimulated contractions of the isolated ileum of the guinea-pig relative to other stimulants; they were effective at low concentrations and the antagonism they produced was surmountable and reversible. These antagonists came to be regarded as acting in competition with histamine for occupation of its specific receptor sites and were used to compare receptors in different tissues and species; for example, mepyramine had similar antagonist potency when tested against histamine on the perfused lung of the guinea-pig, on the isolated ileum and the trachea of the guinea-pig and on human bronchi. The results indicated a homogeneity for the histamine receptors in these tissues.

Several actions of histamine had been noted that could not be specifically antagonized by these drugs: for example, stimulation of gastric acid secretion in the rat, cat or dog; stimulation of isolated atria of the guinea-pig; inhibition of rat uterus contractions. It was also found that these antihistamines reduced the intensity of, but did not completely abolish, vasodilator actions of large doses of histamine, and in 1948 it was suggested by Folkow, Haeger and Kahlson that there may be 'two types of receptor sensitive to histamine only one of which can be blocked by Benadryl^R and related compounds'.

Additional pointers to the differentiation of histamine receptors had been obtained by considering the selectivity of action of agonists on different tissue systems. Thus Ash and Schild (1966) made quantitative estimates of the relative activities of different histamine congeners on the isolated guinea-pig ileum, on the isolated rat uterus and *in vivo* as stimulants of rat gastric acid secretion; they obtained a correlation suggesting that the same type of receptor might be involved in rat gastric acid secretion and inhibition of contraction of rat uterus, and they also proposed that the actions of histamine blocked by the antihistamine drugs characterized one type of histamine receptor, which they named the H_1-receptor. They suggested that the other actions of histamine not specifically antagonized were probably mediated by other histamine receptors, but that the characterization of these receptors awaited the discovery of specific antagonists; in the meantime they were to be regarded as non-H_1. It is worth recalling that at the time receptors were only a concept; there was no physical evidence for their real existence (see Chapter 3).

II.   The Search for a Histamine Antagonist

A.   The Biological Approach

The inability of the antihistamine drugs (H_1-receptor antagonists) to inhibit histamine-stimulated gastric acid secretion has been known for many years and there have been a few published reports of efforts to discover a specific antagonist to this action of histamine. In the 1950s in collaboration with the Eli Lilly Company in the United States, Grossman reported on an extensive study of compounds, chemically related to histamine, that were examined for their action on acid secretion and also tested as possible inhibitors of histamine stimulation, but Grossman did not uncover a histamine antagonist (Grossman *et al.*, 1952).

A similar analysis by Dr J.W. (now Sir James) Black led him to establish at the SK&F Research Laboratories in Welwyn Garden City, in 1964, the test procedures needed to detect antagonists of these other effects of histamine. Compounds were tested for ability to inhibit histamine-stimulated gastric acid secretion in anaesthetized rats. Routinely, the stomach of an anaesthetized rat [starved and pretreated with atropine (anticholinergic) and urethane (as anaesthetic)] was perfused with glucose solution at 37°C. The perfusate was introduced via a tube placed in the oesophagus and collected via a funnel placed in the non-secretory lumen of the stomach; the perfusate was then passed through a micro-flow-type glass electrode system and changes in gastric pH were recorded continuously on a potentiometric pen recorder.

Initially, compounds were administered as a constant intravenous infusion in the middle of a series of fixed-dose histamine responses (given by rapid intravenous injection). Subsequently, the system was modified and a plateau of gastric acid secretion was established by continuous intravenous infusion of histamine (at a dose high enough to produce a near-maximal response) and potential inhibitors were then given by rapid intravenous injection. Since other types of inhibitors of gastric secretion could also be picked up by this test, compounds found to be active were also tested on isolated tissue systems to provide additional criteria for specific antagonism to histamine.

Two *in vitro* test systems were set up, namely, histamine-induced stimulation of the guinea-pig right atrium (which continues to beat spontaneously *in vitro* because it contains the pacemaker and histamine increases the rate of beating) and inhibition by histamine of evoked contractions of the rat uterus.

It is worth reiterating that the atmosphere prevailing in gastroenterological science at that time was far from conducive to the search for a histamine antagonist as a means of controlling gastric acid secretion. In the 1960s many researchers turned their attention to seeking specific inhibitors of gastrin-induced acid secretion. The import of histamine was difficult to prove and there was a widely held view that histamine had no place in the physiological maintenance of gastric acid secretion. The concerted but unsuccessful effort by researchers at Eli Lilly in the 1950s to find an antagonist of histamine-stimulated acid secretion added further to the general feeling that the approach was 'played out'. However, there were still some adherents to the view that histamine played a key role in gastric acid secretion, although they were very much in a minority.

B. The Chemical Approach to an Antagonist

Once the problem of obtaining a competitive antagonist was posed in biological terms it was necessary to then consider how to approach it chemically. How could one obtain such a compound? Where should one start, given no obvious 'lead' compound? Nothing was known chemically about the physiological site of action of histamine.

Returning to first principles, the structure of histamine (Fig. 12.2) was used as a chemical starting point. The simple-minded view was taken that since the search was for a molecule that would compete with histamine for its receptor site, such a molecule would have to be recognized by the receptor and then bind more strongly than histamine, but not trigger the usual response. It therefore seemed worthwhile to retain in potential antagonist structures some chemical features of histamine to aid receptor recognition, and to include chemical groups that might assist binding.

Many different approaches were tried, including the use of analogies derived from known examples of chemical–biological relationships between other types of receptor agonists and antagonists, or enzyme substrates and inhibitors, or antimetabolites. The structure of histamine was modified to alter deliberately its chemical properties, while

TABLE 12.1

Structure and Antagonist Activities[a] of Some Simple Imidazolylalkylisothioureas, -Guanidines and -Amidines

$$(CH_2)_n X - C \underset{NH_2}{\overset{NH}{\diagdown}} \qquad (CH_2)_n NH - C \underset{Z}{\overset{NH}{\diagdown}}$$

Compound	n	Substituent	Activity[a]
3	2	X = NH	+
4	2	X = S	+ +
5	3	X = NH	+ + +
6	3	X = S	±
7	2	Z = SMe	±
8	2	Z = Me	±
9	3	Z = SMe	+ + +
10	3	Z = Me	+ + +

[a]Tested for inhibition of histamine-stimulated gastric acid secretion in the lumen-perfused anaesthetized rat. Results represented semiquantitively as: ±, detectable; +, $ID_{50} > 500\,\mu mol/kg$; + +, $ID_{50} \sim 200\,\mu mol/kg$; + + +, $ID_{50} = 100\text{--}50\,\mu mol/kg$. ID_{50} is the intravenous dose which reduces a near-maximal secretion to 50%.

retaining some definite aspect of its structure or chemistry. Some examples have been discussed elsewhere by Ganellin *et al.* (1976).

Many compounds were made, based on the structure of histamine. In the first 4 years some 200 compounds were synthesized and tested, without providing a blocking drug. The problem for the chemist is that there are too many possible compounds for synthesis. Even small modifications of the natural stimulant histamine introduce many variables for study.

Towards the end of this time many doubts were expressed about whether it really would prove possible to block the action of histamine on gastric acid secretion and there was considerable pressure within the company to abandon the research. Indeed, the American management at the Company headquarters in Philadelphia did eventually order the project to be closed (Spence, 1990). The scientists involved in the project were, however, firmly resolved to continue, and during this period the test system was refined and chemical ideas began to crystallize.

A most important aspect of research is to conduct the work in such a way as to learn from negative results. Even a list of inactive compounds is informative if they have been selected for particular reasons. Having tested many compounds with lipophilic substituents without seeing antagonism, the pharmacologists re-examined some of the early hydrophilic compounds, one of which showed some blocking activity. It was very weak but provided the vital lead. It was missed originally because this compound also acted as a stimulant; in fact it is a partial agonist. The compound is a histamine derivative in which a guanidine group replaces the amino group in the side chain, namely, N^α-guanyl-histamine (**3**) (Table 12.1). It was selected for synthesis by Dr Graham Durant based on an analogy with guanidine structures, which have a high affinity for catecholamine storage sites. Ironically, N^α-guanylhistamine was first synthesized by van der Merwe in 1928 and was reported to be 'devoid of interesting physiological activity'. Clearly,

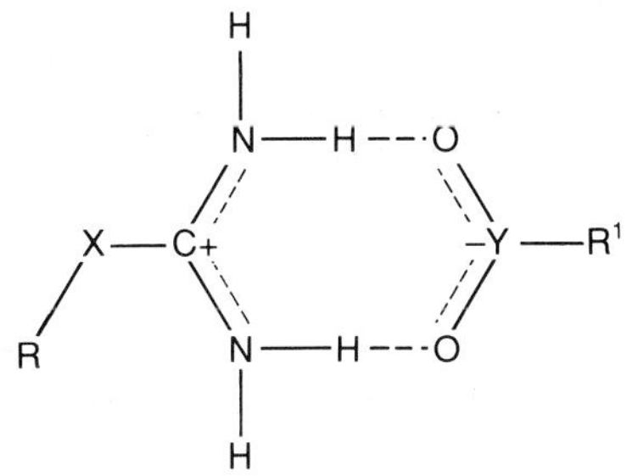

```
            H
            |
      N —— H ——·O
     //            \
X —— C+            —Y —— R¹
    /   \\        //
   R     N —— H ——·O
         |
         H
```

Fig. 12.3 Bidentate hydrogen-bonding envisaged between ion pairs formed between amidinium cations and oxyacid anions.

biological activity may be present, but its discovery depends on the way one looks for it!

C. Developments Based on the First Lead Compounds

The appearance of antagonism, albeit weak, provided a much needed lead and within a few days an analogous compound was retested and found to be more active, namely, (S)-[2-(imidazol-4-yl)ethyl]isothiourea (4) (Table 12.1). However, a much more active compound was required.

An immediate question to be answered was whether activity was due to the presence of the guanidine or isothiourea groups (amidines) *per se* or to the structural resemblance to histamine. Structure–activity studies suggested that for these structures the imidazole ring was important; antagonism did not appear to be a property of amidines in general. It was also necessary to identify the particular chemical properties that conferred antagonist activity in order to make analogues of increased potency.

The amidine groups are strong bases and are protonated and positively charged at physiological pH. Thus the molecules resemble histamine monocation, but also differ in several ways; the amidinium group is planar (whereas the ammonium group of histamine is tetrahedral) and the positive charge is distributed over three hetero atoms. It was noted that the distance between ring and terminal nitrogen is potentially greater than in histamine, and that there are several nitrogen sites for potential interactions instead of one, thereby affording more opportunities for intermolecular bonding.

It was envisaged that amidines might act as antagonists through additional binding being contributed by the amidine group. A type of bidentate hydrogen-bonding between ion pairs can occur with an amidinium cation and an oxyacid anion, for example, carboxylate, phosphate or sulphate, which might be part of the receptor protein (see Fig. 12.3). This was the hypothetical model formulated in 1968 during this research. In 1991, the H_2-receptor was cloned and the presence demonstrated of an aspartic acid residue in the trans-membrane-3 region of the receptor protein (see Chapter 3). The carboxyl group of aspartic acid is postulated to engage the ammonium group of histamine.

Structural variables therefore identified for study initially were the amidine groups, amidino N-substituents, side chain length and alternatives to imidazole. Many analogues of compounds 3 and 4 were made but most turned out to be less active; at that time, unsubstituted imidazole appeared to be the best ring; alkyl substitution on the amidine

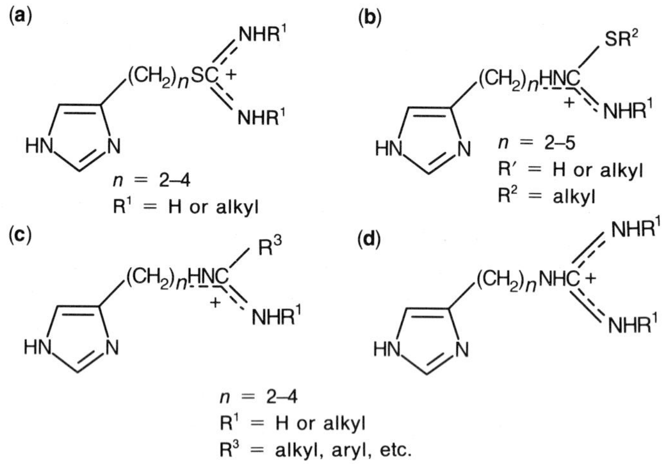

Fig. 12.4 Imidazolylalkylisothioureas, -amidines and -guanidines synthesized and tested as potential antagonists of histamine-stimulated gastric acid secretion. The structures are shown as side chain cations. (a) Isothioureas. (b) 'Reversed' isothioureas. (c) Amidines. (d) Guanidines.

N gave inconsistent results; and lengthening the side chain gave another breakthrough but threw up an apparent contradiction.

For the guanidine structure, increasing the chain length led to a compound (5) showing an increase in antagonist activity. However, for the isothiourea, the reverse result was obtained; i.e. increasing the chain length gave a compound (6) of reduced antagonist activity (Ganellin, 1981).

Thus, although the first two compounds discovered [guanidine (3) and isothiourea (4)] appeared to be closely related in structure in simply being isosteric nitrogen and sulphur analogues, the results with the homologues suggested that the situation was more complex. In an attempt to rationalize these differences, various related amidines were examined (Fig. 12.4). The activities of some of these compounds are expressed semiquantitatively in Table 12.1, reflecting the data available at the time. It was found that the reversed isothioureas (7, 9) (side chain on N instead of S) resembled the guanidines in chain-length requirements, as did carboxamidines (e.g. 8 and 10).

The above results indicated that it was necessary to reappraise the view of amidine interactions. The initial analysis had suggested hydrogen-bonding of the terminal NH groups; in the reversed isothiourea and carboxamidine, hydrogen-bonding would include an NH within the side chain.

The apparent non-additivity between structural change and biological effect posed a typical problem familiar to all practising medicinal chemists: with so many structural variables to study (e.g. ring, side-chain length, amidine system, amidine substituents), there are many millions of structures incorporating different combinations of these variables, and one cannot make and test them all. What then should govern the selection?

An essential feature of the discipline in medicinal chemistry is to find logical ways for defining the boundary conditions for the selection of structures for synthesis. In the case under study there was a continuous search for useful physicochemical models for studying the chemistry of these compounds, and the inconsistencies in the structure–

activity pattern were used to challenge the model or to re-examine the meaning of the biological test results. This dialogue, a search for self-consistency between the chemistry and biology, is vital to a new drug research where no precedent exists.

To explore structure–activity relationships further, it became desirable to increase the side chain length still more, but problems of chemical synthesis were experienced and new synthetic routes were required.

Exploration of amidines and substituents continued but it became clear that progress had become very slow. The problem seemed to be that the compounds had mixed activities, although to varying degrees. In the main they acted both as agonists and as antagonists, that is, they appeared to be partial agonists. This meant that, although the compounds antagonized the action of histamine, they were not sufficiently effective inhibitors of gastric acid secretion because of interference from their inherent stimulatory activity. This appeared to impose a limitation on the potential of this type of structure for providing antagonists and seemed to be hindering progress.

Thus a critical stage was reached in the need for selectivity: it was necessary to achieve a separation between agonist and antagonist activities. It seemed that these compounds might act as agonists by mimicking histamine chemically, since like histamine, they have an imidazole ring and, being basic amidines, the side chain at physiological pH is protonated and carries a positive charge. It also seemed likely that these features would permit receptor recognition and provide binding for a competitive antagonist. This posed a considerable dilemma because the chemical groups that appeared to be required for antagonist activity were the same groups that seemed to confer the agonist effect.

In an attempt to separate these activities, the strongly basic guanidine group was replaced by non-basic groups that, though polar, would not be charged. Such an approach furnished analogues that indeed were not active as agonists; however, the first examples were also not active as antagonists. Eventually, one compound, the thiourea derivative (SK&F 91581, **11**), that did not act as a partial agonist exhibited weak activity as an antagonist. Thioureas are essentially neutral in water because of the electron-withdrawing thiocarbonyl group.

(**11**) SK&F 91581 $n = 3$, $R^2 = H$
(**12**) SK&F 91863 $n = 4$, $R^2 = H$
(**13**) Burimamide $n = 4$, $R^2 = Me$

At this time the higher homologous amine (with the four carbon atom chain length) was synthesized and further exploration revealed that derivatives of this structure possessed a marked increase in antagonist potency. It was not until the side chain had been lengthened that the significance of the result with SK&F 91581 became clear and the desired aim was achieved, that is, a pure competitive antagonist without agonist effects. This compound (**12**, SK&F 91863) paved the way for the synthesis of substituted analogues, and in a short while the *N*-methyl analogue was obtained, which was given the name burimamide (**13**).

A synthesis of burimamide (Scheme 12.1) from lysine ethyl ester dihydrochloride involves reduction with sodium amalgam under carefully controlled conditions to

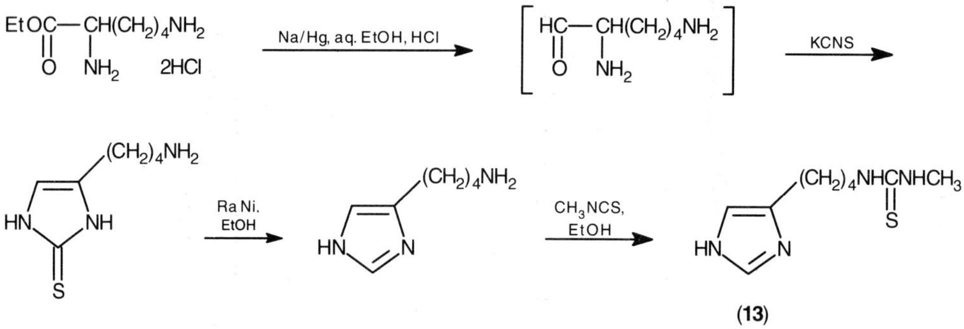

Scheme 12.1   Laboratory synthesis of burimamide (**13**).

generate an α-aminoaldehyde which, on being heated *in situ* with potassium thiocyanate, cyclizes to the imidazole-2-thione. Desulphurization to the imidazole followed by treatment with methyl isothiocyanate gives the required thiourea, burimamide (**13**).

D.   Burimamide, the First Characterized H$_2$-Receptor Histamine Antagonist

Burimamide (**13**) was an extremely important compound, and it provided a vital breakthrough. It was highly selective, showed no agonist activity and antagonized the action of histamine in a competitive manner on the two *in vitro* non-H$_1$ systems, namely, guinea-pig atrium and rat uterus (Table 12.2). It fulfilled the criteria required for characterizing the existence of another set of histamine receptors, namely, the H$_2$-receptors. Thus, it allowed the above tissue systems to be defined as H$_2$-receptor systems, and thereby burimamide was defined as an H$_2$-receptor antagonist. This discovery was announced in 1972; the work had taken 6 years (Black *et al.*, 1972).

Burimamide also antagonized the action of histamine as a stimulant of gastric acid secretion in the rat, cat and dog, and it was the first H$_2$-receptor antagonist to be investigated in man. Given intravenously, it blocked the action of histamine as a stimulant of gastric acid secretion in man, thereby confirming that burimamide behaves in humans as it does in animals. However, its one drawback was that it was not sufficiently active to be given orally. Thus, although burimamide was selective enough to define H$_2$-receptors, it was not active enough to permit proper drug development.

E.   Sulphur–Methylene Isosterism, Imidazole Tautomerism and the Development of Metiamide

Various ways to alter the structure of burimamide were examined in an attempt to increase potency. One approach that proved successful resulted from two lines of exploration that merged. On the one hand, attempts were being made to overcome the problem of synthesizing the side chains by inserting a thioether link. Meanwhile, a study was being made of the pK_a characteristics of burimamide, since it was realized that burimamide in aqueous solution is a mixture of many chemical species in equilibrium. At physiological pH there are three main forms of the imidazole ring, three planar conformations of the thioureido group (a fourth is theoretically possible but is disfavoured by internal steric hindrance), and various trans and gauche rotamer combina-

TABLE 12.2

Structures and H_2-Receptor Histamine Antagonist Activities of Burimamide, Metiamide, Cimetidine and Isosteres

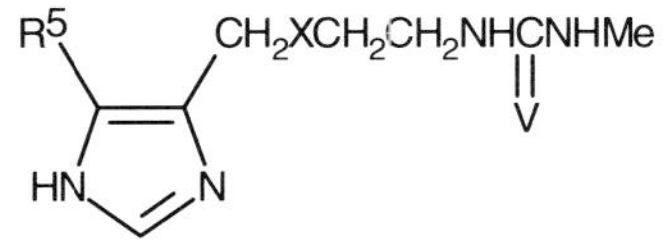

$$R^5 \diagdown \quad CH_2XCH_2CH_2NHCNHMe$$
$$\underset{HN}{\diagup}\diagdown_N \qquad \overset{\|}{V}$$

| | | Structure | | | H₂-Receptor activities | | |
| | | | | | In vitro | | In vivo |
Number	Trivial name	R⁵	X	V	Atrium[a] K_B (95% limits) (μM)	Uterus[b] K_B (95% limits) (μM)	Acid secretion[c] ID_{50} (μmol/kg)
13	Burimamide (thiourea)	H	CH₂	S	7.8 (6.4–8.6)	6.6 (4.9–8.3)	6.1
14	Thiaburimamide	H	S	S	3.2 (2.5–4.5)	3.2 (2.5–4.5)	5[d]
15	Oxaburimamide	H	O	S	28 (13–69)	6.6 (4.9–8.3)	[d]
16	Metiamide (thiourea)	Me	S	S	0.92 (0.74–1.15)	0.75 (0.40–1.36)	1.6
17	Urea isostere	Me	S	O	22 (8.9–65)	7.1 (1.6–30)	27
18	Guanidine isostere	Me	S	NH	16 (8.1–32)	5.5 (2.8–13)	12
19	Nitroguanidine isostere	Me	S	N·NO₂	1.4 (0.79–2.8)	1.4 (0.72–3.2)	2.1
20	Cimetidine (cyanoguanidine)	Me	S	N·CN	0.79 (0.68–0.92)	0.81 (0.54–1.2)	1.4
21	Guanylurea derivative	Me	S	N·CONH₂	7.1 (4.0–14)	6.9 (4.1–12)	7.7

[a] Activities determined against histamine stimulation of guinea-pig right atrium *in vitro*. The dissociation constant (K_B) was calculated from the equation $K_B = B/(x - 1)$, where x is the respective ratio of concentrations of histamine needed to produce half-maximal responses in the presence and absence of different concentrations (B) of antagonist.

[b] Activities determined against histamine inhibition of electrically evoked contractions of rat uterus *in vitro*.

[c] Activities as antagonists of histamine-stimulated gastric acid secretion in the anaesthetized rat as indicated in Table 12.1, footnote *a*.

[d] Not determined.

For burimamide:

R³ = (CH₂)₄NHCNHMe   R⁴ = 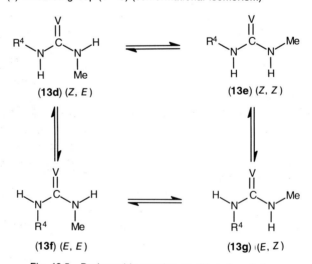   V = S
           ‖
           S

(a) Imidazole ring (ionization and tautomerism)

(b) Alkane chain (C–C bond rotation gives trans and gauche conformers)

(c) Thiourea group (V = S) (conformational isomerism)

Fig. 12.5   Burimamide species equilibria in solution.

tions of the side chain CH_2–CH_2 bonds (Fig. 12.5). This means that at any given instant only a small proportion of the drug molecules would be in a particular form.

The existence of a mixture of species leads one to question which may be biologically active and whether altering drug structure to favour a particular species would alter drug potency [dynamic structure–activity analysis (DSAA)] (Ganellin, 1981). There are substantial energy barriers to interconversion between the species of burimamide so that it is quite likely that a drug molecule, presenting itself to the receptor in a form unfavourable for drug–receptor interaction, might diffuse away again before having time to rearrange into a more favourable form. The relative population of favourable forms might therefore determine the amount of drug required for a given effect.

The various species of burimamide do not interconvert instantaneously, but whereas the rotamers of the side chain and thioureido groups are interconverted simply by internal rotation of a C—C or C—N bond, interconversion of the ring forms probably involves a water-mediated proton transfer. If a molecule presents itself to the receptor with the ring in an unfavourable form it might not readjust, unless there were suitably oriented water molecules (or other hydrogen donor–acceptors) present. The form of the ring therefore merits special consideration.

The above arguments led to a study of the population of imidazole species in burimamide in comparison with histamine. At physiological pH the main species are (Fig. 12.5) the cation (13c) and two uncharged tautomers (13a and 13b), and their populations were estimated qualitatively from the electronic influence of the side chain using pK_a data and the Hammett equation (Charton, 1965b):

$$pK_{a(R)} = pK_{a(H)} + \rho\sigma_m$$

where ρ is the Hammett reaction constant and σ_m the Hammett substituent constant.

For burimamide, the ring pK_a (7.25 at 37°C) is greater than that of unsubstituted imidazole (6.80), indicating that the side chain is mildly electron-releasing. In contrast, for histamine the ammonium ethyl side chain was seen to be electron-withdrawing, since it lowered the pK_a of the imidazole ring (pK_a 5.90). Thus, although both histamine and burimamide are monosubstituted imidazoles, the structural similarity is misleading in that the electronic properties of the respective imidazole rings are different.

If the active form of burimamide were tautomer 13a, the form most preferred for histamine, then increasing its relative population might increase activity; for example, incorporating an electronegative atom into the antagonist side chain should convert it to an electron-withdrawing group and favour species 13a. This would not be the only requirement for activity and it would be necessary to minimize disturbance to other biologically important molecular properties such as stereochemistry and lipid–water interactions. For reasons of synthesis, the first such substitution to be made was the replacement of a methylene group (—CH$_2$—) by the isosteric thioether linkage (—S—) at the carbon atom next but one to the ring, to afford 'thiaburimamide' (14) (Table 12.2), which was found to be more active as an antagonist.

It was argued that a further stabilization of tautomer 13a in thiaburimamide might be obtained by incorporating an electron-releasing substituent in the vacant 4(5) position of the imidazole ring. A methyl group was selected, since it was thought that it should not interfere with receptor interaction, 4-methylhistamine having been shown to be an effective H$_2$-receptor agonist. This approach was successful, and introduction of a methyl group into the ring of the antagonist furnished a more potent drug, which was named metiamide (16) (Black et al., 1974) (Table 12.2). Metiamide is synthesized from methyl isothiocyanate and the requisite amine, 4-[(2-aminoethyl)thiomethyl]-5-methyl-imidazole, the synthesis of which is outlined in Scheme 12.2.

Although the above molecular manipulations were made through consideration of the electronic effects of substituents, evidence has subsequently accrued to suggest that conformational effects are probably more important. Crystal structure studies indicate

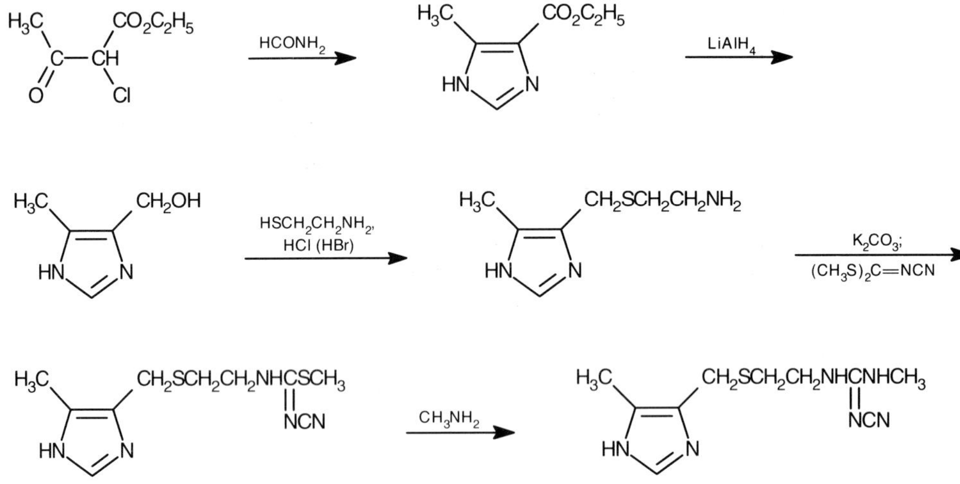

Scheme 12.2 Laboratory synthesis of cimetidine (**20**).

that the thioether linkage may increase molecular flexibility and the ring-methyl group may assist in orientating the imidazole ring (Prout *et al.*, 1977). Furthermore, the oxygen (ether) analogue [oxaburimamide (**15**), Table 12.2] which should fulfil the electronic requirements is less potent than burimamide, possibly by encouraging a different conformation through intramolecular hydrogen-bonding (e.g. Fig. 12.6).

Metiamide represented a major improvement, being 10 times more potent than burimamide *in vitro* and a potent inhibitor of stimulated acid secretion in man. It was investigated for potential use in peptic ulcer therapy and was shown to be effective at increasing the healing of duodenal ulcers and it provided marked symptomatic relief. However, out of 700 patients treated there were a few cases of granulocytopenia (causing a reduction in the number of circulating white cells in the blood and leaving patients open to infection). Although reversible, this severely limited the amount of clinical work. A serious question posed by this finding was whether the granulocytopenia was caused by blockade of H_2-receptors because, if this were so, it would impose a considerable restriction on the therapeutic use of this class of drug.

F.   Isosteres of Thiourea, Guanidine Equilibria and the Discovery of Cimetidine

The possibility existed, however, that the granulocytopenia associated with metiamide was an idiosyncratic reaction to some chemical feature of the compound. The most likely

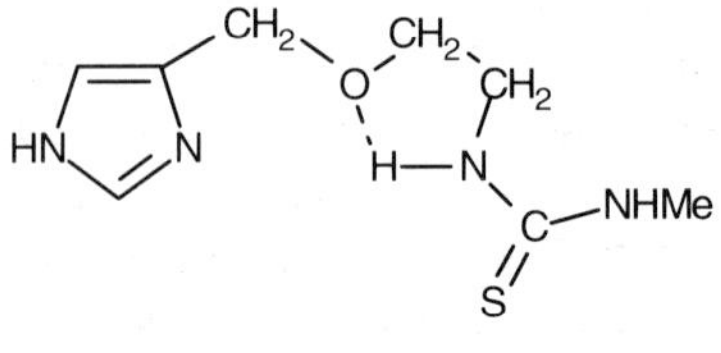

Fig. 12.6   Possible intramolecular hydrogen-bonding by oxaburimamide (**15**) (Table 12.2).

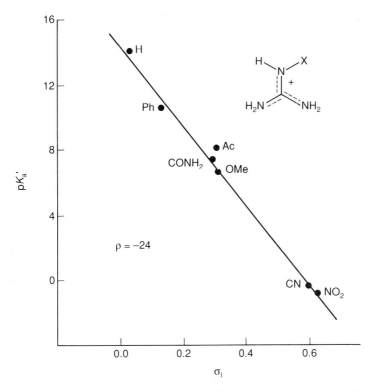

Fig. 12.7 Apparent pK_a values at 25° of N-substituted guanidinium cations versus σ_1 substituent constants. Data from Charton (1965a). The line corresponds to the equation pK'_a = 14.20 − 24.1σ_1. [Reproduced from Burland and Simkins (1977), with permission of the publisher.]

culprit was judged to be the thiourea group in the molecule and this led to the need to examine another compound. Fortunately, exploration had continued with other possible structures and, in particular, with alternatives to the thiourea group. One approach taken was to examine isosteric replacement of the thiourea sulphur atom (=S) of metiamide. Replacement by carbonyl oxygen (=O) gave the urea analogue (**17**), but this was much less active. The idea of guanidine derivatives which had provided the original breakthrough was also re-examined; replacement by imino nitrogen (=NH) afforded the guanidine (**18**) which, interestingly, was not a partial agonist but a fairly active antagonist. However, *in vitro*, the urea and guanidine isosteres were both approximately 20 times less potent than metiamide, and other ways were investigated for removing the positive charge on the guanidine derivative.

An observation that nitroguanidine was not basic led to further investigation and revealed a publication by Charton on the Hammett relationship between σ and pK_a for substituted amidines and guanidines (Charton, 1965a). Guanidine basicity is markedly reduced by electron-withdrawing substituents, and Charton demonstrated a high correlation between the inductive substituent constant σ_1 and pK_a for a series of mono-substituted guanidines (Fig. 12.7). The cyano and nitro groups are sufficiently electron-withdrawing to reduce the pK_a by over 14 units, to values < 0; indeed, the ionization

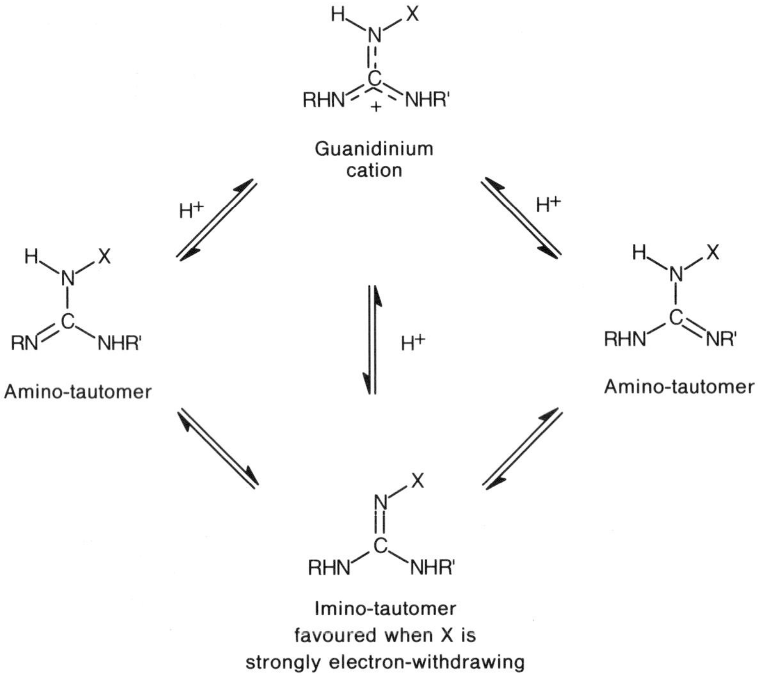

Fig. 12.8   Equilibria between the guanidinium cation and the three conjugate bases.

constants of cyanoguanidine ($pK_a = 0.4$) and nitroguanidine ($pK_a = 0.9$) approach that of thiourea ($pK_a = -1.2$).

The nitroguanidine (19) and cyanoguanidine (20) analogues of metiamide were synthesized and found to be active antagonists comparable with metiamide (Durant et al., 1977). Of these two compounds, the cyanoguanidine (20) is slightly more potent and was selected for development (Brimblecombe et al., 1975, 1978), being given the non-proprietary name 'cimetidine'.

A laboratory synthesis of cimetidine is outlined in Scheme 12.2. 4-Carboethoxy-5-methylimidazole, made from ethyl α-chloroacetoacetate by condensation with formamide in the Bredereck procedure, is reduced by lithium aluminium hydride to give 4-hydroxymethyl-5-methylimidazole. This carbinol is then condensed with cysteamine in the presence of HCl or HBr, and the resulting amine salt is collected and neutralized with aqueous potassium carbonate. The liberated amine is treated with dimethylcyanodithioimidocarbonate and then with methylamine. Many variations of this synthesis have been developed.

The guanidinium cation can lose a proton from each of the three nitrogen atoms to give three different forms of the conjugate base (Fig. 12.8). Powerful electron-withdrawing substituents favour the imino-tautomer over the amino-tautomers, since the proton on the adjacent nitrogen in the cation is more acidic than the protons on the more distant terminal nitrogen atoms. Thus, cyanoguanidines exist predominantly in the cyanoimino form and, in cimetidine, the cyanoimino group ($=NCN$) replaces the thione ($=S$) sulphur atom of metiamide.

Cyanoguanidine and thiourea have many chemical properties in common. They are

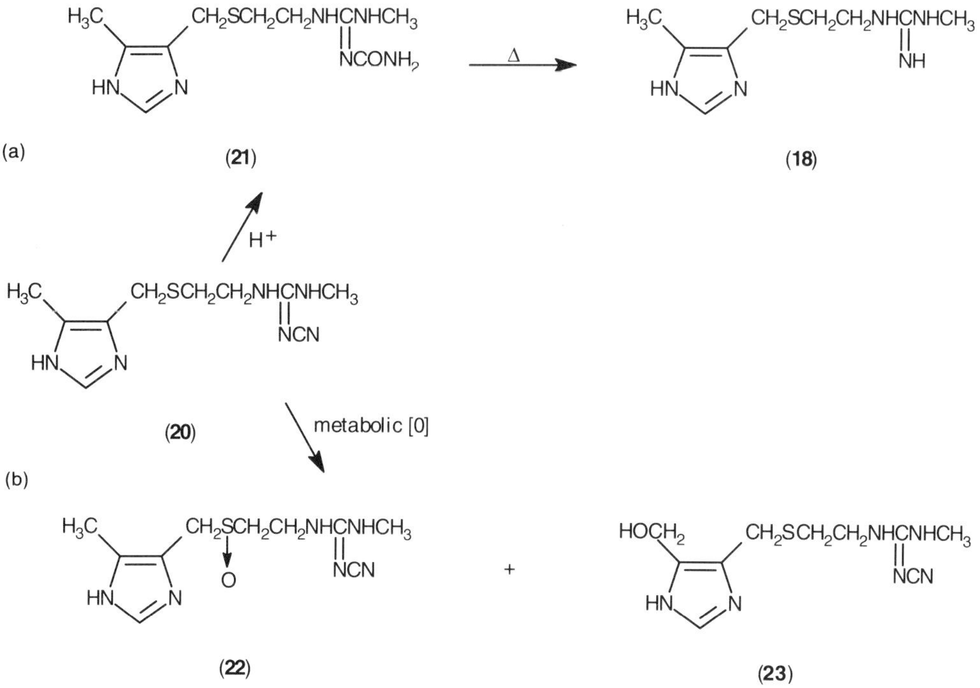

Scheme 12.3 Transformation products of cimetidine (**20**). (a) Hydrolysis gives the guanylurea (**21**) and thence the guanidine (**18**). (b) Metabolic oxidation gives the sulphoxide (**22**) and hydroxymethyl derivative (**23**).

planar structures of similar geometrics; they are weakly amphoteric (very weakly basic and acidic), so that in the pH range 2–12 they are un-ionized; they are very polar and hydrophilic. The similar behaviour of cimetidine and metiamide as histamine H$_2$-receptor antagonists and the close similarity in physicochemical characteristics of thiourea and cyanoguanidine permit the description of the thiourea and cyanoguanidine groups in the present context as bioisosteres (Durant *et al.*, 1977). Nitroguanidine may also be considered to be a bioisostere of thiourea in this series of structures. Bioisosterism is not a universally applicable property, however, and for other biological actions cyanoguanidine and thiourea may be non-equivalent. One important chemical difference between these groups is their conformational behaviour. Three conformations of a disubstituted thiourea group are energetically accessible, namely, **13d**, **13e** and **13g** (Fig. 12.5, V = S) but, owing to steric effects, only two conformations of a disubstituted cyanoguanidine are similarly accessible, namely, **13d** and **13g** (Fig. 12.5, V = NCN).

Cyanoguanidines and thioureas also differ sufficiently in chemical reactivity (e.g. in oxidative and hydrolytic reactions) for differences to be expected in the rates and products of biotransformation of drug molecules containing these groups. In the presence of excess dilute hydrochloric acid, cimetidine is slowly hydrolysed to the guanylurea (**21**, Table 12.2, Scheme 12.3) and on being heated in acid, the latter is further hydrolysed and decarboxylated to give the guanidine (**18**). The guanylurea is also an H$_2$-receptor antagonist but it is less active than cimetidine.

The cyanoguanidine group in cimetidine is metabolically stable and cimetidine is

TABLE 12.3
Inhibition of Gastric Acid Secretion *in vivo* by Intravenous Cimetidine

Animal	Preparation	Stimulant	Intravenous ID_{50} (μmol/kg)
Rat[a]	Lumen-perfused stomach	Histamine	1.37
		Pentagastrin	1.40
Cat[a]	Lumen-perfused stomach	Histamine	0.85
		Pentagastrin	1.45
Dog[b]	Heidenhain pouch	Histamine	1.70
		Pentagastrin	2.00

[a] Anaesthetized.
[b] Conscious.

largely excreted unchanged. The main metabolite is the sulphoxide (**22**) produced by oxidation of the side-chain thioether sulphur atom, together with lesser amounts of the hydroxymethyl oxidation product (**23**) of the ring CH_3 (Scheme 12.3).

III. Cimetidine, a Breakthrough in the Treatment of Peptic Ulcer Disease

Cimetidine has been shown to be a specific competitive antagonist of histamine at H_2-receptors *in vitro*, and to be effective *in vivo* at inhibiting histamine-stimulated gastric acid secretion in the rat, cat and dog (Table 12.3). The ID_{50} values determined in the rat, cat and dog were not significantly different from each other. Cimetidine has also been shown to be active when administered orally, and a dose of 20 μmol/kg (approximately 5 mg/kg) produced a mean inhibition of 90% of maximal histamine-stimulated secretion in the dog.

Cimetidine is also an effective inhibitor of pentagastrin-stimulated acid secretion. (Pentagastrin is a synthetic biologically active analogue of gastrin which contains the terminal four amino acid residues of gastrin, namely, *N-t*-Boc-β-Ala-Trp-Met-Asp-Phe-NH_2.) The ID_{50} values shown in Table 12.3 indicate that the potency of cimetidine against pentagastrin-stimulated secretion is very similar to its potency against histamine-stimulated secretion. Cimetidine is also effective against food-stimulated acid secretion but is less active against cholinergically stimulated secretion.

The finding that H_2-receptor antagonists such as cimetidine effectively inhibit pentagastrin-stimulated secretion clearly indicates that gastrin and histamine are somehow linked in the gastric secretory process. The results with these antagonists firmly establish that histamine has a physiological role in gastric acid secretion.

Cimetidine has been extensively studied in man and its safety and efficacy have been established in the acute treatment of peptic ulcer. Cimetidine given orally as 0.8–1.2 g/day has been shown to relieve symptoms and promote healing of lesions in a majority of patients with peptic ulcer disease. Cimetidine was marketed first in the United Kingdom in November 1976, and was marketed in the United States in August 1977; by 1979 it was sold in over 100 countries under the trademark Tagamet, representing all the major markets with one important exception – it was not granted approval for use in Japan until 1982.

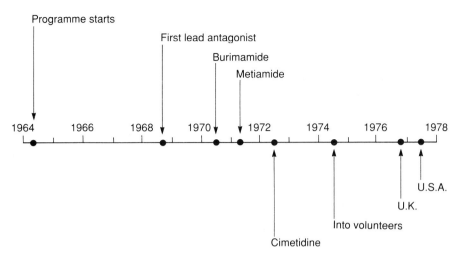

Fig. 12.9 Thirteen years to discover and develop cimetidine and make it generally available for therapeutic use.

Cimetidine changed the medical management of peptic ulcer disease and became a very successful product (Winship, 1978; Freston, 1982). In 1983 its annual worldwide sales reached the level of nearly $1 000 000 000, and in several countries (including Canada and the United States) it was the number one selling prescription product.

The long-term nature of pharmaceutical research and development and the need for tenacity to continue in the face of considerable difficulty and disappointment is well illustrated by this case history of drug discovery (see Fig. 12.9).

The research project was initiated in 1964, and it took 6 years (to 1970) to obtain burimamide, which was used to characterize pharmacologically histamine H_2-receptors and verify the basic concept, and was examined in human volunteers. The next drug, metiamide, a more potent and orally active compound, was investigated clinically, but its use was severely restricted.

Cimetidine followed on from metiamide; it was first synthesized in 1972. Much work had to be done to evaluate the potential of cimetidine in animals, establish its safety in animals and then investigate its behaviour in man. Finally, many clinical studies had to be undertaken to prove its value as a drug therapy before it could be made generally available in 1977. An overall time of 13 years had elapsed!

It is self-evident that any account of drug discovery must be incomplete and certainly only a small proportion of the total studies made have been described in this chapter. Many avenues examined during structure–activity analysis turned out to be ineffective and since, in the main, these are not mentioned here, the net effect may be to make the work appear to be more rational and more perceptive than is warranted. In order to limit the scale of the problem, the early work concentrated on imidazole derivatives. However, it was soon demonstrated that other nitrogen heterocycles such as pyridine and thiazole could be effectively used in place of imidazole in the cimetidine type of structure.

After cimetidine had been selected for development, work continued at SK&F to explore other structures. One moiety which had not been used in a drug before was 1,1-diamino-2-nitroethene. This was designed in an attempt to reduce hydrophilicity since an inverse correlation had been obtained between potency and hydrophilicity (Ganellin, 1981) as indicated by octanol : water partition studies. One aim had been to replace N of cyanoguanidine by CH. This produced complications, however, since the resulting diamino-2-cyanoethene is in tautomeric equilibrium with the cyanoacetamidine (X = CN in Fig. 12.10).

Consideration of CH versus NH acidities (Ganellin, 1981) focused attention on the nitro compound (X = NO$_2$) since the 1,1-diamino-2-nitroethene tautomer predominates, and this led to the synthesis of the 1,1-diamino-2-nitroethene analogue (Fig. 12.10, X = NO$_2$) of cimetidine. This compound was found to be similarly active in animals and it was studied in human volunteers to confirm the findings. Since this compound showed no apparent improvement over cimetidine it was not developed further but the compound was patented. This discovery was taken up later by chemists at Glaxo Research (see below).

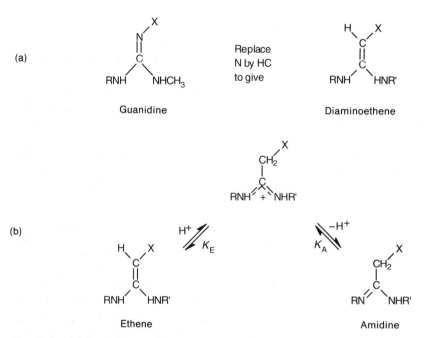

Fig. 12.10 (a) Replacing guanidine N by CH to give a diaminoethene and (b) equilibria between the acetamidinium cation and the diaminoethene and acetamidine conjugate bases. Diaminoethenes are tautomers of acetamidines; the ethene structure is favoured when X = NO$_2$.

IV. Furan in Place of Imidazole; the Discovery of Ranitidine

At Allen and Hanburys Ltd, which was part of the Glaxo Group, a research team had been established in the late 1960s to find improved drugs for treating peptic ulcer disease. Among various approaches the team had investigated the activity of thioamides, e.g. 2-phenyl-2-(2-pyridyl)thioacetamide (SC 15396) (24), which had been reported by Searle

(24) SC 15396
Searle's 'Antigastrin'

scientists to block gastrin (Cook and Bianchi, 1967). In particular, they had synthesized various tetrazole derivatives. When the paper by Black *et al.* (1972) was published defining histamine H$_2$-receptors, the Glaxo chemists turned their attention to the synthesis of tetrazole analogues of burimamide (the first H$_2$-antagonist). The compound where 5-aminotetrazole replaced the imidazole ring (**25**, AH 15475; Table 12.4) was found to be equipotent with burimamide (Bradshaw, 1992). Many other 5-substituted analogues were made but none showed any improvement.

Following the SK&F discovery of active thiourea replacements in H$_2$-antagonists which were published initially in the SK&F patent applications (see above), the Glaxo chemists introduced groups such as cyanoguanidine or diaminonitroethene into the aminotetrazole structure. As for the SK&F compounds, these replacement groups did not increase potency (Bradshaw, 1992).

By 1976, the research at Glaxo had still not been able to improve on AH 15475 and the management argued strongly to abandon H$_2$-antagonists in favour of selective anticholinergics as a target for inhibiting acid secretion. Furthermore, the number of chemists left working on the project was reduced to three (Bradshaw, 1992). One cannot overemphasize the importance that personal experience can make to research. The experience of the senior chemist, Dr John Clitherow, made a crucial contribution. Many years earlier Dr Clitherow had included in his Ph.D. research the synthesis of furan derivatives as intermediates in making analogues of the cholinergic agent muscarine (Beckett *et al.*, 1963). Since aminotetrazole is only weakly basic and yet AH 15475 (in which aminotetrazole functions as an imidazole replacement) was active, the chemists were led to make a furan analogue, i.e. a non-nitrogen heterocycle which was also non-basic. The butyl chain was replaced by CH$_2$SCH$_2$CH$_2$ for ease of synthesis following the lead given by the SK&F work on metiamide; the resulting compound (**26**, AH 18166) showed some activity but had poor solubility in water (Table 12.4). To improve solubility, a Mannich base (from dimethylamine and formaldehyde) was synthesized. This was the breakthrough that the researchers had been waiting for; the product, AH 18665 (**27**) was nearly as active as metiamide (Table 12.5) (Bradshaw *et al.*, 1982).

The chemists also made the cyanoguanidine (**28**, AH 18801) and this proved to be as active as cimetidine (Bradshaw *et al.*, 1982). A problem facing the development chemists was the poor crystallinity and low melting point of this product. There was good reason to expect that the nitrovinyl analogue would be crystalline and so it was synthesized. However, it turned out to be an oil but, fortuitously, it was ten times more potent in the rat than the cyanoguanidine (Table 12.5) (Bradshaw *et al.*, 1982). This compound (**29**, AH 19065) was selected for development and given the name ranitidine.

TABLE 12.4
Some Key Steps in the Discovery of Ranitidine

An aminotetrazole analogue of burimamide is synthesized and found to have activity equipotent with burimamide:

(25)   AH 15475

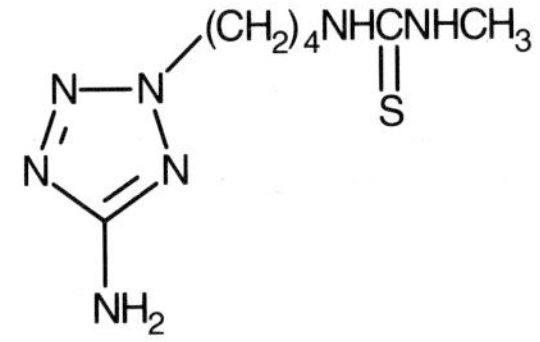

Replacing aminotetrazole by furan and incorporating an —S— link in the chain provides an active compound but it has poor aqueous solubility:

(26)   AH 18166

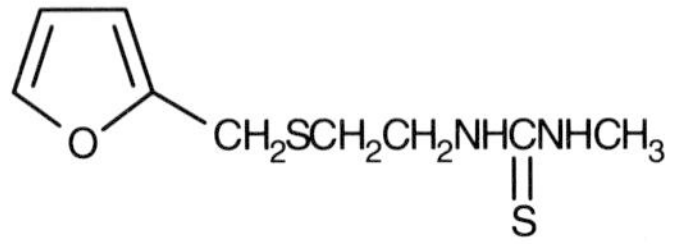

A dimethylaminomethyl group (Mannich base) is introduced to confer solubility and the compound is nearly as potent as metiamide:

(27)   AH 18665

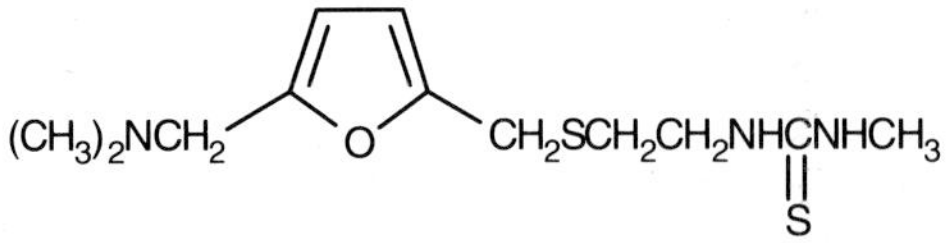

Replacing the thiourea group by a cyanoguanidine group gives a compound equipotent with cimetidine but it has poor crystallinity for pharmaceutical development:

(28)   AH 18801

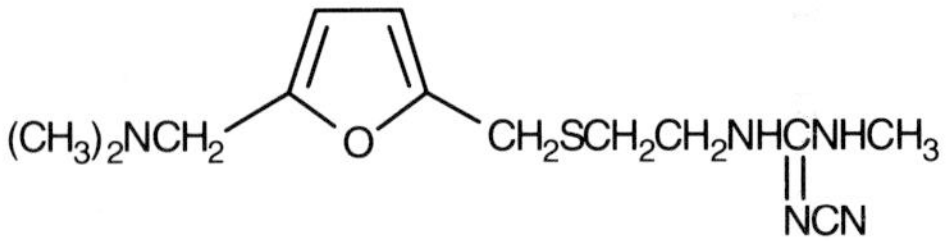

Replacing the cyanoguanidine group by a diaminonitroethene group increases potency by 4–10-fold to give ranitidine

(29)   AH 19065
          Ranitidine

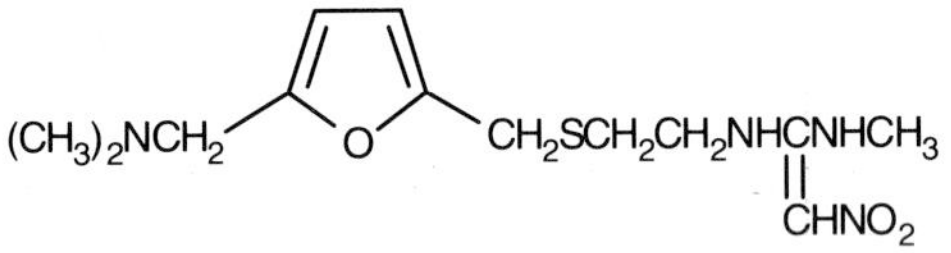

A laboratory synthesis of ranitidine is outlined (Bradshaw *et al.*, 1982) in Scheme 12.4. 2-Dimethylaminomethyl-5-hydroxymethylfuran, made from furfuryl alcohol by a Mannich condensation with formaldehyde and dimethylamine, is condensed with cysteamine in aqueous acid. The resulting amine is liberated with base and treated with 1,1-dimethylthio-2-nitroethene and then with methylamine to furnish ranitidine (29).

Ranitidine is about four or five times more potent than cimetidine in man as an inhibitor of gastric acid secretion (Zeldis *et al.*, 1983; Ireland *et al.*, 1984) and, in 1981 it was marketed as Zantac[R]. Ranitidine is not only more potent than cimetidine but it proved to be more selective. Cimetidine inhibits cytochrome P-450, an import-

TABLE 12.5

Inhibition of Histamine Stimulated Gastric Acid Secretion in the Perfused Stomach Preparation of the Anaesthetized Rat (Bradshaw *et al.*, 1982). Compounds given intravenously; ED$_{50}$ in mg/kg

$$\text{HetCH}_2\text{SCH}_2\text{CH}_2\text{NHCNHCH}_3$$
$$\overset{\parallel}{\text{V}}$$

Het =

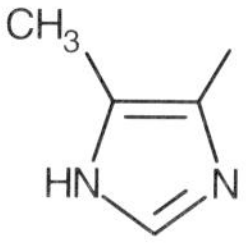

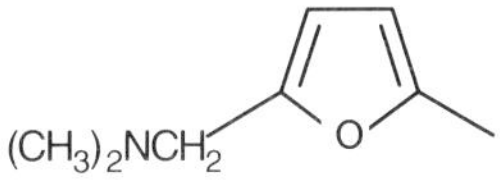

V	ED$_{50}$	ED$_{50}$
S	0.52	2.32
NCN	1.12	1.39
CHNO$_2$	1.75	0.18

ant drug-metabolizing enzyme. This interaction has the effect of inhibiting the metabolism of certain drugs such as propranolol, warfarin, diazepam and theophylline, thus producing effects equivalent to an overdose of these medicines. It is therefore advisable to avoid co-administration. Cimetidine is also a weakly active antiandrogen and high doses can lead, in extreme cases, to gynaecomastia in men. Both these effects, i.e. cytochrome P-450 inhibition and binding to androgen receptors, are avoided by ranitidine.

Clinically therefore ranitidine offered some advantages; eventually, ranitidine built up its own safety record for treating peptic ulcer disease and by 1987 its world-wide sales had overtaken those of cimetidine and it became the best-selling prescription product.

V. Discovery of Tiotidine and Famotidine

At ICI Pharmaceuticals, antiulcer research had been aimed at peptide analogues of gastrin as potential gastrin antagonists but one was not obtained. Following the publication of Black *et al.* (1974) they turned to histamine as a target and began screening compounds which they already had on file. From this emerged 2-guanidino-4-methyl-thiazole (**30**) which was shown to be an H$_2$-receptor histamine antagonist on the isolated guinea-pig right atrium, with an apparent dissociation constant (K_B) of 3.6×10^{-6} M (Gilman *et al.*, 1982).

Compound (**30**) served as a new lead which was outside the scope of the SK&F patents, since SK&F had not covered guanidine as a substituent in the heterocyclic part of their molecules.

This was followed up by the synthesis of compounds in which the appropriate side chain replaced the methyl group. Structure (**31**), ICI 125211, was approxi-

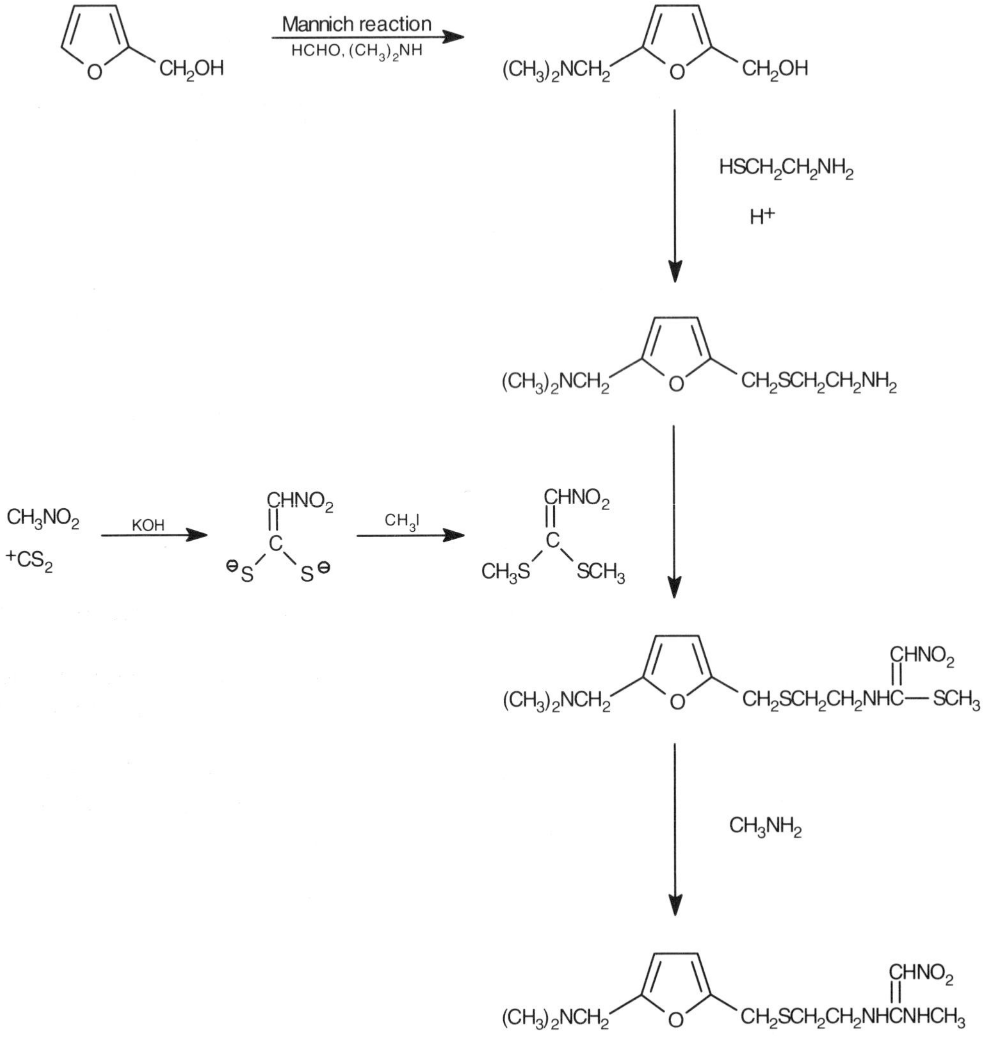

(**29**) Ranitidine

Scheme 12.4 Laboratory synthesis of ranitidine (**29**).

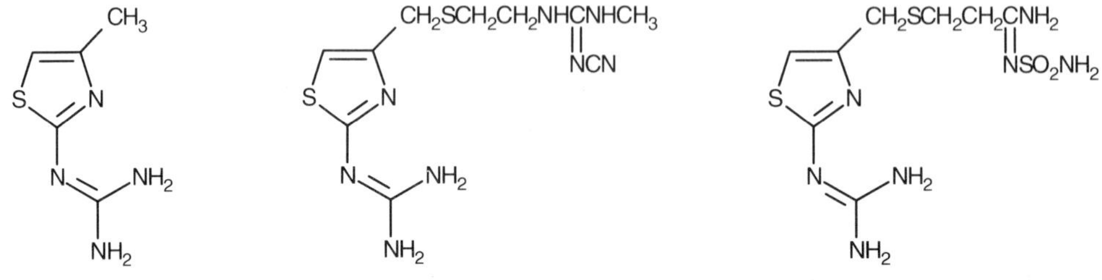

(**30**) (**31**) Tiotidine (**32**) Famotidine

mately 40 times more potent than cimetidine as an H$_2$-receptor antagonist (Gilman et al., 1982). This compound, given the name tiotidine, was selected for drug development.

Tiotidine was eight times more potent than cimetidine in man with a relatively slow onset and duration of action. It looked to be a very promising compound as a potential therapeutic agent but, unfortunately, after prolonged high dose safety studies in rats a problem was discovered: tiotidine was found to have a toxic effect on the cells lining the stomach and so clinical investigations had to be suspended.

The guanidinothiazole group appears to confer high affinity at the H$_2$-receptor and many active derivatives are now known. Only one of these, however, has become a therapeutically useful drug; this is a sulphamidoamidine analogue discovered in the laboratories of the Japanese Company, Yamanouchi. This compound, famotidine (YM 11170, **32**) (Yanagisawa et al., 1987), is about thirty times more potent than cimetidine in man and is marketed in Western countries by Merck, Sharp and Dohme (under licence from Yamanouchi).

VI. Other Compounds

The discovery and development of cimetidine and ranitidine provided a revolution in the medical treatment and management of peptic ulcer disease.

Many pharmaceutical companies became involved in research programmes to discover additional compounds as H$_2$-receptor histamine antagonists. A very wide range of chemical structures now exists for this class of drug; see Cooper et al. (1990) for a review. Many of these compounds have been investigated in human studies but very few are marketed. Apart from famotidine, there is a thiazole analogue nizatidine (**33**) and a piperidinomethylphenoxypropyl derivative roxatidine (**34**).

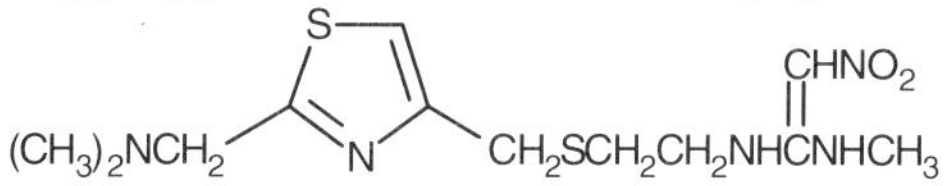

(**33**) Nizatidine

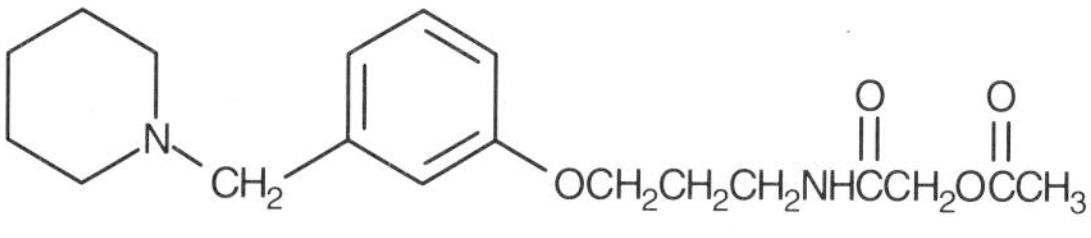

(**34**) Roxatidine

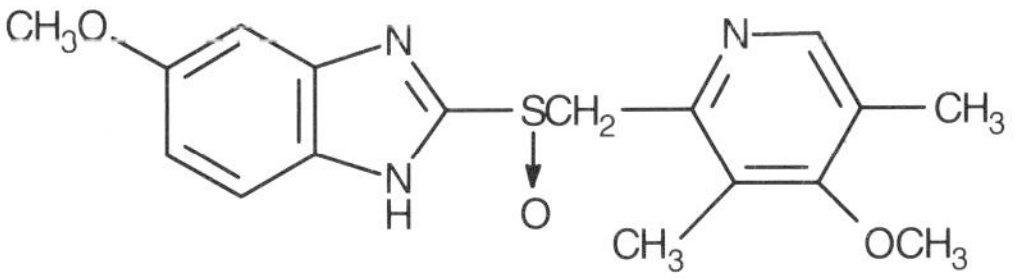

(**35**) Omeprazole

A further development is the discovery of a new class of drug which acts beyond the histamine receptor and inhibits the enzyme H$^+$/K$^+$-ATPase in the parietal cell.

This enzyme is responsible for effecting an exchange between potassium ions and hydrogen ions and is known as the 'proton pump'. It appears to be the final stage in the complex sequence of biochemical events that leads to the secretion of hydrochloric acid by the parietal cell. The first compound of this class to be marketed is omeprazole (35) which was launched in 1988. Omeprazole is a very potent and long-lasting inhibitor.

As the above events demonstrate, the situation in drug therapy is continuously evolving. A treatment that is heralded as a major breakthrough in one decade may later be surpassed by another drug discovery and, potentially, even displaced by an alternative approach to intervention in a physiological mechanism that controls a pathological condition. Whichever treatment prevails, however, the result has been to bring relief to millions of patients who suffered from peptic ulcers.

References

Ash, A.S.F. and Schild, H.O. (1966). *Br. J. Pharmacol. Chemother.* **27**, 427.

Beckett, A.H., Harper, N.J. and Clitherow, J.W. (1963). *J. Pharm. Pharmacol.* **15**, 362.

Black, J.W., Duncan, W.A.M., Durant, G.J., Ganellin, C.R. and Parsons, M.E. (1972). *Nature (London)* **236**, 385.

Black, J.W., Durant, G.J., Emmett, J.C. and Ganellin, C.R. (1974). *Nature (London)* **248**, 65.

Bovet, D. (1950). *Ann. N.Y. Acad. Sci.* **50**, Art. 9, 1089.

Bradshaw, J., Butcher, M.E., Clitherow, J.W., Dowle, M.D., Hayes, R., Judd, D.B., McKinnon, J.M. and Price, B.J. (1982). *In* 'The Chemical Regulation of Biological Mechanisms' (A.M. Creighton and S. Turner, eds), pp. 45–57, Special publications No. 42. Royal Society of Chemistry, London.

Bradshaw, J. (1992). *In* 'Chronicles of Drug Design' (D. Lednicer, ed.) American Chemical Society, New York.

Brimblecombe, R.W., Duncan, W.A.M., Durant, G.J., Emmett, J.C., Ganellin, C.R. and Parsons, M.E. (1975). *J. Int. Med. Res.* **3**, 86.

Brimblecombe, R.W., Duncan, W.A.M., Durant, G.J., Emmett, J.C., Ganellin, C.R., Leslie, G.B. and Parsons, M.E. (1978). *Gastroenterology* **74**, 339.

Charton, M. (1965a). *J. Org. Chem.* **30**, 369.

Charton, M. (1965b). *J. Org. Chem.* **30**, 3346.

Cook, D.L. and Bianchi, R.G. (1967). *Life Sci.* **6**, 1381.

Cooper, D.G., Young, R.C., Durant, G.J. and Ganellin, C.R. (1990). *In* 'Comprehensive Medicinal Chemistry' (J.C. Emmett, ed.), Vol. 3, pp. 323–421, Pergamon Press, Oxford.

Dale, H.H. and Laidlaw, P.P. (1910). *J. Physiol. London* **41**, 318.

Durant, G.J., Emmett, J.C., Ganellin, C.R., Miles, P.D., Prain, H.D., Parsons, M.E. and White, G.R. (1977). *J. Med. Chem.* **20**, 901.

Folkow, B., Haeger, K. and Kahlson, G. (1948). *Acta Physiol. Scand.* **15**, 264.

Freston, J.W. (1982). *Ann. Intern. Med.* **97**, 573.

Ganellin, C.R. (1981). *J. Med. Chem.* **24**, 913.

Ganellin, C.R., Durant, G.J. and Emmett, J.C. (1976). *Fed. Proc., Fed. Am. Soc. Exp. Biol.* **35**, 1924.

Gilman, D.J., Jones, D.F., Oldham, K., Wardleworth, J.M. and Yellin, T.O. (1982). *In* 'The Chemical Regulation of Biological Mechanisms' (A.M. Creighton and S. Turner, eds), pp. 58–76, Special publications No. 42. Royal Society of Chemistry, London.

Grossman, M.I., Robertson, C. and Rosiere, C.E. (1952). *J. Pharmacol. Exp. Ther.* **104**, 277.

Ireland, A., Colin-Jones, O.G., Gear, P., Golding, P.L., Ramage, J.K., Williams, J.G., Leicester, R.J., Smith, C.L., Ross, G., Bamforth, J., De Gara, C.J., Gledhill, T. and Hunt, R.H. (1984). *Lancet* **ii**, 274.

Prout, K., Critchley, S.R., Ganellin, C.R. and Mitchell, R.C. (1977). *J. Chem. Soc., Perkin Trans.* **2**, 68.

Spence, C. (1990). *Pharm. Market.* **13**.

Winship, D.H. (1978). *Gastroenterology* **74**, 402.

Yanagisawa, I., Hirata, Y. and Ishii, Y. (1987). *J. Med. Chem.* **30**, 1787.

Zeldis, J.B., Friedman, L.S. and Isselbacher, K.J. (1983). *New Engl. J. Med.* **309**, 1368.

–13–

Fluconazole, An Orally Active Antifungal Agent

K. RICHARDSON
Discovery Chemistry,
Central Research,
Pfizer Limited,
Sandwich, Kent, U.K.

Dr Ken Richardson joined Boots Research Department in Nottingham at the age of 16. He subsequently obtained a G.R.I.C. at the local college followed by a Ph.D. at Nottingham University, supervised by Professor A.W. Johnson. Post-doctoral research on B_{12} synthesis at Harvard (supervised by Professor R.B. Woodward) was followed by 4 years' research at Pfizer in the U.S.A. before his move to Pfizer Central Research in Sandwich, Kent. He is currently a Director in Discovery Chemistry with responsibilities for research in infectious diseases, pulmonary, gastrointestinal and urogenital areas.

I. Introduction

The therapy of fungal infections, particularly those in immune-compromised individuals, is an increasingly difficult medical problem (Stahel *et al.*, 1982; Robinson *et al.*, 1990). Fungi that once were innocuous, merely interfering with the freshness of one's bread or asymptomatically colonizing most human gastrointestinal tracts, are now often the cause of life-threatening infections in immune-compromised patients (Heeres, 1985). Many factors have contributed to an increased number of such susceptible patients. Chemotherapy and radiation used in cancer patients may damage mucosal barriers, which are part of our natural defence against pathogens (Slavin *et al.*, 1978). Similarly, complicated surgical procedures such as the use of in-dwelling catheters or long-term corticosteroid treatment may predispose patients to the development of fungal infection. Broadspectrum antibacterial treatment may promote fungal growth and in the presence of a deficient immune system, as in cancer and organ-transplant patients, this can lead to fungaemia and disseminated infection.

MEDICINAL CHEMISTRY 2nd Edition
ISBN 0-12-274120-X

Accordingly, in 1978 a group of scientists were brought together at Pfizer Central Research in Sandwich with the objective of discovering a novel agent for the treatment of serious systemic infections. At that time, the incidence of such infections was quite low but we reasoned that in the future, with the advances in medical technology, there would be increasing numbers of patients at risk to such infections. Major advances were being made in the search for successful therapies for disorders that had previously been fatal. Organ and bone-marrow transplantation, cytotoxic chemotherapy for malignancy and the use of highly potent antibacterial agents were all being used with increasing frequency with the result that there was a slow but sure increase in the numbers of reported systemic fungal infections (Hart *et al.*, 1969; Young *et al.*, 1974; Eilard and Norrby, 1978). In addition, because of the difficulties in diagnosis of invasive fungal infections, the existence of a systemic fungal infection was often not proven until autopsy.

The available drugs at that time had marked deficiences. The most frequently administered agent was amphotericin B, a polyene antibiotic which binds to the fungal sterol ergosterol, in the cell membrane, and promotes the loss of vital ions from the cell. However, amphotericin also has some affinity for the mammalian sterol cholesterol and this lack of selectivity results in a number of side effects in man, including shaking, chills, fever, cramps, nausea, vomiting and renal failure (Maddux and Barriere, 1980). These side effects, together with the fact that amphotericin has to be given by slow intravenous infusion, made clinicians reluctant to use it and frequently amphotericin was used as a last resort, when it was too late. The only other widely available systemically effective drug was 5-fluorocytosine (flucytosine), which had the major advantages of good oral bioavailability and low toxicity but had the disadvantage of being active against only a limited range of fungi and resistance frequently arose during long-term treatment (Block *et al.*, 1973).

Amphotericin B   Flucytosine

There was clearly a need for a safe effective agent that could be given on suspicion of a fungal infection and we felt that demand would increase in the future. We believed that the ideal agent would be one that could be administered both orally and intravenously. The oral route would be the most frequently used but certain cancer patients have difficulty with oral dosing and therefore the availability of an intravenous dosage form would be a considerable advantage. We also wanted our agent to be effective against a wide range of fungal species. Our belief was that an agent satisfying all of these criteria would represent a considerable advance in antifungal chemotherapy.

II. Approach and Mechanism of Action

A number of possible approaches were considered but the one adopted was based on *imidazole* derivatives. These compounds possessed several of the properties that we felt should be in our target antifungal agent. For example, compounds such a clotrimazole, miconazole and tioconazole (Jevons *et al.*, 1979), possessed very potent activity *in vitro*

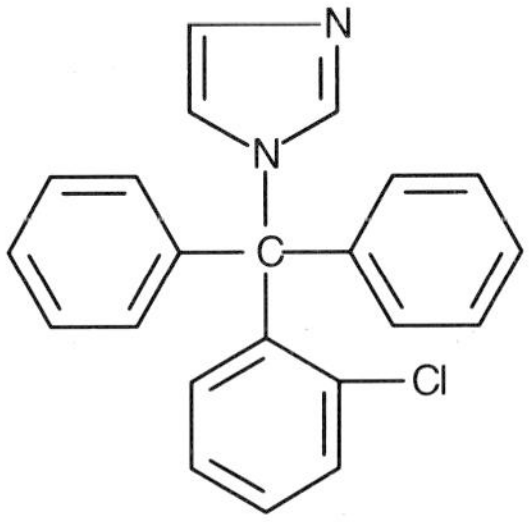

Clotrimazole

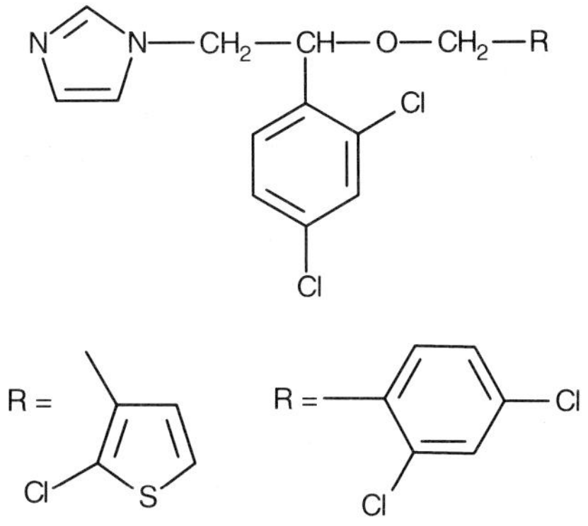

Tioconazole   Miconazole

against a broad range of fungal pathogens. Their mechanism of antifungal action was known to be due to inhibition of fungal C-14 demethylase, a cytochrome *P*-450-containing enzyme essential for the production of the principal fungal sterol ergosterol (Van den Bossche *et al.*, 1978), as shown in Fig. 13.1. It was also known that these imidazole derivatives could be remarkably selective for the fungal C-14 demethylase, rather than the closely related mammalian enzyme, suggesting that these compounds were inherently safe. The imidazole antifungals were also known to be very effective clinically, when administered topically, against fungal infections of the skin and vagina. However, these compounds were poorly effective in animal models of fungal infections when administered by either the oral or intravenous routes. Our studies in Sandwich suggested that these imidazole antifungals were susceptible to metabolic inactivation, resulting in low oral bioavailability and low poorly sustained plasma levels. In addition, they were very lipophilic (octanol : water log *P* ~ 5) and highly bound to plasma proteins resulting in very low circulating levels of the unbound active form. Therefore, the approach that we decided to adopt was to make the compounds as metabolically stable as possible whilst minimizing their overall lipophilicity. Our belief was that metabolic stability would lead

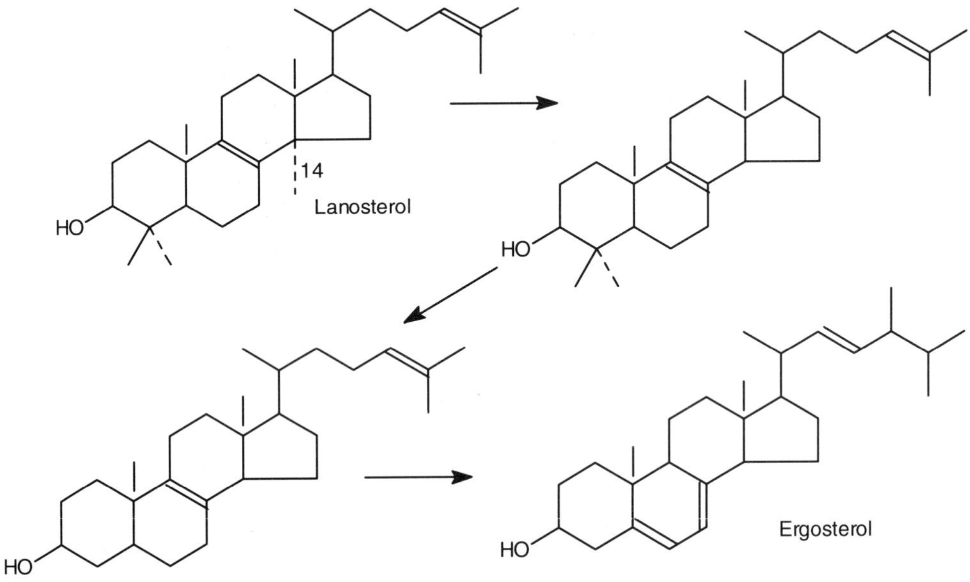

Fig. 13.1   C-14 demethylation, a key step in the synthesis of ergosterol.

to improved oral bioavailability and sustained plasma levels while reduced lipophilicity would result in lower protein binding with the overall effect being to produce high plasma levels of unbound drug.

III.   Chemistry and Structure–Activity Relationships

Our initial investigations involved modifications of tioconazole in which we introduced a range of polar-substituted alkyl, phenyl and heterocyclic groups in place of the chlorothiophene moiety, as an approach to compounds with reduced lipophilicity. Some progress was being made but our plans were modified in view of the report from Thienpont *et al.* (1979) that orally administered ketoconazole was active in several

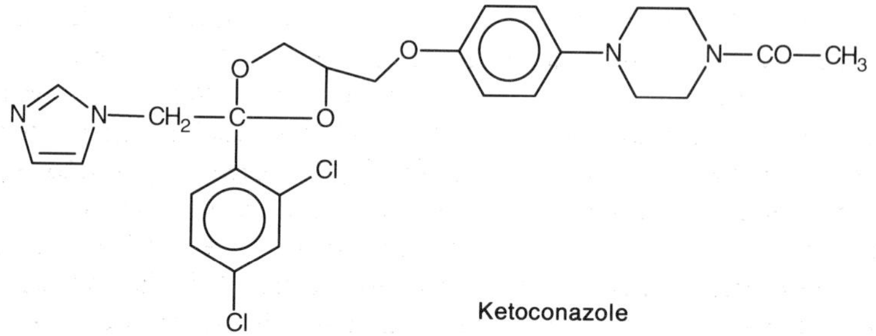

Ketoconazole

animal models of fungal infection. We confirmed these results in our fungal infection models and also showed that, although ketoconazole was less metabolically vulnerable than earlier imidazole antifungal derivatives, resulting in good oral bioavailability, it was metabolized and < 1% of unchanged drug was excreted in the urine. Ketoconazole was

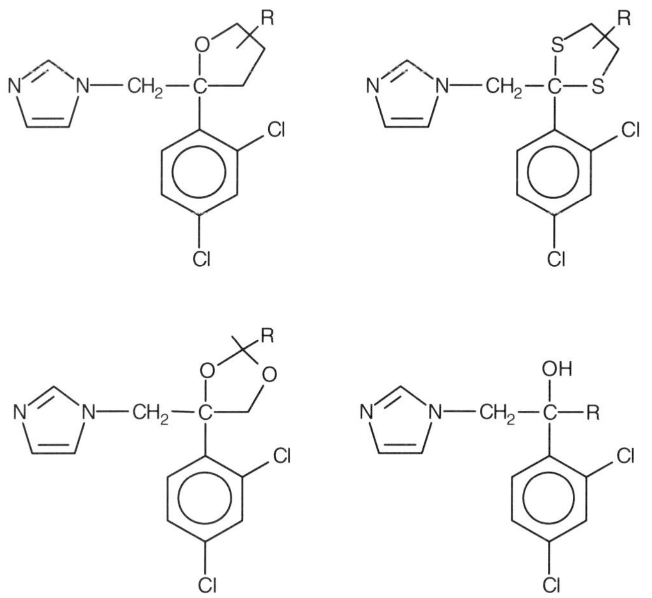

Fig. 13.2 Examples of the imidazole series prepared.

also less lipophilic than earlier imidazole derivatives resulting in somewhat higher blood levels, but it was still highly bound to plasma proteins, with < 1% being in the unbound form. Thus, although ketoconazole was clearly an important advance from the earlier imidazole antifungal derivatives (Heeres, 1985), it left considerable scope for improvement (Holt, 1980) and clearly fell well short of the target that we had set ourselves.

Ketoconazole did, however, represent an important structure–activity step and our plans were adjusted to take account of it. We prepared a range of structural types which were influenced by the structure of ketoconazole, with the major emphasis being placed on the dioxolanes, dithiolanes, tetrahydrofurans and hydroxy derivatives shown in Fig. 13.2. In each of these series, we investigated a wide range of modifications as we sought to reduce lipophilicity and increase metabolic stability. The lipophilicity problem was tackled by introduction of a variety of polar groups at different positions on the structures. However, we found that, although polar functionality such as carboxamide, sulphone, nitrile and polar heterocycles were usually well tolerated, the introduction of ionized groups such as amino and carboxylic acid resulted in a marked loss of both *in vitro* and *in vivo* antifungal activity. The assessment of *in vitro* activity was carried out on agar plates (Kobayashi and Medoff, 1983) against a panel of organisms representing all of the major fungal pathogens (*Candida*, *Cryptococcus*, *Aspergillus* and dermatophytes). Initial *in vivo* evaluation used a systemic *Candida* infection in mice where animals were given a potentially lethal infection then treated with test compounds, or saline, by oral or intravenous dosing (Richardson *et al.*, 1985). Efficacy assessment was carried out at 2 days post-infection when all saline-treated mice would have succumbed to the infection. The dose of compound resulting in the survival of 50% of the treated animals is the ED_{50}. Compounds showing ketoconazole-like anti-*Candida* activity *in vivo* were progressed to evaluation in other animal models, including vaginal candidosis in mice, dermatophytosis in guinea-pigs and studies in immune-compromised animals.

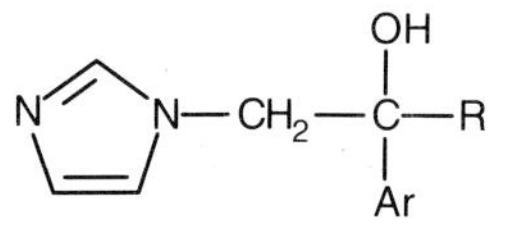

$$\underset{\displaystyle Ar}{\overset{\displaystyle OH}{N{\diagdown}\!\!\diagup N - CH_2 - \underset{|}{\overset{|}{C}} - R}}$$

R = Alkyl,
    Aryl,
    $-(CH_2)_n$–Het,
    $-(CH_2)_n$–$S(O)_x$–R′
    $-(CH_2)_n$–O–R′, etc.

Fig. 13.3 Series of substituted 2-hydroxyethyl imidazole derivatives.

These studies gave a rapid indication of activity against several fungal species, at different infection sites, in two animal species. In addition, all interesting compounds underwent pharmacokinetic evaluation in mice.

In each of the different structural series shown in Fig. 13.2, it proved possible to obtain several compounds showing *in vivo* efficacy similar to that of ketoconazole, but none of these derivatives showed the clear advantage that we were seeking. In order to increase our rate of progress we decided to concentrate our efforts on one series, the hydroxy series. We chose to emphasize this series for several reasons. 1. Owing to the presence of the hydroxy group, this series tended to be less lipophilic. 2. This was the only non-cyclic series and was therefore most structurally distinct from ketoconazole. 3. Compounds of this type were readily synthesized and hence ideas could be most rapidly evaluated.

However, despite the synthesis of 200–300 compounds (Fig. 13.3), and the production of a number of compounds with ketoconazole-like activity, we were unable to obtain any examples which showed *in vivo* activity superior to that of ketoconazole. Pharmacokinetic evaluation of 30 of these derivatives showed that, despite the wide range of structural variations, they all remained metabolically vulnerable and had, at best, a ketoconazole-like pharmacokinetic profile, with a half-life of ~ 1 h in mice with $< 1\%$ being excreted unchanged in the urine.

The only consistent structural feature of all the compounds that we had synthesized was the presence of the imidazole moiety and we therefore concluded that it was possible that this group could be one of the reasons for the metabolic vulnerability of these compounds. This view received support from our discovery that in tioconazole the imidazole group was a site for metabolic inactivation, as was later reported by Macrae *et al.* (1990). We therefore came to the conclusion that the imidazole had to be replaced and we decided to introduce a range of groups (Fig. 13.4) which we believed might interact with cytochrome *P*-450, in place of the imidazole unit. This particular series was chosen to evaluate potential imidazole replacements because compounds were readily synthesized and the corresponding imidazole derivative demonstrated *in vitro* and *in vivo* activity equivalent to that of ketoconazole. Therefore, any improvement in metabolic stability was expected to lead to *in vivo* efficacy superior to that of keto-conazole.

A total of twenty groups were examined as potential replacements, including a range of substituted imidazole derivatives, several other heterocycles and a number of basic groups. However, the only group offering encouragement was 1,2,4-triazole. Replacement of imidazole by a 1,2,4-triazol-1-yl unit gave UK-46,245 which was twice as potent

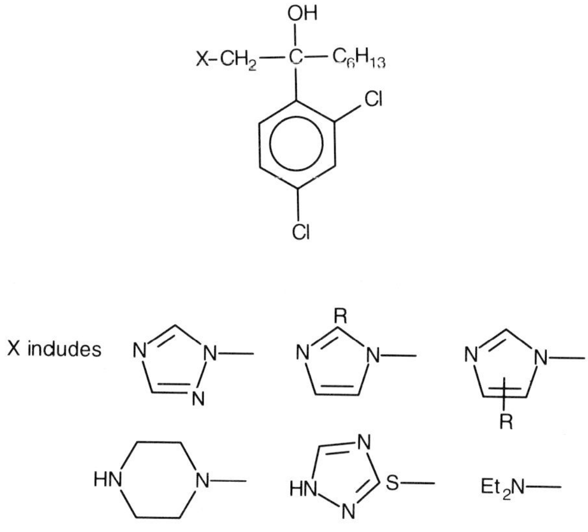

X includes

Fig. 13.4   Preparation of non-imidazole *t*-alcohol derivatives.

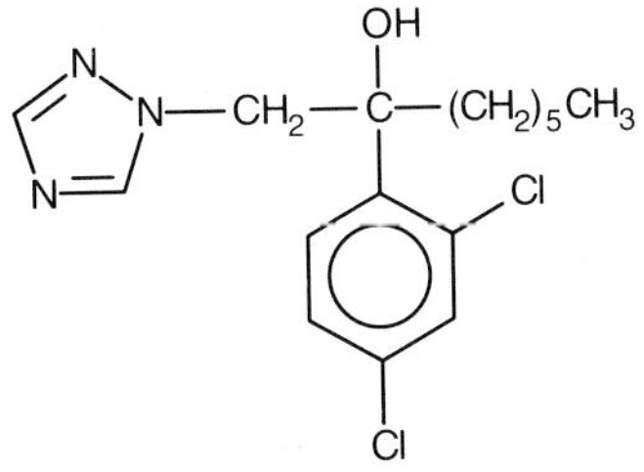

UK-46, 245

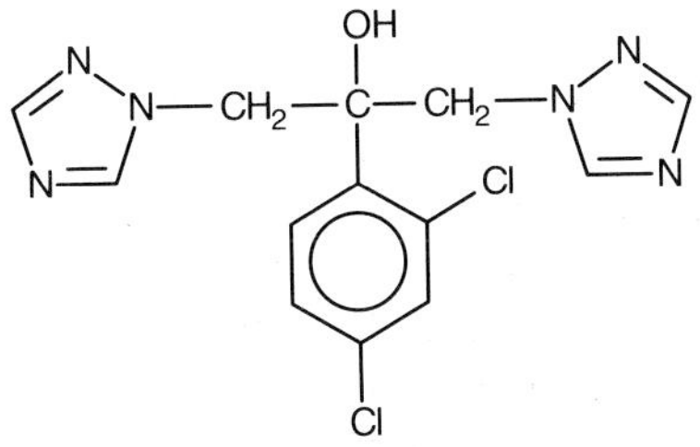

UK-47, 265

in the standard murine systemic candidosis model as the corresponding imidazole analogue, despite being approximately six times less potent against *Candida in vitro*. We interpreted these data as indicating that the triazole group was less susceptible to metabolic inactivation than an imidazole moiety, but yet it retained the ability to interact with the cytochrome *P*-450 unit in the fungal C-14 demethylase enzyme. However, in UK-46,245 there remained a lipophilic metabolically vulnerable hexyl group and therefore we considered how we could replace this moiety and achieve our original objectives of a compound which would combine good metabolic stability with low lipophilicity. A number of potential replacements were considered but the first to be examined was a 1,2,4-triazole-1-yl unit to yield the symmetrical *bis*-triazole compound UK-47,265.

This was a most remarkable compound! When examined in a systemic candidosis model in mice, UK-47,265 was virtually 100 times more potent than ketoconazole following administration by either oral or intravenous dosing. This level of activity was completely unprecedented. Further evaluation against vaginal candidosis in mice and rats and against dermatophytosis in mice and guinea-pigs showed highly impressive levels of efficacy, as did examination against systemic candidosis in immune-compromised mice and rats. These findings were even more remarkable when it was shown that UK-47,265 had only modest activity against fungi when examined using standard *in vitro* assay methods (Troke *et al.*, 1990). Pharmacokinetic studies in rodents indicated that the *in vivo* activity was not due to the formation of active metabolites since UK-47,265 was very stable, showing high and persistent levels of unchanged drug with approximately 30% being excreted intact in the urine, in marked contrast to the < 1% observed with ketoconazole (Richardson *et al.*, 1990).

The poor correlation between *in vitro* activity and *in vivo* efficacy was well recognized with antifungal compounds, although UK-47,265 appeared to represent a rather extreme example. However, it was considered that *in vitro* assessment of antifungal activity would be essential for the development of UK-47,265 since it would be necessary to examine its activity against many clinical isolates and against examples of all important species of fungal pathogens. We therefore initiated a major investigation of the possible reasons for the very low levels of *in vitro* antifungal activity seen with this compound. The activity of UK-47,265 in over 30 different commercially available agar and liquid-based growth media was evaluated. It became clear that complex media, especially those containing peptones, antagonized the activity of UK-47,265 whereas good *in vitro* activity was observed when a tissue culture-based medium similar to the SAAMF medium of Hoeprich (1972) was used.

UK-47,265 showed excellent activity in a wide range of systemic and superficial infection models. The systemic infections were due to *Candida, Cryptococcus* and *Aspergillus* species, and were in a range of infection sites including the kidney, liver, brain, gut and lung. The animals used were both normal and those which had been immune-compromised due to treatment with agents such as cyclophosphamide, dexamethasone and cortisone. The superficial infections were dermatophytosis in mice and guinea-pigs and vaginal candidosis in mice and rats. Pharmacokinetic evaluation in several species (mouse, rat, guinea-pig, rabbit, dog) showed excellent oral bioavailability together with a long plasma half-life and therefore UK-47,265 was progressed into pre-clinical safety evaluation. The results were extremely disappointing since UK-47,265 proved to be hepatotoxic in mice and dogs, and teratogenic in rats. These findings clearly precluded any further progression and we therefore turned our attention to the search for a replacement.

While UK-47,265 was being evaluated in antifungal, pharmacokinetic and safety studies, an intensive follow-up programme had been in progress resulting in the synthesis of over 100 *bis*-triazole compounds. Two principal routes were used for the construction of these derivatives (Fig. 13.5). In the first approach, the chloroacetyl compound (**1**) was

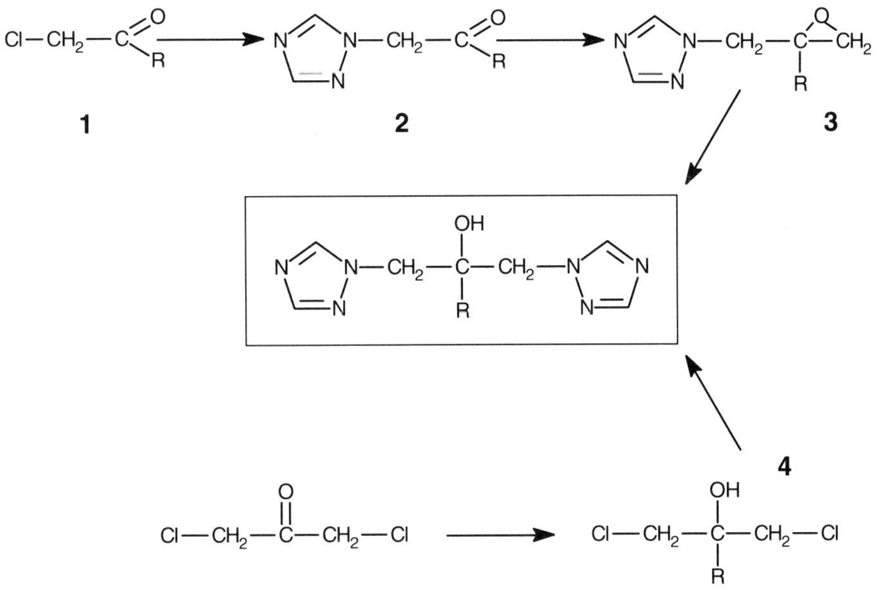

Fig. 13.5   Synthesis of *bis*-triazole derivatives.

reacted with 1,2,4-triazole to yield ketone (**2**) which was converted to epoxide (**3**). Further treatment with 1,2,4-triazole opened the epoxide to yield the *bis*-triazole derivative. In the second approach 1,3-dichloroacetone was converted to the 1,3-dichloropropan-2-ol derivative (**4**) which was then reacted with 1,2,4-triazole, as shown.

All of the *bis*-triazole derivatives were examined in the mouse model of systemic candidosis and a particularly interesting series of compounds resulted from replacement of the dichlorophenyl moiety by a range of aryl and heteroaryl groups (Table 13.1). Many of these derivatives showed very good *in vivo* anti-*Candida* activity and the best of these were progressed to evaluation in mouse models of vaginal candidosis and dermatophytosis (Table 13.2). The best three compounds, the 2,4-difluorophenyl, 2-chloro-4-fluorophenyl and 4-chlorophenyl analogues were progressed to pharmacokinetic evaluation in the mouse (Table 13.3). The outstanding compound was the 2,4-difluorophenyl analogue which had a plasma half-life of 5.1 h with 75% of drug being excreted unchanged in the urine. In addition, this compound was water-soluble (8 mg/ml at room temperature), a property that would allow it to be formulated for intravenous administration. This derivative, UK-49,858 (Fig. 13.6) showed outstanding activity in our full range of fungal infection models in both animals with normal immune function and those with suppressed immune function (Richardson *et al.*, 1985; Troke *et al.*, 1985). Safety evaluation showed that UK-49,858, now known as fluconazole, was not teratogenic or hepatotoxic and it was therefore progressed to studies in man. The pharmacokinetic and antifungal properties of fluconazole are summarized in Table 13.4.

IV.   Clinical Studies

In healthy human volunteers, fluconazole showed excellent (> 90%) oral absorption,

TABLE 13.1
Evaluation of 1,3-*Bis*-triazol-1-ylpropan-2-ol Derivatives vs Systemic Candidosis in Mice

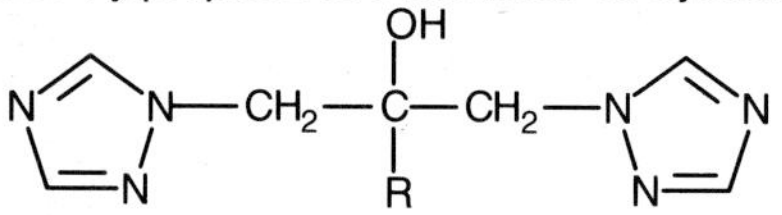

R	ED_{50} (mg/kg)	R	ED_{50} (mg/kg)	R	ED_{50} (mg/kg)	R	ED_{50} (mg/kg)
2,4-difluorophenyl	0.1	2,4-dichlorophenyl	> 10	chloropyridinyl	0.4	phenyl	5.7
3,4-difluorophenyl	> 10	4-chlorophenyl	0.1	2,4,5-trifluorophenyl	0.6	Et	> 10
2,6-difluorophenyl	> 10	2-chlorophenyl	0.2	2-methyl-4-chlorophenyl	> 10	4-chlorobenzyl (CH_2)	> 10
2,4-difluorophenyl	0.1	4-bromophenyl	> 10	4-chlorophenyl	> 10	cyclohexyl (H)	> 10
2,3-difluorophenyl	> 10	4-fluorophenyl	0.3	2-chloro-4-fluorophenyl	0.1	2-chlorothiophenyl (S, Cl)	> 10
2,6-difluorophenyl	> 10	2,4-dichlorophenyl	2.0	2,4,6-trifluorophenyl	0.8	2-bromothiophenyl (S, Br)	1.9
2,4-dichlorophenyl	0.1	4-(CF_3)phenyl	0.1	2,4-dimethylphenyl	> 10	pyridin-2-yl	> 10
4-bromophenyl	0.1	4-iodophenyl	0.1	4-methoxyphenyl (OMe)	> 10	pyridin-4-yl	> 10

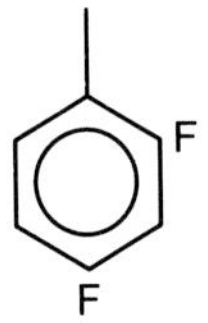

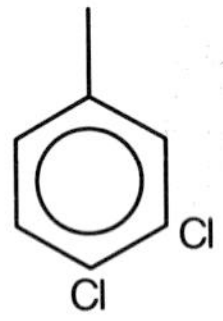

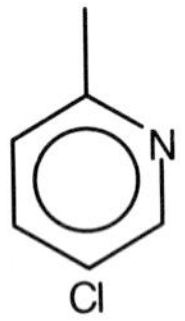

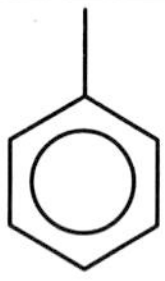

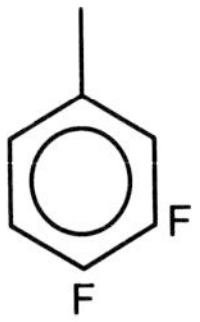

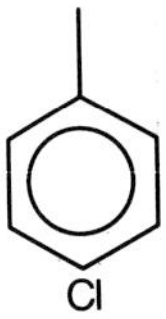

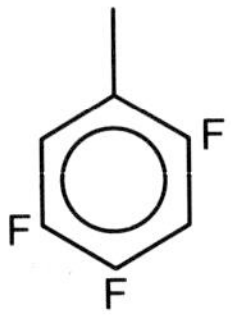

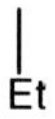

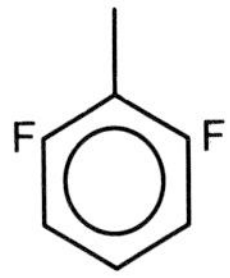

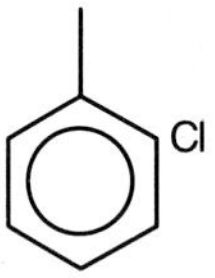

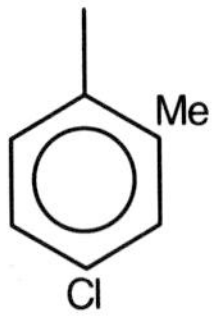

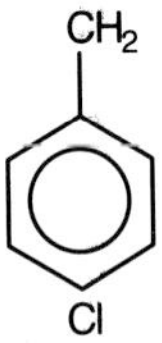

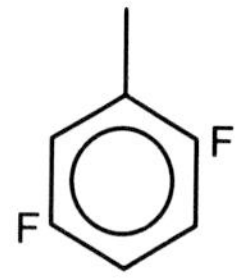

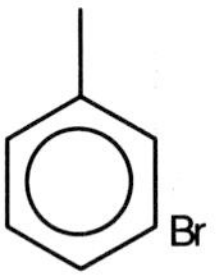

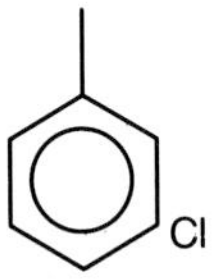

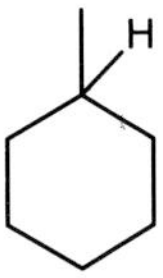

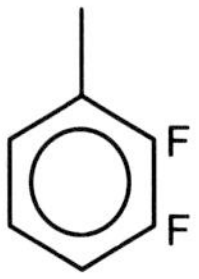

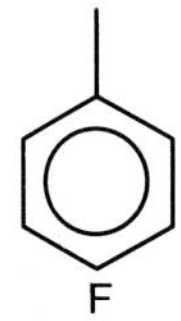

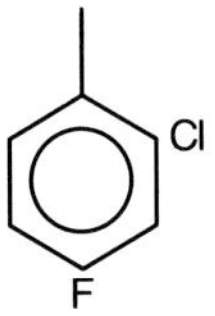

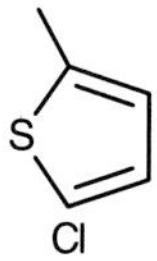

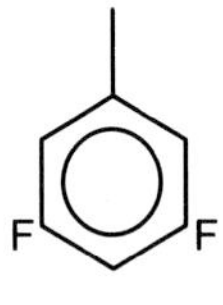

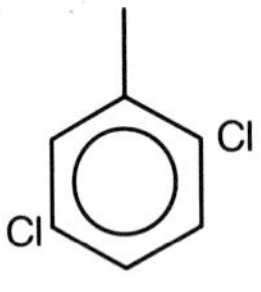

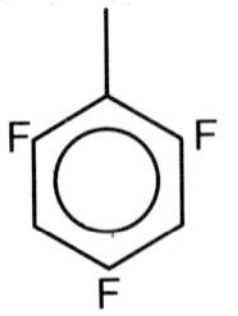

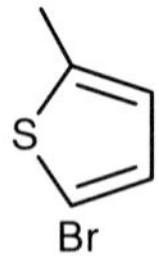

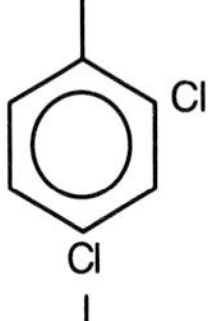

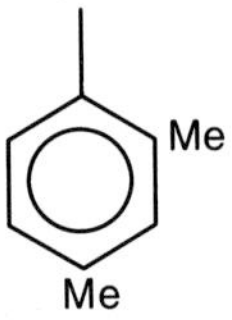

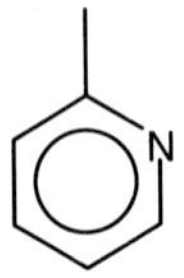

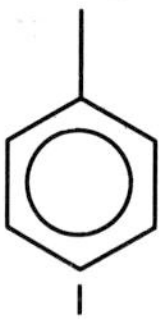

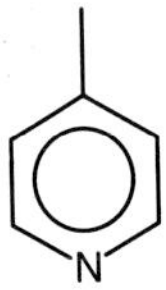

TABLE 13.2
Evaluation vs Vaginal Candidosis and Dermatophytosis in Mice

R	Dermatophytosis (% clinical cure)[a]	Vaginal candidosis (% clinical cure)[a]
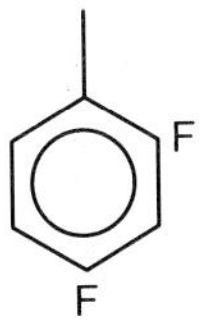 2,4-difluorophenyl	79 (5)	80 (10)
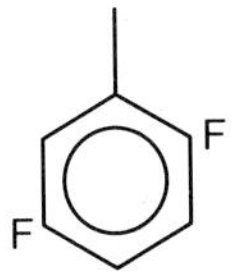 2,4-difluorophenyl (other isomer)	20 (20)	N.D.[b]
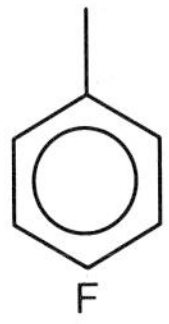 4-fluorophenyl	30 (10)	14 (10)
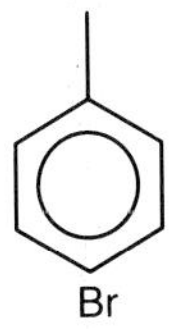 4-bromophenyl	49 (20)	90 (10)
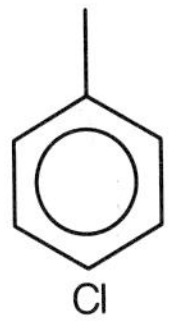 4-chlorophenyl	76 (20)	80 (10)
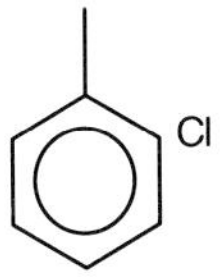 2-chlorophenyl	35 (10)	64 (10)
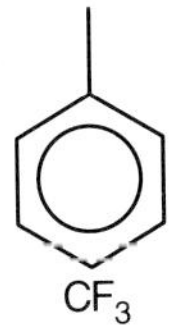 4-trifluoromethylphenyl	0 (20)	34 (5)
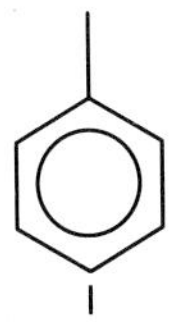 4-iodophenyl	70 (10)	30 (5)

TABLE 13.2
Continued

R	Dermatophytosis (% clinical cure)[a]	Vaginal candidosis (% clinical cure)[a]
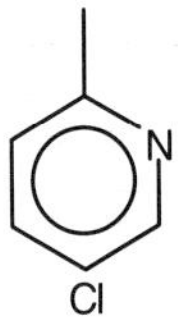	24 (20)	N.D.
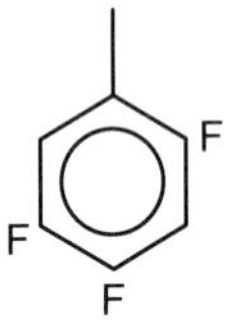	59 (20)	N.D.
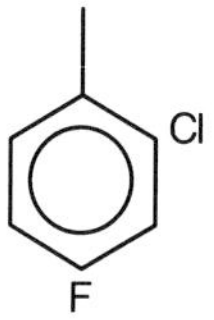	85 (20)	84 (10)
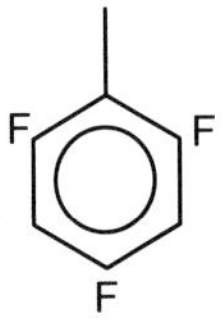	0 (20)	N.D.

[a] Dermatophytosis and vaginal candidosis are given as % clinical cure (dose in mg/kg) (Richardson et al., 1985).
[b] N.D. = Not done.

both from solution and solid capsule formulations, with peak plasma concentrations occurring rapidly. Studies comparing peak concentrations and systemic exposure following oral administration to both fed and fasted volunteers showed that food did not affect the absorption of fluconazole. The plasma half-life was approximately 30 h leading to predictable accumulation following daily dosing, with a steady rate being reached within 4–5 days (Brammer et al., 1990). Approximately 80% of orally administered fluconazole was excreted unchanged in the urine, as was predicted from studies in animals. Importantly, there were no adverse side effects observed in volunteers and therefore fluconazole was progressed to evaluation against fungal infections in man.

Initial efficacy studies were carried out in patients with acute vulvovaginal candidosis. Two hundred patients were enrolled in this study with half being given a 150 mg single oral dose of fluconazole and half were dosed intravaginally (200 mg daily for 3 consecutive days) with commercial Canesten (clotrimazole) vaginal tablets. Both drugs produced excellent clinical responses (fluconazole 100%; Canesten 97%) with no side effects (Brammer and Lees, 1987). Following these very encouraging data, fluconazole was evaluated for the treatment of fungal skin infections. In an initial study, 43 patients received fluconazole as a 50 mg tablet daily. Therapy lasted for up to 6 weeks, depending

TABLE 13.3
Solubility and Pharmacokinetic Properties in Mice

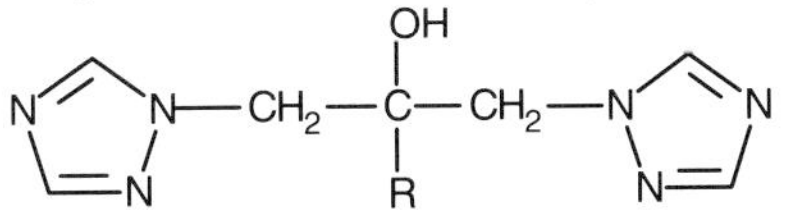

R	$T_{1/2}$ (h)	Percentage in urine	Water solubility (mg/ml)
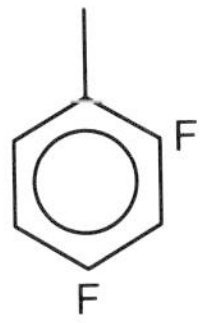	5.1	75	8
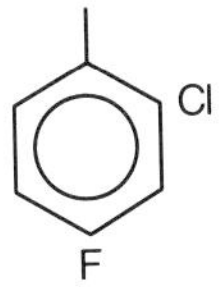	3.6	29.4	< 1
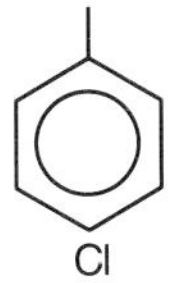	4.0	12.5	< 1

on the severity of the infection, and all patients were clinically cured or improved at the end of treatment (Naeyaert et al., 1987).

Evaluation in immune-compromised patients commenced in patients with oral Candida infections. The large majority of these patients were human immunodeficiency virus (HIV)-positive and the authors (Dupont and Drouhet, 1988) reported 'spectacular' clinical efficacy with a 100% success rate after 5–7 days' treatment. Tolerance to fluconazole was difficult to assess because of the patient's underlying disease but Dupont and Drouhet could detect no clinically significant side effects. De Wit et al. (1989) compared the activities of fluconazole and ketoconazole, both given orally, for the treatment of oropharyngeal candidosis in patients with acquired immune deficiency syndrome (AIDS). Fluconazole had a 100% cure rate compared with 75% for keto-conazole. The authors suggested that the superiority of fluconazole, despite the greater

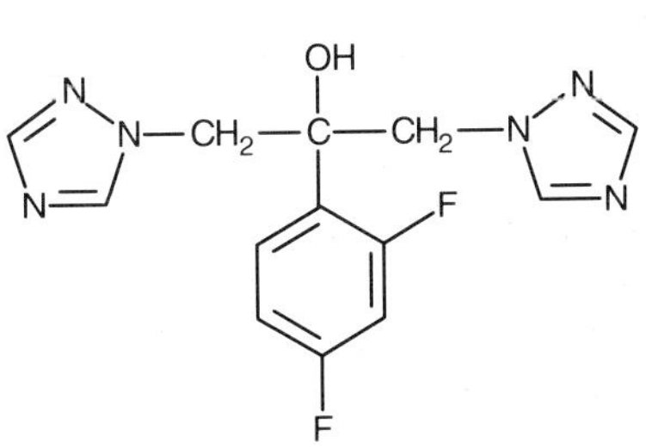

Fig. 13.6   Structure of fluconazole (UK-49,858).

TABLE 13.4
Pharmacokinetic and Antifungal Properties of Fluconazole

- Potent *in vivo* activity against *Candida, Cryptococcus, Aspergillus* and dermatophytes.
- Active in immune-normal and immune-compromisd animals.
- Excellent oral bioavailability in several animal species.
- Water-solubility facilitates formulation for i.v. administration.
- Good safety profile in animals.

in vitro potency of ketoconazole, could arise because fluconazole did not adversely affect the proliferative responses of lymphocytes.

These extremely promising results encouraged progression to studies of potentially life-threatening infections. Cohen (1989) reported the treatment of *Candida* infections in five patients who were immune-suppressed because of chemotherapy for cancer or after organ transplantation. Doses of 50–400 mg/day were given intravenously for up to 15 days and all five patients were cured of their infections. Further studies showed very good efficacy against *Candida* infections at a wide range of body sites in AIDS, cancer, leukaemia and organ-transplant patients (Van't Wout *et al.*, 1988; Kujath and Lerch, 1989; Gritti *et al.*, 1990; Kauffman *et al.*, 1991).

Cryptococcal meningitis is a life-threatening fungal infection in up to 30% of AIDS patients, and the results of treatment with amphotericin, alone or combined with flucytosine, have been disappointing (Diamond, 1991). Dupont and Drouhet (1987) achieved remarkable clinical success with fluconazole and a much larger study (Dismukes *et al.*, 1989) using 200 mg a day has confirmed that fluconazole can achieve dramatic success against this life-threatening infection. Patients with AIDS will require some form of maintenance antifungal therapy to prevent relapse/reinfection and a recent study (Diamond, 1991) showed that prophylaxis with fluconazole was highly effective, and fluconazole is now the drug of choice for these patients. Fluconazole is also being used successfully for the treatment of infections due to Coccidioides, Blastomyces, Histoplasma, Fusarium and Penicillium species.

V. Conclusions

In summary, fluconazole is a safe effective broad-spectrum antifungal agent that may be given by both the oral and intravenous routes, in a once-daily dosing schedule. It therefore satisfies all of the original objectives of our research programme. The safety appears to be due to fluconazole's remarkable selectivity for the fungal C-14 demethylase, rather than related enzymes present in man. This selectivity appears to be due to fluconazole's polarity and the fact that it is a triazole derivative. Our studies have shown that triazole derivatives are more selective for the fungal system than their imidazole analogues while reductions in lipophilicity aid selectivity. Therefore, fluconazole benefits from both being more polar than other azole antifungals and being a triazole derivative.

Fluconazole's low lipophilicity also results in an even distribution throughout the body, including the cerebrospinal fluid, and in its very low protein binding (12%) which together ensure high levels of unbound drug at sites of infection. The low lipophilicity results in fluconazole being water-soluble leading to excellent oral absorption in fed and fasted patients, and in those receiving H_2-antagonists such as cimetidine. It also facilitates ready formulation for intravenous administration which is unique among azole antifungal agents.

Fluconazole's metabolic stability results in excellent oral bioavailability, by preventing first-pass metabolism in the liver and leads to a long half-life (30 h) and a high percentage of drug eliminated, unchanged, in the urine.

Fluconazole was marketed in 1988 under the trade name of Diflucan and it is now on the market in over 30 countries and has already been used by approximately four million patients. Over the next few years Diflucan will be marketed all around the world and, if the initial promise is maintained, clearly, it is destined to become a major breakthrough for the treatment and prevention of fungal infections.

Acknowledgements

The author wishes to acknowledge scientific colleagues and collaborators – too many to mention individually – for their invaluable contributions to the fluconazole programme in all its phases.

Additional Reading

'Recent Trends in the Discovery, Development and Evaluation of Antifungal Agents' (R.A. Fromtling ed.), pp. 75–174. J.R. Prous, Barcelona, 1987.
'Reviews of Infectious Diseases', Vol. 12, Supplement 3, March–April 1990.
'Scrip's New Product Review', No. 41 Fluconazole. PJB Publications Ltd, Richmond, Surrey, 1989.

References

Block, E.R., Jennings, A.E. and Bennett, J.S. (1973). *Antimicrob. Agents Chemother.* **3**, 649.
Brammer, K.W. and Lees, L.J. (1987). In 'Recent Trends in the Discovery, Development and Evaluation of Antifungal Agents' (R.A. Fromtling, ed.), p.151, J.R. Prous, Barcelona.
Brammer, K.W., Farrow, P.R. and Faulkner, J.K. (1990). *Rev. Infect. Dis.* **12** (Suppl. 3), S318.
Cohen, J. (1989). *J. Antimicrob. Chemother.* **23**, 294.
De Wit, S., Weerts, D., Goossens, H. and Clumeck, N. (1989). *Lancet i* 746.
Diamond, R.D. (1991). *Rev. Infect. Dis.* **13**, 480.
Dismukes, W., Cloud, G., Thompson, S., Sugar, A. and Tuazon, C. (1989). '1989 ICAAC' in Houston, Abstr. 1065.
Dupont, B. and Drouhet, E. (1987). *Ann. Intern. Med.* **106**, 778.
Dupont, B. and Drouhet, E. (1988). *J. Med. Vet. Mycol.* **26**, 67.
Eilard, T. and Norrby, R. (1978). *Scand. J. Infect. Dis.* Suppl. 16, 15.
Gritti, F.M., Raise, E., Bonazzi, L., Vannin, V., Di Giandomenico, G., Lanzoni, G. and Cucci, A.M. (1990). *Curr. Ther. Res.* **47**, 1049.
Hart, P., Russel, E. and Remington, J. (1969). *J. Infect. Dis.* **120**, 169.
Heeres, J. (1985). In 'Medicinal Chemistry. The Role of Organic Chemistry in Drug Research' (S.M. Roberts and B.J. Price, eds.), p. 249, Academic Press, London.
Hoeprich, P.D. and Finn, P.D. (1972). *J. Infect. Dis.* **126**, 353.
Holt, R.J. (1980). In 'Antifungal Chemotherapy' (D.C.E. Speller, ed.), p.107, John Wiley, New York.
Jevons, S., Gymer, G.E., Brammer, K.W., Cox, D.A. and Leeming, M.R.G. (1979). *Antimicrob. Agents Chemother.* **15**, 597.

Kauffman, C.A., Bradley, S.F., Ross, S.C. and Weber, D.R. (1991). *Am. J. Med.* **91**, 137.

Kobayashi, G.S. and Medoff, G. (1983). *In* 'Fungi Pathogenic for Humans and Animals' (D.H. Howard, ed.), p. 357, Marcel Dekker Inc., New York.

Kujath, P. and Lerch, K. (1989). *Infection* **17**, 111.

Macrae, P.V., Kinns, M., Pullen, F.S. and Tarbit, M.H. (1990). *Drug Metab. Dispos.* **18**, 1100.

Maddux, M.S. and Barriere, S.L. (1980). *Drug Intell. Clin. Pharm.* **14**, 177.

Naeyaert, J.M., de Bersaques, J., de Cuyper, C., Hindryckx, P., van Landuyt, H. and Gordts, B. (1987). *In* 'Recent Trends in the Discovery, Development and Evaluation of Antifungal Agents' (R.A. Fromtling, ed.), p.157, J.R. Prous, Barcelona.

Richardson, K., Brammer, K.W., Marriott, M.S. and Troke, P.F. (1985). *Antimicrob. Agents Chemother.* **27**, 832.

Richardson, K., Cooper, K., Marriott, M.S., Tarbit, M.H., Troke, P.F. and Whittle, P.J. (1990). *Rev. Infect. Dis.* **12** (Suppl. 3), S267.

Robinson, P.A. Knirsch, A.K. and Joseph, J.A. (1990). *Rev. Infect. Dis.* **12** (Suppl. 3), S349.

Slavin, R.E., Dions, M.A. and Saral, R. (1978). *Cancer* **42**, 1747.

Stahel, R.A., Vogt, P., Schüler, G., Rüttner, J.R., Frick, P. and Oelz, O. (1982). *J. Infect.* **5**, 269.

Thienpont, D., Van Cutsem, J., Van Gerven, F., Heeres, J. and Janssen, P.A.J. (1979). *Experientia* **35**, 606.

Troke, P.F., Andrews, R.J., Brammer, K.W., Marriott, M.S. and Richardson, K. (1985). *Antimicrob. Agents Chemother.* **28**, 815.

Troke, P.F., Andrews, R.J., Pye, G.W. and Richardson, K. (1990). *Rev. Infect. Dis.* **12** (Suppl. 3), S276.

Van den Bossche, H., Willemsens, G., Cools, W., Lauwers, W.F.J. and LeJeune, L. (1978). *Chem. Biol. Interact.* **21**, 59.

Van't Wout, J.W., Mattie, H. and van Furth, R. (1988). *J. Antimicrob. Chemother.* **21**, 655.

Young, R.C., Bennett, J.E., Geelhoed, G. and Levine, A.S. (1974). *Ann. Intern. Med.* **80**, 605.

–14–

Clavulanic Acid and Related Compounds: Inhibitors of β-Lactamase Enzymes

A.G. BROWN
SmithKline Beecham Pharmaceuticals,
Brockham Park,
Betchworth,
Surrey RH3 7AJ, U.K.

I. FRANÇOIS
SmithKline Beecham Pharmaceuticals,
Brockham Park,
Betchworth,
Surrey RH3 7AJ, U.K.

Dr Allan Brown received a B.Sc. degree from the University of Strathclyde and Ph.D. and D.Sc. degrees from the University of Aberdeen. Following further study at the University of Sheffield he joined Beecham Pharmaceuticals 1966. Since then he has been a Project Manager in areas of research aimed at the discovery of new microbial metabolites and antiinfective agents. Currently he is Director of the Antiviral Chemotherapy Programme within SmithKline Beecham Pharmaceuticals, U.K. In 1981 he received the Royal Society of Chemistry, Medicinal Chemistry Award. His interests include new approaches to antiinfective agents and the interaction of these with related medicinal chemistry topics.

Dr Irene François obtained her B.Sc. and her Doctorate, for research on specific enzyme inhibitors in dihydrofolate biosynthesis, from the University of Strathclyde. Since joining Beecham Pharmaceuticals in 1972 she continued her interest in mechanism-based enzyme inhibitors. Her major involvement has been in the field of β-lactamase inhibitors, particularly the naturally occurring clavulanic acid and more recently in totally synthetic alkylidene penem derivatives. Her research interests at SmithKline Beecham also include penem antibacterials and synthetic chemistry associated with β-lactam mimics.

MEDICINAL CHEMISTRY 2nd Edition
ISBN 0-12-274120-X

I. Introduction

In the last 35 years and following the discovery of penicillin by Fleming over 60 years ago, various β-lactam antibiotics have been developed, either as microbial metabolites, or as semisynthetic antibacterials. Today the most widely used β-lactam antibiotics are derivatives of the penicillin (1) and cephalosporin (2) families. Newer entities, such as clavulanic acid (3) and thienamycin (4) have, however, gained considerable interest in recent years (Rolinson, 1979; Morin and Gorman, 1982). In this chapter the discovery, development and clinical application of clavulanic acid, an inhibitor of β-lactamase, is described.

During the early studies on the purification and evaluation of penicillin, Abraham and Chain (1940) recognized that bacteria could produce an enzyme capable of inactivating the antibiotic. The enzyme was called penicillinase, which is one of a class of hydrolytic enzymes referred to as β-lactamases. β-Lactamases destroy penicillins and cephalo-sporins by hydrolysing the β-lactam ring to yield a penicilloic acid derivative (5) in the case of penicillins or a variety of products (e.g. 6 and 7) with cephalosporins, depending upon the nature of the R_2 substituent (Scheme 14.1), thereby preventing the antibiotics from exerting their therapeutic effect.

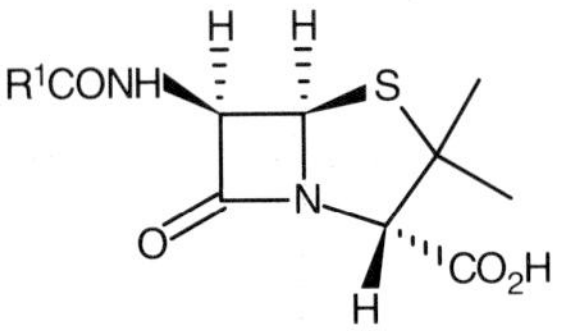

(1)

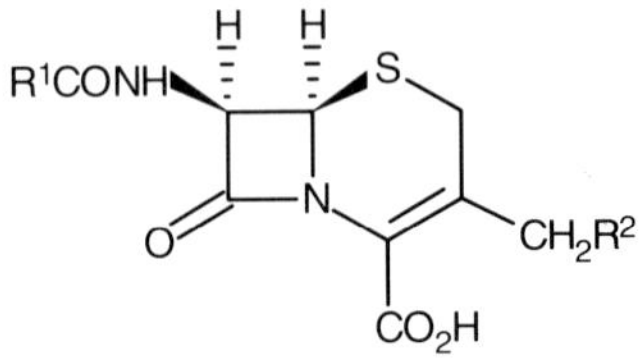

(2)

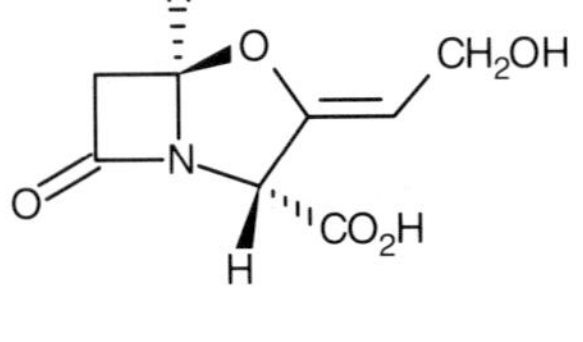

(3) Clavulanic acid

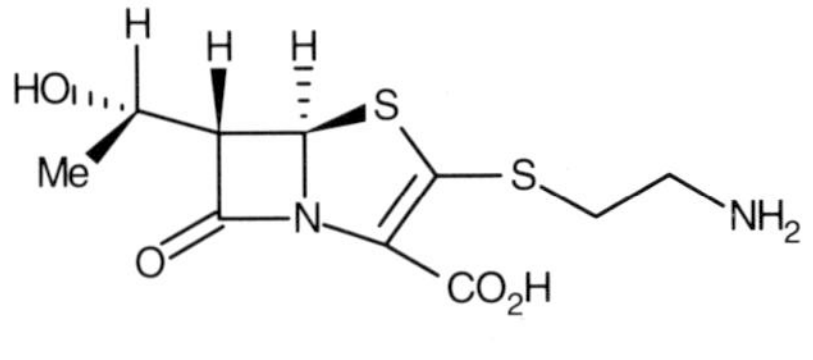

(4) Thienamycin

The ability of bacteria to produce a variety of β-lactamases has proved over the years to be the principal mechanism by which resistance has developed to antibiotic therapy with penicillins and cephalosporins. Initial attempts in the late 1950s and early 1960s to overcome resistance due to β-lactamase production (particularly by *Staphylococcus aureus*) led to the successful development of penicillins and cephalosporins with a high degree of stability towards this enzyme; such compounds were, for example, the penicil-lins methicillin (8) and flucloxacillin (9) and semisynthetic cephalosporins like cefuroxime (10).

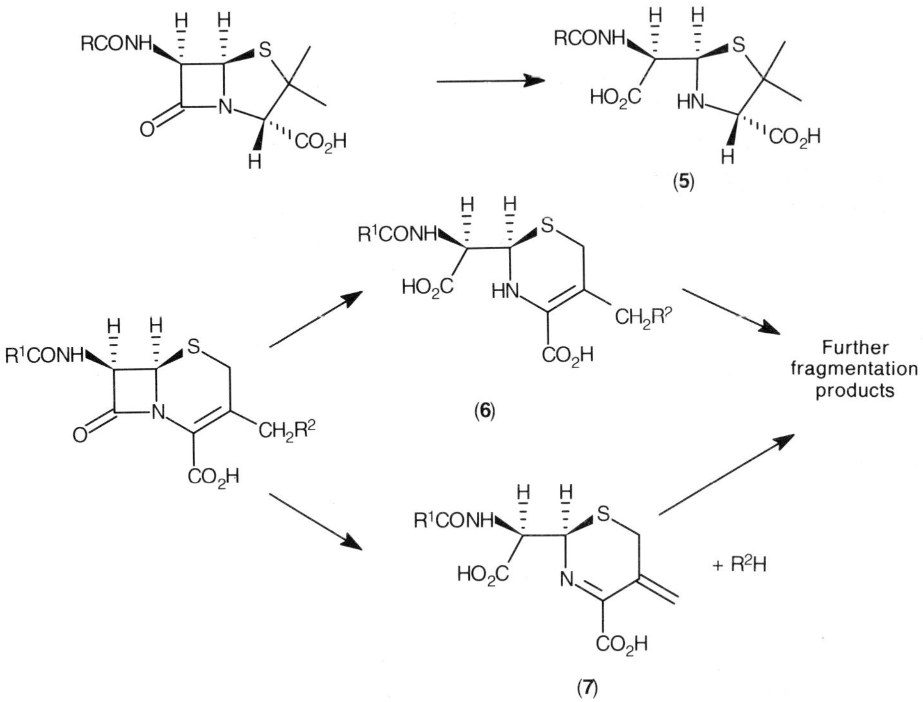

Scheme 14.1 General reaction scheme for the hydrolysis of penicillins and cephalosporins by β-lactamase.

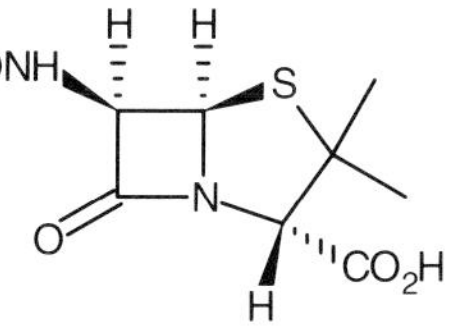

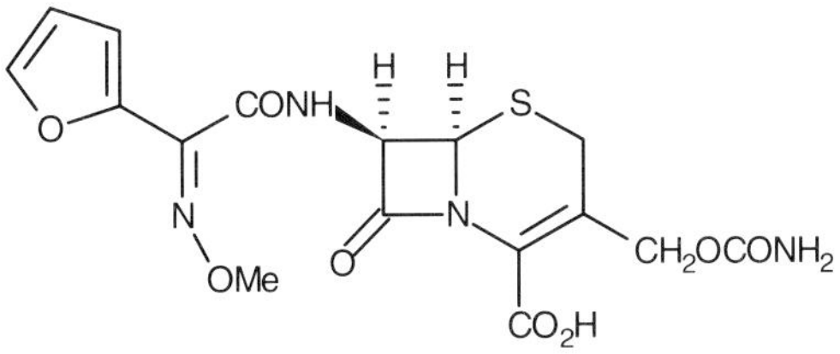

(8) Methicillin R =

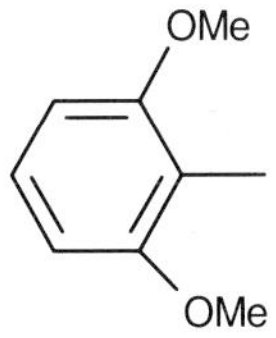

(9) Flucloxacillin R =

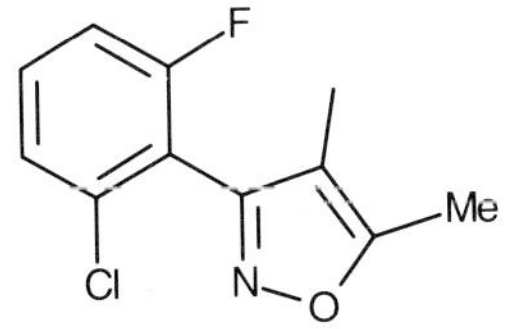

(10) Cefuroxime

Since their discovery a large amount of work has been published on the production, biochemistry and chemistry of β-lactamases (Hamilton-Miller and Smith, 1979). They have been classified and divided according to whether they are plasmid-mediated or chromosomally mediated,* and on the basis of their physical properties, substrate specificity (e.g. penicillin or cephalosporin) and inhibition profiles (Bush, 1989; Möellering, 1991) (Table 14.1).

II. Isolation of Clavulanic Acid

As a result of studying the production of β-lactamases and the microbiological properties of various semisynthetic penicillins, Dr George Rolinson at Beecham Pharmaceuticals considered the possibility of screening microorganisms for naturally occurring β-lactamase inhibitors. In a microbiological screen (Brown et al., 1976), based on β-lactamase inhibition and the traditional hole-in-plate antibacterial assay system, culture filtrates derived from a variety of fungi and actinomycetes were tested for the production of a β-lactamase inhibitor. In the screen, agar containing a fixed amount of penicillin G is inoculated with the β-lactamase-producing strain Klebsiella aerogenes A (now designated K. pneumoniae). On incubation, if no β-lactamase inhibitor is present, the K. aerogenes A grows as an opaque lawn, because the penicillin G is destroyed. If, however, a solution containing an inhibitor is placed in a well cut in the plate, diffusion occurs into the agar, and now on incubation and as a result of inactivation of the β-lactamase, a clear zone of inhibition results around the well. Application of this assay (KAG assay) to various actinomycetes and fungi ($\sim$1000 isolates), obtained from a variety of soil samples collected from different parts of the world, led to the identification of a number of strains of Streptomyces olivaceus which gave potent β-lactamase inhibitory activity. The metabolites responsible for the β-lactamase inhibition produced by these strains of S. olivaceus were isolated, characterized and designated as the olivanic acids (Brown et al., 1977; Corbett et al., 1977; Brown et al., 1990) and are members of the carbapenem family of antibiotic β-lactamase inhibitors. While this work was in progress, other Streptomyces species were reported to produce the cephamycin group of β-lactams. These cultures were obtained and examined for the production of β-lactamase-inhibitory activity. One culture in particular, S. clavuligerus, gave pronounced activity in the above assay and the metabolite responsible was isolated and characterized as clavulanic acid (3) (Brown et al., 1976; Howarth et al., 1976; Reading and Cole, 1977).

Clavulanic acid (3) can be isolated as a crystalline alkali metal salt from culture filtrate and its structure was established via infrared, ^{1}H and ^{13}C NMR and mass spectroscopy studies on its methyl ester. An alternative non-β-lactam structure (11) which appeared to comply with certain spectroscopic data was discounted on examination of the product after catalytic reduction. Separation of the resulting epimers was accomplished by HPLC, and scrutiny of the ^{1}H NMR of the pure major isomer (12) confirmed the β-lactam structure. The C-5 proton was observed to have moved upfield by 0.47 p.p.m. consistent with the reduction of the double bond. Also, the multiplicity of this proton remained unchanged, i.e. a double doublet, which is the same as that of the starting material, whereas, in the alternative structure (13) the multiplicity would have been different. Finally the chemical shift of the newly appointed methine proton is consistent

---

*Chromosomally mediated β-lactamases are carried on the chromosomes (DNA) of the bacterium and are therefore always present in varying amounts in the species producing them. Plasmids are small pieces of circular bacterial DNA which may be passed from one bacterial strain to another; this represents an important source of transmissible bacterial resistance.

TABLE 14.1
General Classification Scheme for Bacterial β-Lactamases[a]

Group	Subtitle	Preferred substrates	Inhibited by:[b]		Representative enzyme(s)
			CA	EDTA	
1	CEP-N	Cephalosporins	No	No	Chromsomal enzymes from Gram-negative bacteria
2a	PEN-Y	Penicillins	Yes	No	Penicillinases from Gram-positive bacteria
2b	BDS-Y	Cephalosporins, penicillins	Yes	No	TEM-1, TEM-2, SHV-1, HMS-1, CEP-2
2b'	EBS-Y	Cephalosporins, penicillins, cefotaxime	Yes	No	TEM-3, TEM-5, TEM-7, TEM-8, TLE-1, RHH-1, SHV-2, SHV-3
2c	CAR-Y	Penicillins, carbenicillin	Yes	No	PSE-1, PSE-3, PSE-4, CARB-3, CARB-4
2d	CLX-Y	Penicillins, carbenicillin	Yes	No	OXA-1, OXA-2, OXA-3, OXA-7, PSE-2, *Bacteriodes fragilis*
2e	CEP-Y	Cephalosporins	Yes	No	*Proteus vulgaris, B. fragilis*
3	MET-N	Variable	No	Yes	*Bacillus cereus II, Pseudomonas maltophilia*
4	PEN-N	Penicillins	No	?[c]	*Pseudomonas cepacia, B. fragilis*

[a] Adapted from Bush (1989) and Möellering (1991).
[b] CA = clavulanic acid; EDTA = ethylenediaminetetraacetic acid.
[c] Variable.

with it being adjacent to a carbon-bearing oxygen atom. The structure of clavulanic acid was finally confirmed by X-ray analysis of the p-nitro-, p-bromobenzyl esters and sodium and potassium salts which revealed the absolute configuration as (R) at C-2 and C-5; these studies also confirmed the Z configuration of the exocyclic double bond. Thus clavulanic acid was defined as (Z)-(2R,5R)-3-(β-hydroxyethylidene)-7-oxo-4-oxa-1-azabicyclo[3.2.0]heptane-2-carboxylic acid. Trivial nomenclature, similar to that used for penicillin and cephalosporin derivatives, is usually preferred so that the parent ring system is called clavam and is numbered starting from oxygen as shown in structure (14). This numbering system will be used henceforth.

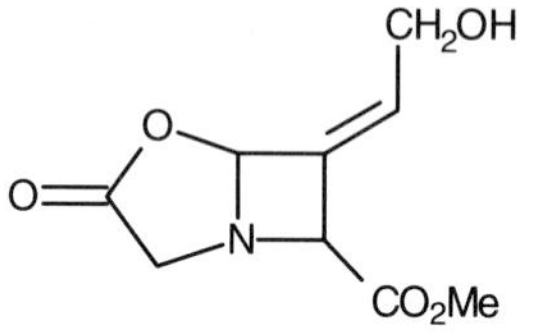

(11) 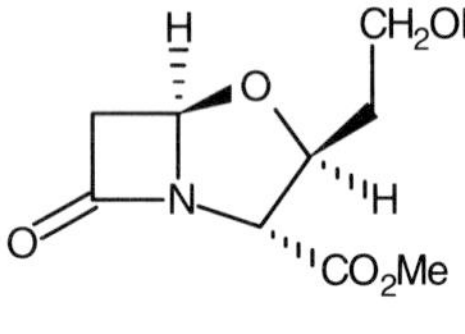   (12)

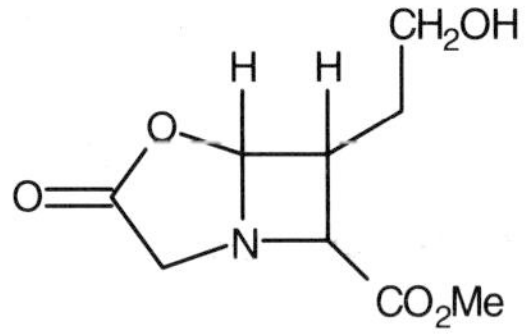

(13) 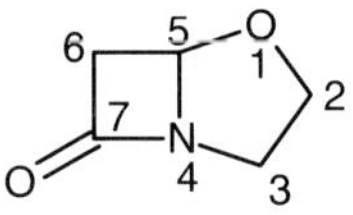   (14) Clavam

Clavulanic acid was the first example of a 'non-traditional' fused β-lactam to be isolated from a natural source. In contrast with the traditional examples of penicillins and cephalosporins, it contains an oxazolidine ring rather than a thiazolidine or dihydro-thiazine ring; it lacks an acylamino substituent at C-6 on the β-lactam ring and contains a β-hydroxyethylidene function at C-2.

Since the discovery of the production of clavulanic acid by S. clavuligerus other groups have isolated it from other species of Streptomyces. Other clavam derivatives have also been reported, e.g. (15)–(19) from S. clavuligerus and other Streptomyces sp. (Southgate and Elson, 1985). In general, these clavam derivatives possess the opposite stereochemistry at C-5 to clavulanic acid.

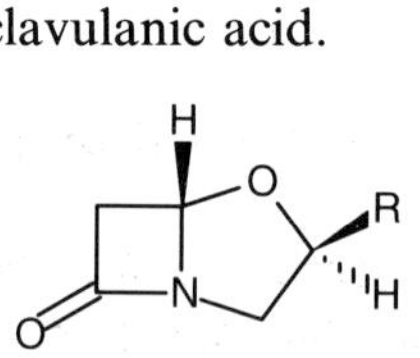

(15) R = CO_2H (18) R = CH_2OH
(16) R = CH_2CH_2OH (19) R = CH_2CHO
(17) R = CH_2CO_2H
              |
            NH_2

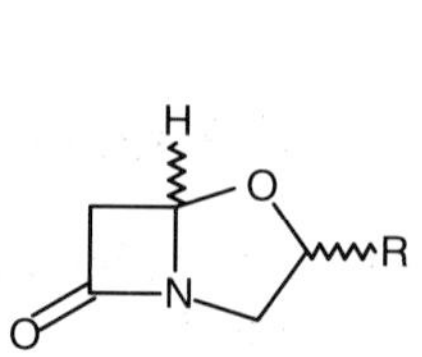

The biosynthesis of clavulanic acid has been examined in some depth (Baggaley et al., 1987; Elson et al., 1987; Baldwin et al., 1991; Salowe et al., 1991). Initial studies which involved feeding labelled precursors to various strains of S. clavuligerus indicated that

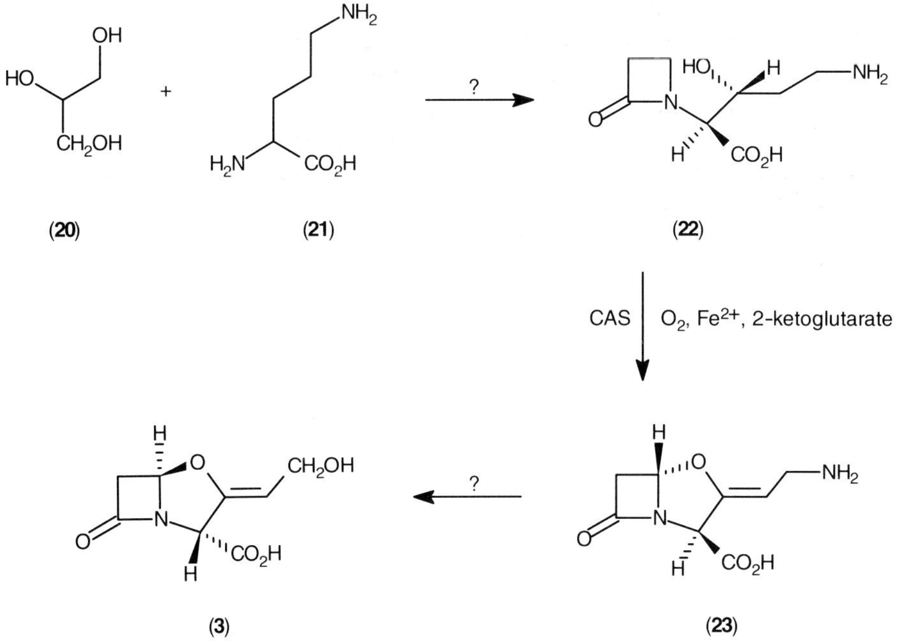

Scheme 14.2   Biosynthesis of clavulanic acid.

carbon atoms C-5, C-6 and C-7 of the β-lactam ring were derived from glycerol (20), while those making up the oxazolidine ring originated from L-ornithine (21). Some of the key synthetic steps and intermediates involved in the pathway from these two precursors to clavulanic acid (3) have been determined (Scheme 14.2). The monocyclic β-lactam proclavaminic acid (22) is produced at an early stage and is converted into clavaminic acid (23) via an oxidative cyclization involving a 2-oxoglutarate and an Fe^{2+}-dependent dioxygenase called clavaminic acid synthase (CAS). Thus the oxygen atom of the oxazolidine ring of clavulanic acid is obtained from the 3-OH function of proclavaminic acid (22). Further studies to determine the nature of the other biosynthetic steps in the pathway are in progress.

III. Biological Activity of Clavulanic Acid

Clavulanic acid is a potent inhibitor of a wide variety of β-lactamases (Coleman *et al.*, 1989) (Table 14.2) as defined by an IC_{50} value, that is, the concentration of clavulanic acid required to result in a 50% inhibition of the rate of hydrolysis of a standard penicillin or cephalosporin substrate. Such IC_{50} values for an inhibitor are determined with and without preincubation of enzyme and inhibitor. Analysis of these results, as well as others from detailed kinetic studies, can indicate if an agent is a reversible or irreversible inhibitor and whether it is exerting competitive or non-competitive inhibition. Against the β-lactamases listed in Tables 14.1 and 14.2, clavulanic acid is a relatively poor inhibitor of the Group 1 enzymes (referred to as cephalosporinases), which are produced by a variety of genera such as *Enterobacter*, *Citrobacter*, *Serratia*,

TABLE 14.2
Intrinsic β-Lactamase Inhibitory Activity of Clavulanic Acid[a]

Organism	Enzyme group[d]	IC$_{50}$ values (μg/ml)[b] with (+) or without (−) 5 min preincubation	
		−	+
Enterobacter cloacae P99	1	> 50	> 50
Escherichia coli JT410	1	> 50	> 50
Bacteroides fragilis 11295/BC4	2e	1.4	0.006
Proteus vulgaris H	2e	0.84	0.017
Pseudomonas aeruginosa A	1	> 50	> 50
Proteus mirabilis C889	2c	3.6	0.021
Escherichia coli JT4 (TEM-1)[c]	2b	0.88	0.055
Escherichia coli K-12 R1010 (SHV-1)	2b	2.4	0.035
Klebsiella pneumoniae E70	2b′	1.0	0.011
Klebsiella oxytoca 1082 (K1)	2b′	3.2	0.047
Escherichia coli K-12 RGN 238 (OXA-1)	2d	> 50	0.71
Escherichia coli K-12 pMG19 (PSE-4)	2c	2.0	0.022
Staphylococcus aureus NCTC 11561	2a	> 50	0.063

[a] Adapted from Coleman *et al.* (1989).
[b] IC$_{50}$ = conc. of inhibitor required to reduce the initial rate of hydrolysis of nitrocefin by 50%.
[c] Enzymes in parentheses were produced particularly by the strains shown.
[d] Enzyme type based on the Bush classification.

Providentia, Pseudomonas aeruginosa and certain strains of *Escherichia coli*. However, it is an excellent inhibitor of the Group 2 β-lactamases; this group encompasses the majority of clinically important enzymes including the TEM type which are most frequently encountered in Gram-negative populations. The Group 2a β-lactamase produced by the Gram-positive organism *Staphylococcus aureus* is very susceptible to clavulanic acid.

Clavulanic acid, when combined with a β-lactamase-sensitive β-lactam antibiotic such as amoxycillin (24) or ampicillin (25), will protect the antibiotic from inactivation by bacterial β-lactamases (Hunter *et al.*, 1978, 1979, 1980). The antibacterial spectrum of amoxycillin plus clavulanic acid (in a 2 : 1 formulation) compared with amoxycillin alone and clavulanic acid alone against representative β-lactamase-producing Gram-positive and Gram-negative bacteria is shown in Table 14.3. From the data it can be seen that the formulation of amoxycillin with clavulanic acid is markedly more effective than amoxycillin alone against amoxycillin-resistant strains of *E. coli, K. pneumoniae, Proteus* species, *Y. enterocolitica, A. hydrophila, Bacteroides* species, *H. influenzae, N. gonorrhoeae, M. catarrhalis* and *Staphylococcus* species at clinically achievable levels (Rolinson and Watson, 1981; Leigh and Robinson, 1982; Slocombe *et al.*, 1984).

TABLE 14.3

Antibacterial Spectrum of Amoxycillin in the Presence of Clavulanic Acid Compared with Amoxycillin Alone and Clavulanic Acid Alone Against β-Lactamase-Producing Organisms[a]

| Organism | MIC values ($\mu g/ml$)[b] | | |
	Amoxycillin	Amoxycillin + clavulanic acid	Clavulanic acid
Escherichia coli NCTC 11560	> 512	8.0	16
Escherichia coli ATCC 35218[c]	> 512	4.0	16
Klebsiella pneumoniae 12	64	2.0	32
Proteus mirabilis 889	> 512	4.0	32
Proteus vulgaris T510	512	2.0	32
Yersinia enterocolitica 10723	32	8.0	32
Aeromonas hydrophila U53	256	32	16
Bacteroides fragilis B3	32	0.5	32
Bacteroides melaninogenicus 109	16	0.1	32
Bacteroides vulgatus 940	16	0.1	32
Bacteroides thetaiotaomicron 4873	32	0.5	16
Haemophilus influenzae NEMC1[c]	128	0.5	32
Neisseria gonorrhoeae AX 1729[c]	16	1.0	4.0
Moraxella catarrhalis 2001E[c]	16	0.25	8.0
Staphylococcus aureus ATCC 29213	8.0	0.25	16
Staphylococcus aureus NCTC 11561[c]	256	1.0	16
Staphylococcus aureus V532[+c,d]	256	16	512
Staphylococcus epidermidis 810	128	1.0	8.0
Staphylococcus epidermidis 254[+c,d]	256	4.0	64

[a] Adapted from Slocombe *et al.* (1984).
[b] Expressed as a concentration of amoxycillin.
[c] Strains producing a plasmid-mediated β-lactamase.
[d] β-Lactamase-producing methicillin-resistant.
MIC, minimum inhibiting concentration.

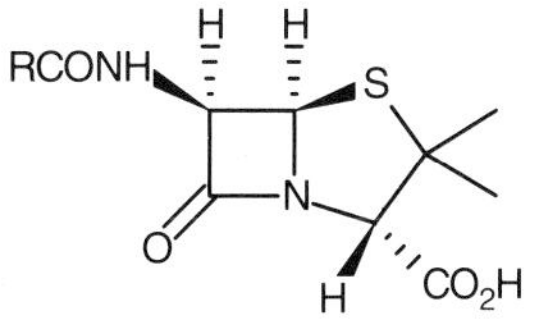

(**24**) Amoxycillin

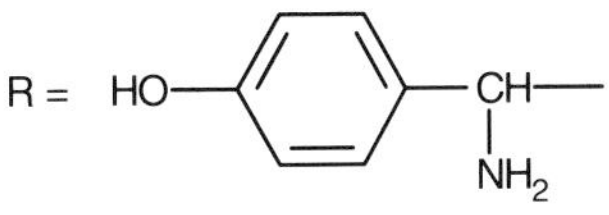

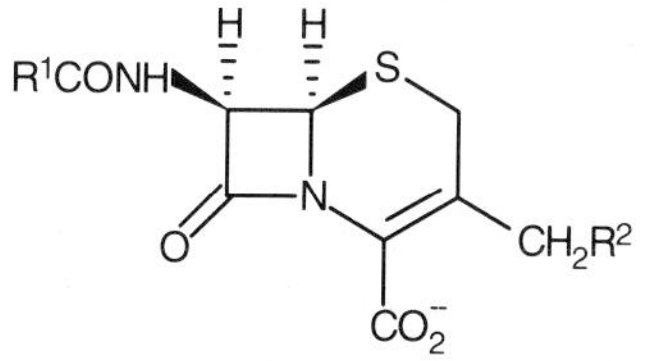

(**29**) Cephalexin

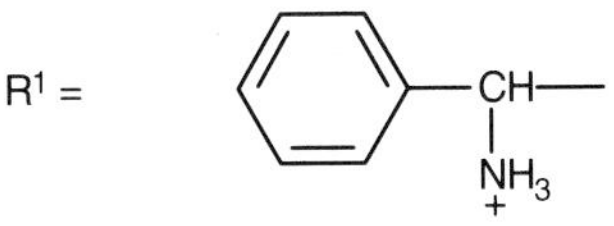

$R^2 = H$

continued over

(25) Ampicillin

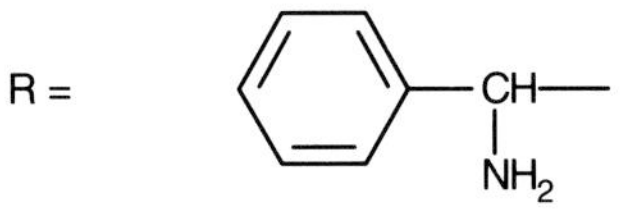

R =

(26) Benzylpenicillin

R =

(27) Ticarcillin

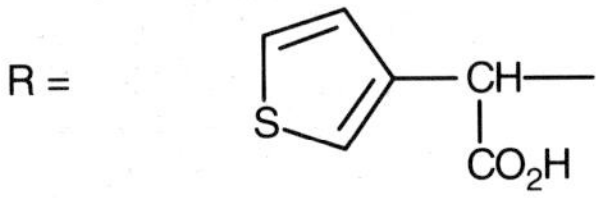

R =

(28) Pipericillin

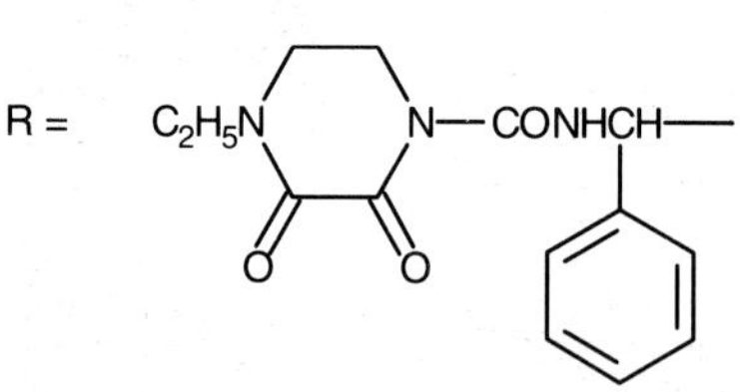

R = C_2H_5N N—CONHCH—

(30) Cephaloridine

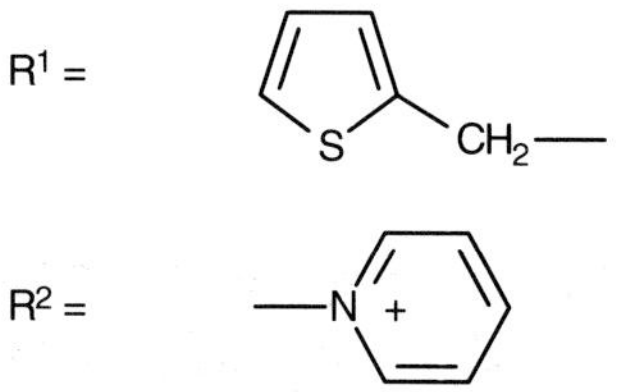

R^1 =

R^2 =

(31) Ceftazidime

R^1 = H_2N—

$(CH_3)_2C—CO_2H$

R^2 =   —N +

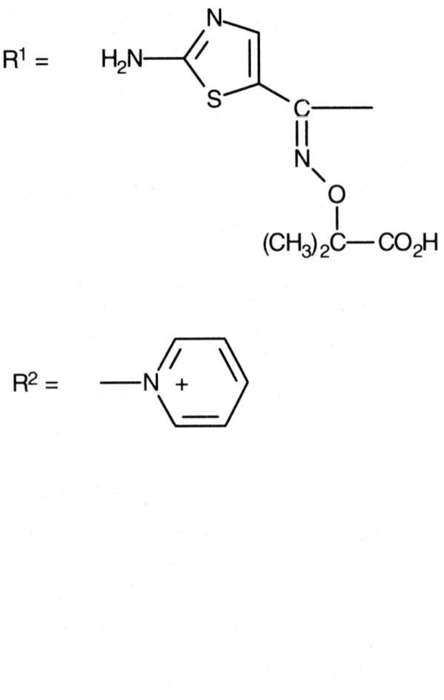

Similarly striking protection can be obtained with other penicillins such as benzylpenicillin (26), ticarcillin (27) and pipericillin (28) as with first-generation cephalosporins such as cephalexin (29) and cephaloridine (30). Second-generation and third-generation cephalosporins such as cefuroxime (10) and ceftazidime (31), which have pronounced stability to β-lactamases, show little potentiation of their antibacterial activity in the presence of clavulanic acid. At the concentrations generally used to potentiate the activity of various β-lactamase-susceptible β-lactam antibiotics, clavulanic acid alone exhibits a low level of antibacterial activity (Table 14.3, MIC range 4–64 μg/ml).

The morphological effects produced by amoxycillin (Fig. 14.1), compared with those

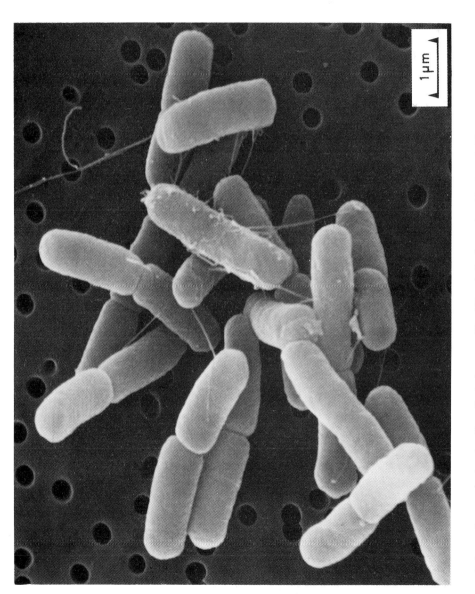

Fig. 14.1 Scanning electron micrograph showing the effect of amoxycillin (10 mg/ml) alone after 60 min at 37°C against a plasmid-mediated β-lactamase-producing strain of *E.coli* JT39.

produced by the formulation of amoxycillin plus clavulanic acid (Fig. 14.2), on β-lacta-mase-producing cultures can be seen using the scanning electron micrograph. The protective β-lactamase-inhibitory effect of clavulanic acid allows amoxycillin to exert its normal characteristic action on the bacterial cell wall which results in spheroplast formation and ultimately cell lysis.

Clavulanic acid is well absorbed by a number of animal species (e.g. mouse, dog, squirrel, monkey) when dosed by the subcutaneous, intramuscular and oral routes. Against experimental infections in mice caused by β-lactamase-producing bacteria, clavulanic acid plus amoxycillin was very effective. Extensive toxicological and metabolic studies on clavulanic acid allowed it to be cleared for detailed pharmacological investigations in man. Orally it is well absorbed in man, a dose of 125 mg resulting in a peak serum level of $\sim 3.5 \,\mu g/ml$ 1 h after administration, with a urinary recovery of 30–40%. Both amoxycillin and clavulanic acid are well absorbed when administered together by mouth to man, and the two compounds show similar serum profiles and half-lives (Fig. 14.3) (Jackson et al., 1984).

IV. Chemistry of Clavulanic Acid and Biological Activity of Some Derivatives

Clavulanic acid contains a number of functional groups which can be readily modified leading to a large variety of derivatives (Brown et al., 1984, 1990). The acid moiety is easily esterified using a diazoalkane or by alkylation of the corresponding alkali metal salt in a suitable solvent, or via an alcohol or thiol with a carbodiimide; the benzyl (**32**) and p-nitrobenzyl (**33**) esters are the most useful in that these ester functions can be conveniently removed via catalytic hydrogenolysis. Catalytic hydrogenation of methyl (**34**) or benzyl clavulanate (**32**) gave methyl or benzyl dihydroclavulanate (**35**) as a mixture of C-2 epimers. Hydrogenolysis of benzyl clavulanate in aqueous ethanol, containing an equivalent amount of sodium hydrogen carbonate, at room temperature

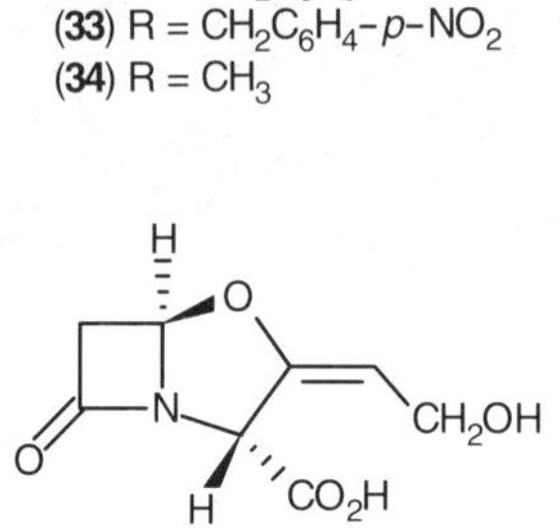

(**32**) R = CH₂C₆H₅
(**33**) R = CH₂C₆H₄-p-NO₂
(**34**) R = CH₃

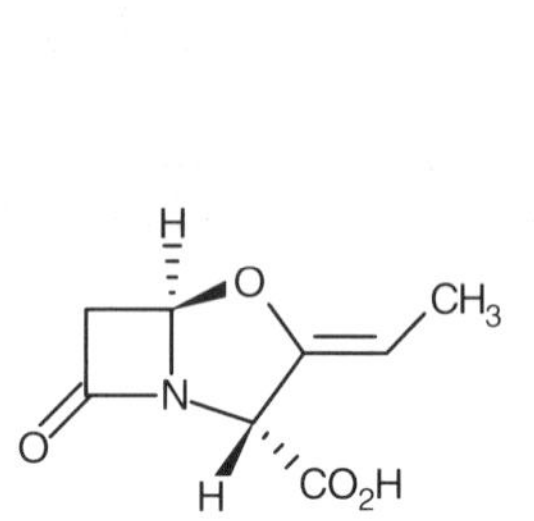

(**35**) R = CH₃ or CH₂C₆H₅

(**36**)

(**37**)

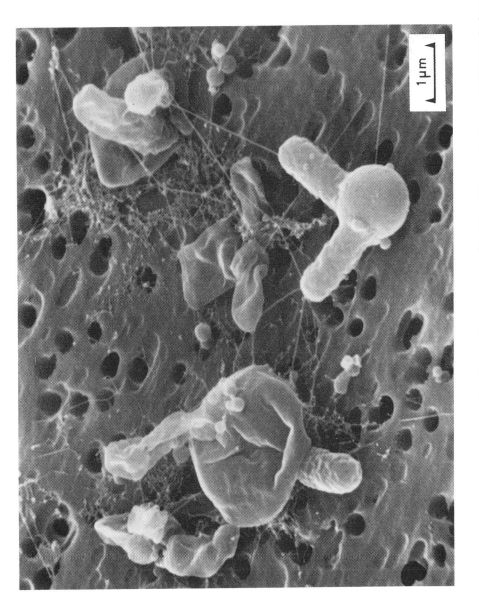

Fig. 14.2 Scanning electron micrograph showing bacteriolytic effects produced by amoxycillin (10 mg/ml) plus clavulanic acid (2.5 mg/ml) after 60 min at 37°C against a β-lactamase-producing strain of *E.coli* JT39.

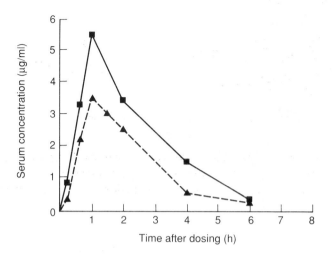

Fig. 14.3   Amoxycillin (250 mg, p.o., ■)/clavulanic acid (125 mg, p.o., ▲) serum concentrations in man. Adapted from Jackson *et al.* (1984).

and atmospheric pressure with uptake of 1 mol of hydrogen afforded sodium clavulanate. Under more rigorous hydrogenolysis conditions, benzyl clavulanate (**32**) could be converted into the isoclavulanic acid derivative (**36**), which is the *E*-isomer of (**3**), and into the C-9-deoxyclavulanic acid analogue (**37**).

The allylic hydroxy function can be acylated, etherified, sulphated, replaced by halogen (and hence azido leading to amino and acylamino functions), or by amino and sulphur nucleophiles and also oxidized to an aldehyde. The allylic double bond can be isomerized photolytically as well as catalytically. An interesting product from the photolysis of phenacyl clavulanate was the tetracyclic β-lactam oxetane (**38**). The alkene moiety in clavulanic acid can also be epoxidized and cleaved by ozonolysis. Certain derivatives undergo elimination reactions with the formation of a diene (a clavem), while alkylation can be carried out at C-3 and C-6. The β-lactam carbonyl group reacts with certain Wittig reagents and the parent acid can be decarboxylated. The degradation of clavulanic acid or its esters under a variety of conditions produces a number of non-β-lactam species as well as rearrangement products. The pyrroles (**39**) and (**40**) can be obtained via azetidinone (**41**) on treatment of ester (**32**) with triethylamine and acetic acid, while hydrolysis of clavulanic acid gave the amino ketone (**42**) (Brown *et al.*, 1990).

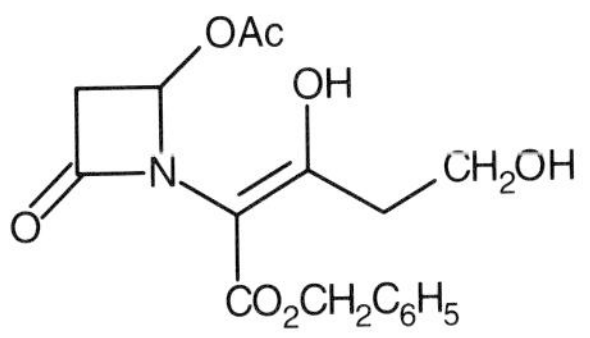

(41)

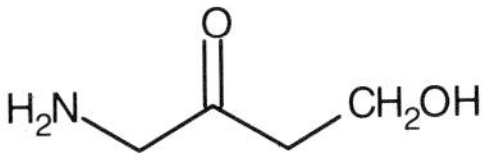

(42)

The clavam nucleus is an interesting target for total synthesis, and the structure of clavulanic acid has been confirmed by the preparation of racemic methyl clavulanate (34) from the azetidinone (43) by the sequence of reactions shown in Scheme 14.3; (±)-

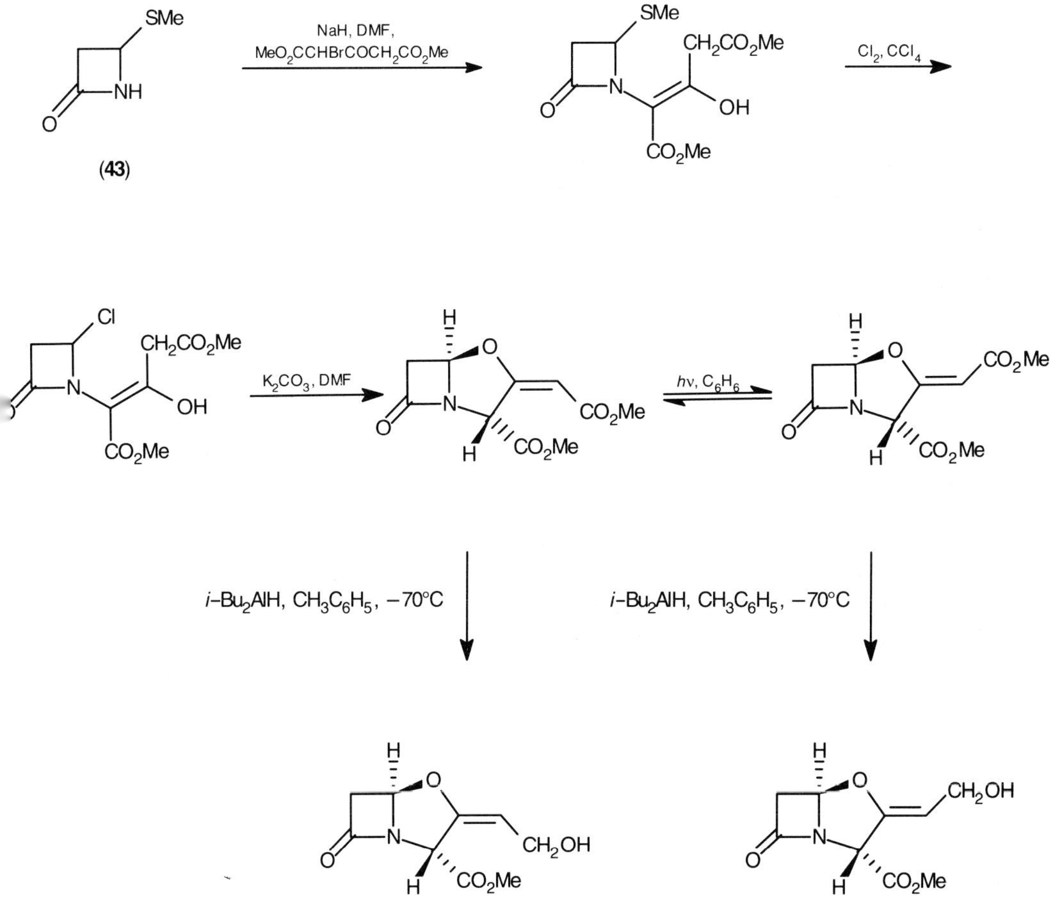

Scheme 14.3   Synthesis of (±)-methyl clavulanate.

methyl isoclavulanate (**44**) was also prepared as a consequence of this work (Bentley *et al.*, 1977, 1979).

The β-lactamase-inhibitory (IC_{50}) data for a number of representative clavam derivatives obtained from clavulanic acid are given in Table 14.4. From these (Brown, 1981) and other data one can identify a number of empirical structure–activity features.

1. Potent β-lactamase-inhibitory activity appears to require a close structural analogy to clavulanic acid, only variation of the C-9 hydroxy group being allowed; any modification resulting in removal of the double bond or destruction of the β-lactam moiety gives poor inhibitors, as does C-6 modification, C-5 inversion [i.e. (5*R*) stereochemistry is required], C-3 decarboxylation or C-3 amide formation.
2. Isoclavulanic acid (the *E*-isomer) derivatives are less active than the corresponding *Z*-isomers.
3. Clavem derivatives have limited interest.
4. In general terms, β-lactamase-inhibitory activity is closely related to the size and hydrophilicity of the substituent replacing the hydroxy function, with small hydrophilic groups yielding the most active compounds in any series.

V. Mechanism of Action of Clavulanic Acid

The mode of action of clavulanic acid as a β-lactamase inhibitor has been examined in detail (Reading and Hepburn, 1979; Fisher *et al.*, 1980; Reading and Farmer, 1981). The interaction with the TEM β-lactamase from *E.coli* is illustrated in Fig. 14.4. An acyl-enzyme intermediate (formed by cleavage of the β-lactam ring by a hydroxy group of a serine residue and concomitant ring opening of the unstable oxazolidine ring) is derived from the Michaelis complex; this intermediate now breaks down in three possible ways. It can (a) undergo hydrolysis to yield original enzyme plus degraded clavulanic acid, (b) form a transiently stable complex which ultimately hydrolyses to enzyme and degradation product(s) or (c) form irreversibly inactivated complexes. A turnover of 115 clavulanic acid molecules occurs for every irreversible inactivation of enzyme. A similar mechanism is proposed for the inactivation of the chromosomally mediated β-lactamase from *K. pneumoniae*. With the *Staph. aureus* β-lactamase the interaction, which proceeds in an approximately 1 : 1 stoichiometry, yields an acyl-enzyme intermediate which reacts further to give irreversibly inactivated enzyme, though hydrolysis of the intermediate can occur in the absence of an excess of clavulanic acid to give enzyme-degraded clavulanic acid and regeneration of enzyme activity.

These and other studies have confirmed clavulanic acid as a progressive irreversible inhibitor (or 'active-site-directed suicide inhibitor') of β-lactamases.

Several other non-classical β-lactams have also been found to be 'mechanism-based inactivators' of β-lactamases. Prominent amongst these are certain carbapenems such as the olivanic acids (see next section), asparenomycins [e.g. (**45**) asparenomycin A] (Brown *et al.*, 1990) and SF-2103A (**46**) (Niwa *et al.*, 1986), also 6β-bromopenicillanic acid (**47**; bromobactam) (Pratt and Loosemore, 1978; Moore *et al.*, 1981; Wise *et al.*, 1981), the penam sulphones (**48**; sulbactam) (English *et al.*, 1978), (**49**; tazobactam) (Micetich *et al.*, 1987) and (**50**) (Keith *et al.*, 1983), and the C-6-alkylidene penams (**51**) (Chen *et al.*, 1986), (**52**) and (**53**) (Brown *et al.*, 1990). More recently, in these laboratories, we have identified the C-6 triazolylmethylene penem (**54**; BRL 42715) (Bennett *et al.*, 1988; Osborne *et al.*,

TABLE 14.4

β-Lactamase-inhibitory Properties of Some Derivatives of Clavulanic Acid[a]

	IC_{50} (µg/ml)				
	Staphylococcus aureus Russel, 2a β-lactamase	*Escherichia coli* JT4, 2b (TEM-1) β-lactamase	*Klebsiella pneumoniae* E70, 2c β-lactamase	*Proteus mirabilis* C889, 2c β-lactamase	*Citrobacter freundii* Manti, 1 β-lactamase
Clavulanic acid; $R = OH$	0.06	0.07	0.03	0.03	10
Deoxyclavulanic acid; $R = H$	0.12	0.09	0.05	–	5
Isoclavulanic acid (*E*-geometry)	0.6	1.0	0.45	–	5
Acetate; $R = OCOCH_3$	0.04	–	>0.4	–	0.4
Carbamate; $R = OCONHCH_3$	1.5	2.5	2.5	–	0.45
Methyl ether; $R = OMe$	0.05	0.18	0.07	0.01	8.5
Benzyl ether; $R = OCH_2Ph$	0.005	0.1	0.04	0.02	4.4
Thioether; $R = SMe$	0.11	0.04	0.13	0.01	$\gg 10$[b]
Amine; $R = N(CH_2Ph)_2$	0.002	0.04	0.08	0.01	0.62

[a] Adapted from Brown (1981).
[b] β-lactamase from *Enterobacter cloacae* P99.

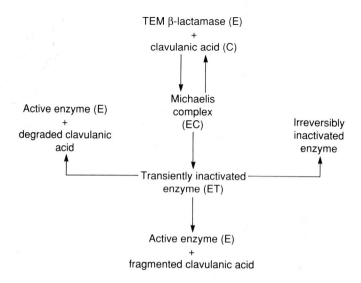

TEM β-lactamase (E)
+
clavulanic acid (C)

Michaelis
complex
(EC)

Active enzyme (E)
+
degraded clavulanic
acid

Irreversibly
inactivated
enzyme

Transiently inactivated
enzyme (ET)

Active enzyme (E)
+
fragmented clavulanic acid

Fig. 14.4   Mechanism of inactivation of TEM β-lactamase by clavulanic acid.

1989) as an extremely potent broad-spectrum inhibitor, effecting rapid and progressive inactivation of a wide range of β-lactamases including the Group 1 cephalosporinases.

(45)

(46)

(47)

(48)

(49)

(50)

(51)

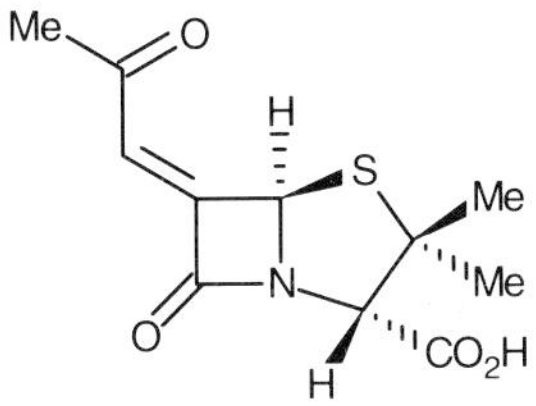

(52) 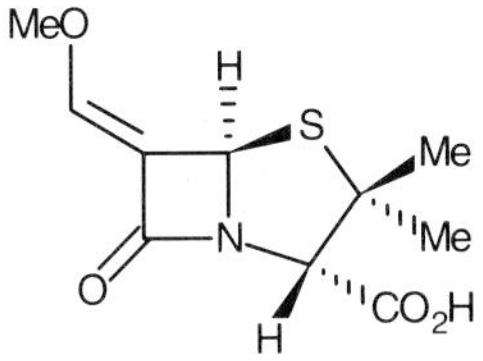 (53) 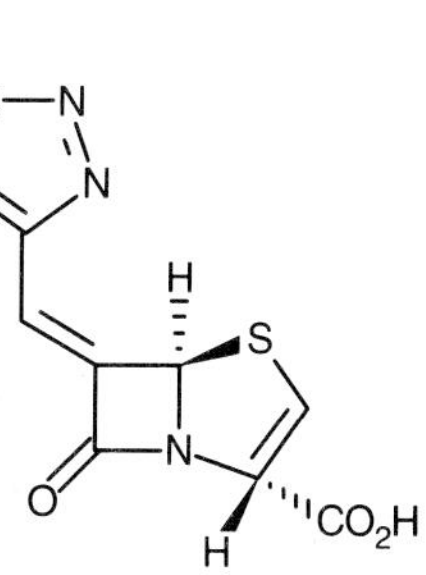 (54)

VI. The Olivanic Acids and Thienamycins

The assay which led to the discovery of clavulanic acid also identified a series of *S. olivaceus* strains which were shown to produce a family of antibiotic β-lactamase inhibitors called the olivanic acids (carbapenems) characterized as the sulphate esters (**55**)–(**57**) and the related hydroxy compounds (**58**)–(**61**). At this time (early 1970s) Merck workers had isolated the thienamycins from *S. cattleya* (**4**), (**62**) and (**63**) and the epithienamycins (**58**)–(**61**) from *S. flavogriseus* while screening for inhibitors of peptidoglycan synthesis. Over 40 carbapenems have now been derived from *Streptomyces* sp. (Southgate and Elson, 1985; Brown *et al.*, 1990). Structurally they differ considerably from the penicillins and clavulanic acid. They have a CH$_2$ unit instead of S or O in the five-membered ring and contain an endocyclic double bond; the C-2 substituent is usually a substituted alkyl or alkenylthio unit, while alkyl, hydroxyalkyl or alkylidene functions are present at C-6; the relative stereochemistry of the β-lactam ring protons can be either *cis* or *trans*.

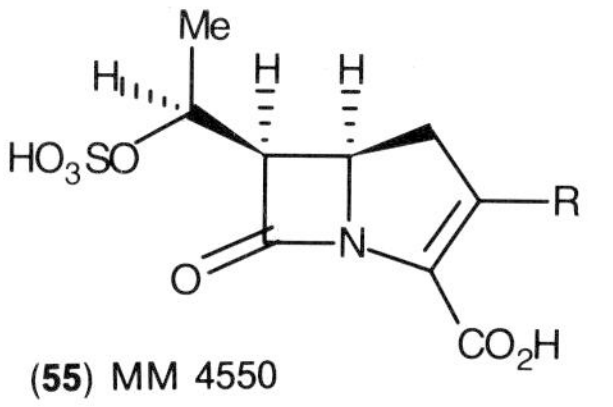

(**55**) MM 4550

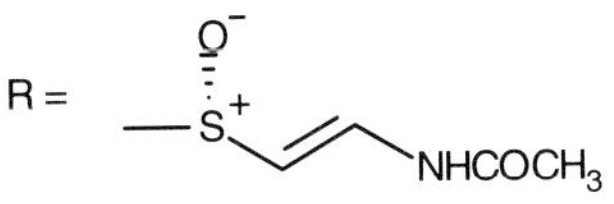

R =

(**56**) MM 13902

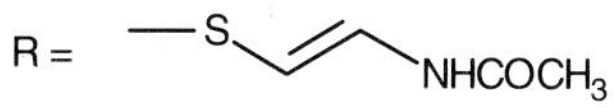

R =

(**57**) MM 17880

R =

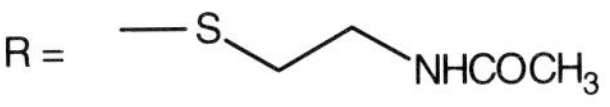

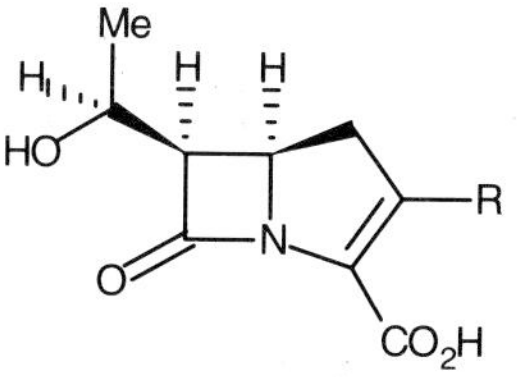

(**58**) MM 22380 Epithienamycin A

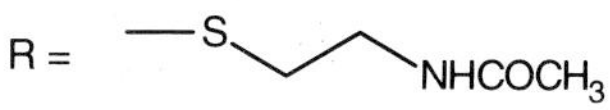

R =

(**59**) MM 22381 Epithienamycin C

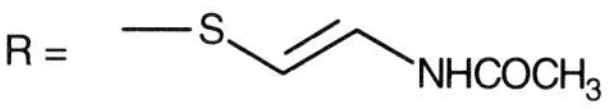

R =

(60) MM 22382 Epithienamycin B

R = —S⟍⟍⟍NHCOCH₃

(61) MM 22383 Epithienamycin D

R = —S⟍⟍NHCOCH₃

(62) R = ⟍⟍⟍NHCOCH₃

(63) R = ⟍⟍⟍NHCOCH₃

(64) R = ⟍⟍⟍N=CHNH₂

The olivanic acids are potent β-lactamase inhibitors (up to 1000 times more active than clavulanic acid) and highly effective inhibitors of the cephalosporinases (Group 1 type β-lactamases) (Brown, 1981). Like clavulanic acid the potentiation of the activity of certain penicillins and cephalosporins can be demonstrated in vitro, but evaluation is complicated by the pronounced antibacterial activity of the carbapenem inhibitor (Basker et al., 1980). Thienamycin (4) and a chemically more stable derivative N-formimidoyl derivative (64; imipenem) are potent broad-spectrum antibiotics. Carbapenems, however, have poor metabolic stability leading to low urinary recovery in man, the enzyme renal dehydropeptidase-1 present in human kidney being responsible for their degradation. In clinical practice, imipenem is used in conjunction with cilastatin, an inhibitor of renal dehydropeptidase (Brown et al., 1990).

VII. Conclusion

Currently an oral formulation containing potassium clavulanate (125 mg) plus amoxycillin trihydrate (250 mg) called Augmentin (Trade Mark of SmithKline Beecham) is marketed in a number of countries following satisfactory toxicological examination and detailed clinical trials (Rolinson and Watson, 1981; Leigh and Robinson, 1982). Augmentin is highly efficacious against a wide range of urinary tract, skin and soft tissue, upper and lower respiratory tract and gynaecological infections found in hospital and in general practice.

References

Abraham, E.P. and Chain, E. (1940). Nature (London) 149, 837.
Baggaley, K.H., Nicholson, N.H. and Sime, J.T. (1987). J. Chem. Soc. Chem. Commun., 567.
Baldwin, J.E., Adlington, R.M., Bryans, J.S., Bringham, A.O., Coates, J.B., Crouch, N.P., Lloyd, M.D., Schofield, C.J., Elson, S.W., Baggaley, K.H., Cassels, R. and Nicholson, N. (1991). Tetrahedron 47, 4089.
Basker, M.J., Boon, R.J. and Hunter, P.A. (1980). J. Antibiot., 33, 878.
Bennett, I.S., Brooks, G., Broom, N.J.P., Coleman, K., Coulton, S., Edmondson, R.A., Griffin, D.R., Harbridge, J.B., Osborne, N.F., Stirling-François, I. and Walker, G. (1988). In 'Proceedings of the 28th Interscience Conference on Antimicrobial Agents and Chemotherapy, Los Angeles', Abstr. 118.
Bentley, P.H., Berry, P.D., Brooks, G., Gilpin, M.L., Hunt, E. and Zomaya, I.I. (1977). J. Chem. Soc. Chem. Commun., 748.
Bentley, P.H., Brooks, G., Gilpin, M.L. and Hunt, E. (1979). Tetrahedron Lett. 20, 1889.
Brown, A.G. (1981). J. Antibiot. Chemother. 7, 15.

Brown, A.G., Butterworth, D., Cole, M., Hanscombe, G., Hood, J.D., Reading, C. and Rolinson, G.N. (1976). *J. Antibiot.* **29**, 668.

Brown, A.G., Corbett, D.F., Eglington, A.J. and Howarth, T.T. (1977). *J. Chem. Soc. Chem. Commun.,* 523.

Brown, A.G., Corbett, D.F., Goodacre, J., Harbridge, J.B., King, T.J., Ponsford, R.J. and Stirling, I. (1984). *J. Chem. Soc. Perkin Trans. 1,* 635.

Brown, A.G., Pearson, M.J. and Southgate, R. (1990). *In* 'Comprehensive Medicinal Chemistry' (C. Hansch, P.G. Sammes and J.B. Taylor, eds), p. 655, Pergamon Press, Oxford.

Bush, K. (1989). *Antimicrob. Agents Chemother.* **33**, 259.

Chen, Y.L., Chang, C-W., and Hedberg, K. (1986). *Tetrahedron Lett.* **27**, 3449.

Coleman, K., Griffin, D.R.J., Page, J.W.J. and Upshon, P.A. (1989). *Antimicrob. Agents. Chemother..* **33**, 1580.

Corbett, D.F., Eglington, A.J. and Howarth, T.T. (1977). *J. Chem. Soc. Chem. Commun.,* 953.

Elson, S.W., Baggaley, K.H., Gillett, J., Holland, S., Nicholson, N.H., Sime, J.T. and Woronieki, S.R. (1987). *J. Chem. Soc. Chem. Commun.,* 1736 and 1739.

English, A.R., Retsema, J.A., Girard, A.E., Lynch, J.E. and Barth, W.E. (1978). *Antimicrob. Agents Chemother.* **14**, 414.

Fisher, J., Belasco, J.G., Charnas, R.L., Khosla, S. and Knowles, J.R. (1980). *Philos. Trans. R. Soc. London, Ser.B* **289**, 309.

Hamilton-Miller, J.M.T. and Smith, J.T. (eds) (1979). 'Beta-Lactamases', Academic Press, London.

Howarth, T.T., Brown, A.G. and King, T.J. (1976). *J. Chem. Soc., Chem. Commun.,* 266.

Hunter, P.A., Reading, C. and Witting, D.A. (1978). *Current Chemother. Proc. Int. Congr. Chemother.,* 10th, **1**, 478.

Hunter, P.A., Coleman, K., Fisher, J., Taylor, D. and Taylor, E. (1979). *Future Trends Chemother., 4th, Proc. Int. Symp.* **5**, 1.

Hunter, P.A., Coleman, K., Fisher, J., and Taylor, D. (1980). *J. Antimicrob. Chemother.* **6**, 455.

Jackson, D., Cooper, D.L., Filer, C.W. and Langley, P.F. (1984). 'Postgraduate Medicine: Custom Communications', p. 51, New York.

Keith, D.D., Tengi, J., Rossman, P., Todaro, L. and Weigle, M. (1983). *Tetrahedron* **39**, 2445.

Leigh, D.A. and Robinson, O.P.W. (eds). (1982). 'Augmentin'. Excerpta Medica, Amsterdam.

Micetich, R.G., Hall, T.W., Maiti, S.N., Spevak, P., Yamabe, S., Ishida, N., Tanaka, M., Yamazaki, T., Nakai, A. and Ogawa (1987). *J. Med. Chem.* **30**, 1469.

Möellering Jr., R.C. (1991). *Rev. Infect. Dis.* **13** (Suppl. 9), S723.

Moore, B.A. and Brammer, K.W. (1981). *Antimicrob. Agents Chemother.* **29**, 327.

Morin, R.B. and Gorman, M. (eds). (1982). 'Chemistry and Biology of β-Lactam Antibiotics', Vols. 1, 2 and 3, Academic Press, New York.

Niwa, T., Yoshida, T., Tamura, A., Kazuno, Y., Inouye, S., Ito, T. and Kojima, M. (1986). *J. Antibiot.* **39**, 943.

Osborne, N.F., Broom, N.J.P., Coulton, S., Harbridge, J.B., Harris, M.A., Stirling-François, I. and Walker, G. (1989). *J. Chem. Soc. Chem. Commun.,* 371.

Pratt, R.F. and Loosemore, M.J. (1978). *Proc. Natl. Acad. Sci. U.S.A.* **75**, 4145.

Reading, C. and Cole, M. (1977). *Antimicrob. Agents Chemother.* **11**, 852.

Reading, C. and Farmer, T. (1981). *Biochem. J.* **199**, 779.

Reading, C. and Hepburn, P. (1979). *Biochem. J.* **179**, 67.

Rolinson, G.N. (1979). *J. Antimicrob. Chemother.* **5**, 7.

Rolinson, G.N. and Watson, A. (eds) (1981). 'Augmentin'. Excerpta Medica, Amsterdam.

Salowe, S.P., Krol, W.J., Iwata-Reuyl, D. and Townsend, C.A. (1991). *Biochemistry* **30**, 2281.

Slocombe, B., Beale, A.S., Boon, R.J., Griffin, K.E., Masters, P.J., Sutherland, R. and White, A.R. (1984). 'Postgraduate Medicine: Custom Communications', p. 29, New York.

Southgate, R. and Elson, S. (1985). *In* 'Progress in the Chemistry of Organic Natural Products' (W. Herz, H. Grisebach, G.W. Kirby and Ch. Tamm, eds), p. 1. Springer-Verlag Wien, New York.

Wise, R., Andrew, J.M. and Patel, N. (1981). *J. Antimicrob. Chemother.* **7**, 531.

Index